Springer Proceedings in Physics **427**

Indexed by Scopus

The series Springer Proceedings in Physics, founded in 1984, is devoted to timely reports of state-of-the-art developments in physics and related sciences. Typically based on material presented at conferences, workshops and similar scientific meetings, volumes published in this series will constitute a comprehensive up to date source of reference on a field or subfield of relevance in contemporary physics. Proposals must include the following:

- Name, place and date of the scientific meeting
- A link to the committees (local organization, international advisors etc.)
- Scientific description of the meeting
- List of invited/plenary speakers
- An estimate of the planned proceedings book parameters (number of pages/articles, requested number of bulk copies, submission deadline).

Please contact:

For Americas and Europe: Dr. Zachary Evenson; zachary.evenson@springer.com
For Asia, Australia and New Zealand: Dr. Loyola DSilva; loyola.dsilva@springer.com

Phang Chang · Kavikumar Jacob ·
Logenthiran Machap · Siti Amira Othman ·
Shakila Abdullah · Nurul Nadia Adnan
Editors

Proceedings of the 9th International Conference on the Applications of Science and Mathematics

SCIEMATHIC 2024, 03–04 October, Putrajaya, Malaysia

Springer

Editors
Phang Chang
Department of Mathematics and Statistics
Universiti Tun Hussein Onn Malaysia
Parit Raja, Johor, Malaysia

Kavikumar Jacob
Department of Mathematics and Statistics
Universiti Tun Hussein Onn Malaysia
Parit Raja, Johor, Malaysia

Logenthiran Machap
Department of Mathematics and Statistics
Universiti Tun Hussein Onn Malaysia
Parit Raja, Johor, Malaysia

Siti Amira Othman
Department of Physics and Chemistry
Universiti Tun Hussein Onn Malaysia
Parit Raja, Johor, Malaysia

Shakila Abdullah
Department of Physics and Chemistry
Universiti Tun Hussein Onn Malaysia
Parit Raja, Johor, Malaysia

Nurul Nadia Adnan
Department of Physics and Chemistry
Universiti Tun Hussein Onn Malaysia
Johor, Malaysia

ISSN 0930-8989 ISSN 1867-4941 (electronic)
Springer Proceedings in Physics
ISBN 978-981-96-7678-1 ISBN 978-981-96-7679-8 (eBook)
https://doi.org/10.1007/978-981-96-7679-8

This Springer imprint is published by the registered company Springer Nature Singapore Pte Ltd.
The registered company address is: 152 Beach Road, #21-01/04 Gateway East, Singapore 189721, Singapore

Preface

We are honoured to present the Proceedings of the 9th International Conference on the Applications of Science and Mathematics (SCIEMATHIC) 2024, held at the Everly Hotel, Putrajaya, Malaysia, on October 3–4, 2024. We believe all participants gained valuable insights and experiences in science and mathematics throughout the conference.

The conference's primary objective is to promote discussion and knowledge exchange between experts in science, mathematics and biodiversity for a sustainable future. Other than that, to empower scientists and professionals to deliver science, mathematics and nature in a way that promotes harmony and holistic understanding for future generations and, lastly, to showcase and celebrate successful examples of integrating science, mathematics and biodiversity for global sustainability initiatives. Renowned experts in the field, including esteemed professors from Malaysia and Indonesia, will be invited as keynote speakers to share the latest advancements in their respective areas of expertise. This conference serves as a valuable platform for students and researchers to share knowledge and expertise, indirectly allowing the university to gain recognition among participants from within and outside the country.

This proceeding features a selection of 25 papers submitted to the conference by contributors from universities, research institutes and industries. Each paper underwent a rigorous peer-review process by the conference committee and external reviewers. The papers were selected based on their quality and suitability to the conference themes. The proceedings are organised into six key topics: mathematics and statistics, physics, chemistry, biology, engineering sciences and artificial intelligence.

We sincerely thank all the authors for their valuable contributions to these proceedings. We also thank the organising committee, reviewers, speakers, chairpersons and all conference participants for their unwavering support of SCIEMATHIC 2024.

Contents

Mathematics and Statistics

Solving Influenza Infection Model Using Runge-Kutta Fourth Order Method and Euler Method

Nurul Farina Rosli and Norzuria Ibrahim(✉)

Department of Mathematics and Statistics, Faculty of Applied Sciences and Technology, UTHM Kampus Cawangan Pagoh, Hab Pendidikan Tinggi Pagoh, KM 1, Jalan Panchor, 84600 Pagoh, Muar, Johor, Malaysia

norzuria@uthm.edu.my

Abstract. This project discussed the numerical solutions of first-order nonlinear ordinary differential equations (ODEs) that established the epidemic model. The application under consideration is a SEIR model of the influenza outbreak in Australia in 1919. Two proposed methods, the Runge-Kutta fourth order (RK4) and the Euler method, were used, and they proved to be good approximations of the behaviour of each SEIR variable and practically well-suited for solving this problem. The analytical solution was hard to obtain, so this study used ode45 in MATLAB to solve the simulation of the SEIR model. The outcomes of the RK4 and Euler methods were acquired through MATLAB software implementation. Additionally, we intended to evaluate two different ways to determine a good approximation to the behaviour of each of the SEIR variables. To verify good approximation to the behaviour of each SEIR variable, approximation methods implementing different step sizes, which were $h = 0.01$, 0.1 and 0.25, were compared to the ode45 in MATLAB provided by the SEIR model. The step size had to be smaller for the solution to approximate each SEIR variable's behaviour better. The RK4 order and Euler methods were compared numerically and illustrated graphically. To examine errors, we concluded by analyzing and computing the errors of the two proposed methods for various step sizes. The RK4 order method was observed to be a good approximation to the behaviour of four systems SEIR in solving the SEIR model using ode45 in MATLAB.

Keywords: Influenza · Ordinary Differential Equation · Runge-Kutta Fourth Order Method

1 Introduction

Influenza, commonly known as the flu, is a highly contagious respiratory illness caused by influenza viruses [9]. Influenza is highly contagious and quickly transferred by contact with droplets from an infected person's coughing and sneezing nose and throat. The infection affected the nose, throat, and lungs. According [4] several nations reported human swine influenza A (H1N1) virus infection. Cases of new influenza (H1N1) were

P. Chang et al. (Eds.): SCIEMATHIC 2024, SPPHY 427, pp. 3–12, 2025.
https://doi.org/10.1007/978-981-96-7679-8_1

initially reported in California in mid-April 2009, followed by Texas and Mexico. [7] examined a mathematical model for the remedy and stability of Influenza. This study aimed to convert the continuous model's essential properties to discrete strategies to obtain unconditionally stable procedures.

The population consisted of four compartments represented as individuals which are species of individuals, $S(t)$, E(t), $I(t)$, and $R(t)$. The first compartment, represented by (S), consists of individuals who haven't been exposed to the illness yet. The second compartment, denoted by (E), comprises infected individuals who do not transmit the disease during the incubation period. The third compartment, represented by (I), contains infected individuals who develop the disease after the incubation period. Lastly, the fourth compartment, denoted by (R), comprises those who either recover or die away. The SEIR compartment model was used to examine the spreading dynamics of the new pandemic Influenza A (H1N1) virus. Influenza could be avoided by being vaccinated every year. A one-year vaccination design might be useless, followed by another year as the virus mutates quickly. [16] used the likelihood-based method to analyze a mathematical model of yearly influenza epidemics in Australia. The presence of multiple parameters and multivariate variables characterizes this model.

Several investigations used the Runge-Kutta method to solve models of influenza. [5] stated that the Runge-Kutta method was extensively employed to solve ODEs due to its accuracy and simplicity. Carl David, Tolme Runge and Martin Wilhelm Kutta were the German mathematicians who developed the Runge-Kutta method. [13] investigated the enhanced Runge-Kutta method that solves ODEs. Runge-Kutta was an iterative technique that employed a weighted average of function values to estimate the function's value at the subsequent time step, where we were given a general form of RK4 order method.

[8] explained that the Euler technique was a first-order approach, which indicated that the local error was proportional to the square root of the step size. In contrast, the global error was proportional to the step size directly. [11] stated that the Euler techniques were the most well-known numerical methods for solving ordinary differential equations. The Euler method had a slow convergence. [2] studied that the accuracy of Euler's method is significantly lower than that of the Runge-Kutta method.

The SEIR model is used to examine influenza transmission dynamics using two approaches that comprised analytical and numerical methods. The problem statement for this study was that an analytical solution is often used to solve the SEIR model. However, analytical solutions are hard to use for complex problems, and some results cannot be obtained from analytical solutions, so this study used ode45 in MATLAB to solve the SEIR model. Ode45 is defined as a general-purpose single step ODE solver in MATLAB. This study aims to compare the solution of the influenza infection model using ode45 in MATLAB with the RK4 and Euler methods with different step sizes.

2 Materials and Method

This paper discussed solving the Influenza Infection model using ode45 in MATLAB. This model was compared with numerical methods like the RK4 and Euler. The comparison between these two methods is conducted to observe a good approximation of the behaviour of each of the SEIR variables.

2.1 Influenza Infection Model

The outbreak of Influenza in Australia was investigated to anticipate the rise of infection levels. The compartment population encountered four species of individuals: $S(t)$, $E(t)$, $I(t)$, and $R(t)$. The Influenza model was a given as first order nonlinear ordinary differential equation as follows (Samsuzzoha Singh et al. 2013):

$$S'(t) = -\beta\frac{IS}{N} - \mu S + rN + \delta R \tag{1}$$

$$E'(t) = \beta\frac{IS}{N} - (\mu + \sigma + \kappa)E \tag{2}$$

$$I'(t) = \sigma E - (\mu + \alpha + \gamma)I \tag{3}$$

$$R'(t) = \kappa E + \gamma I - \mu R - \delta R \tag{4}$$

where Table 1 described the variables of the influenza model, $S(t)$, $E(t)$, $I(t)$, $R(t)$ and Table 2 showed the parameters of influenza model [6, 14]. The initial conditions were taken from [6] which are $S(0) = 0.4865$, $E(0) = 0.0000009$, $I(0) = 0.000068$, $R(0) = 0$ with total population, $N = S(0) + E(0) + I(0) + R(0)$ and time interval, $0 \le t \le 70$.

Table 1. Describing the variables of the influenza model [14]

Variables	Description
$S(t)$	Proportion of susceptible population
$E(t)$	Proportion of exposed population
$I(t)$	Proportion of infected population
$R(t)$	Proportion of recovered population

Table 2. Parameter value for influenza infection model [6, 14].

Parameter	Description	Value of Parameters
β	Contact rate	0.6300000
μ	Natural mortality rate	0.0003671
r	Birth rate	0.0006762
δ	Duration of immunity loss	0.0027400
σ	Mean duration of latency	0.6990000
κ	Recovery rate of latent	0.0001500
α	Flu induced mortality rate	0.1800000
γ	Mean recovery time for clinically ill	0.1000000

2.2 Approximation Method

Runge-Kutta Fourth Order Method

The RK4 methods were developed by two German Mathematicians Carl Runge (1856–1927) and Wilhelm Kutta (1867–1944). The Runge-Kutta method was widely regarded as the most dependable and accurate method for solving initial value problems involving linear and nonlinear ordinary differential equations, producing the best outcomes. This study required four steps to solve the step at the final shape. Runge-Kutta can be expressed generally as below [12, 15],

$$y_{i+1} = y_i + \frac{1}{6}(k_1 + 2k_2 + 2k_3 + k_4) \tag{5}$$

The general formula of k_1, k_2, k_3, k_4 for the RK_4 method [15] is:

$$k_1 = hf_1(t_i, y_i) \tag{6}$$

$$k_2 = hf_2(t_i + \frac{h}{2}, y_i + \frac{k_1}{2}) \tag{7}$$

$$k_3 = hf_3(t_i + \frac{h}{2}, y_i + \frac{k_2}{2}) \tag{8}$$

$$k_4 = hf_4(t_i + \frac{h}{2}, y_i + \frac{k_3}{2}) \tag{9}$$

where f_1, f_2, f_3, f_4 were unknown functions, t is a time and h is a step size which $h =$ 0.01, 0.1 and 0.25 [3].

The general formula for RK4 for SEIR as shown below. The general formula of RK4 was used to solve the system at Eq. (5) and produce the numerical results. S_{i+1}, E_{i+1}, I_{i+1} and R_{i+1} for the RK4 method were:

$$S_{i+1} = S_i + \frac{1}{6}(k_{S1} + 2k_{S2} + 2k_{S3} + k_{S4}) \tag{10}$$

$$\mathrm{E_{i+1}} = \mathrm{E_i} + \frac{1}{6}(\mathrm{k_{E1}} + 2\mathrm{k_{E2}} + 2\mathrm{k_{E3}} + \mathrm{k_{E4}}) \tag{11}$$

$$\mathrm{I_{i+1}} = \mathrm{I_i} + \frac{1}{6}(\mathrm{k_{I1}} + 2\mathrm{k_{I2}} + 2\mathrm{k_{I3}} + \mathrm{k_{I4}}) \tag{12}$$

$$R_{i+1} = R_i + \frac{1}{6}(k_{R1} + 2k_{R2} + 2k_{R3} + k_{R4}) \tag{13}$$

where $i = 0,1, ..., n - 1$.

Euler Method

Euler's method is the most straightforward one-step method. The explicit numerical integration of ordinary differential equations is simple. In 1768, Euler developed his method for the initial value problem (IVP). The general formula for Euler approximation is [1]:

$$S_{i+1} = S_i + \Delta t\left(\frac{-\beta I_i S_i}{N}\right) - \mu S_i + rN + \delta R_i \tag{14}$$

$$E_{i+1} = E_i + \Delta t(\frac{-\beta I_i S_i}{N}) - (\mu + \sigma + \kappa)E_i \tag{15}$$

$$I_{i+1} = I_i + \Delta t(\sigma E_i - (\mu + \alpha + \gamma I_i)) \tag{16}$$

$$R_{i+1} = R_i + \Delta t(\kappa E_i + \gamma I_i + \mu R_i + \delta R_i) \tag{17}$$

where:

S_i, E_i, I_i, R_i= susceptible, exposed, infected, and recovered population.
$S_{i+1}, E_{i+1}, I_{i+1}, R_{i+1}$= sequence of susceptible, exposed, infected, and recovered population.

We use Δt as step size in the time interval which h = 0.01,0.1,0.25 [3].

3 Results and Discussion

Numerical solutions for the nonlinear influenza model were discussed in this section, as well as the simulation of the SEIR Model for the spread of the Influenza Infection model with initial conditions. Next, the Influenza Infection model was solved using ode45 in MATLAB with RK4 and Euler methods with different step sizes, *h*. The results obtained were compared with the SEIR model using ode45 in MATLAB with approximate methods, the RK4 and Euler methods, to assess the good approximation to the behaviour of each of the SEIR variable for each method.

3.1 Simulation of SEIR Model of Influenza

The analysis for spreading of Influenza disease for Australia was done with the simulation of SEIR Model using ode45 in MATLAB software. For the initial value problem and the value of parameter is referred to Sect. 2.1 and Table 2. The population epidemic in Australia 1919 was studied to predict the progression of disease levels.

Figure 1 demonstrated that the curve was sharp before it began to fall after 30 days in the proportion that is susceptible to population *S(t)*. According to the [10], Australia was involved in World War I early in 1919. While the curve was low, it began to rise after

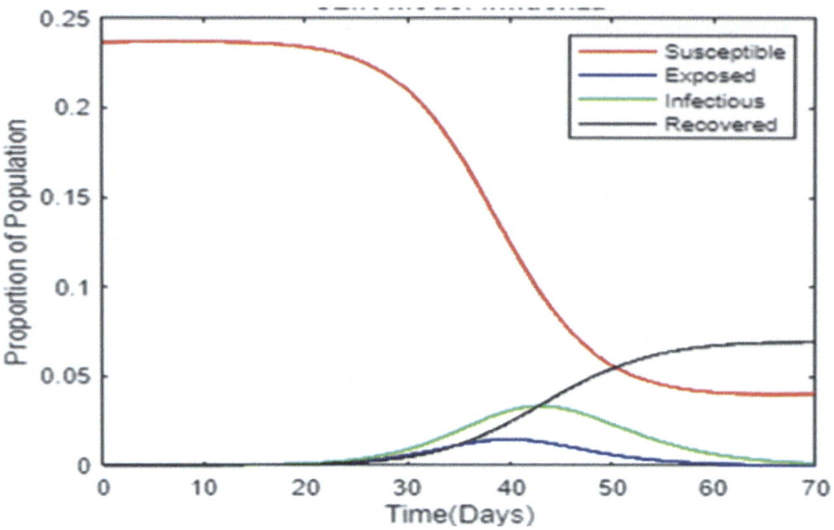

Fig. 1. Time series of SEIR model

20 days for the proportion of the exposed population, $E(t)$, and the proportion of the infected population, $I(t)$. At the time, this disease affected and exposed most troopers and soldiers quickly spread throughout military camps. While the curve for infected disease was higher, the curve for the proportion of recovered, $R(t)$ grew 30 days after the proportion of exposed to prevent the illness before starting in numerous numbers of infections by the disease. During World War I, Commonwealth Serum Laboratories were founded to reduce Australia's dependence on imported vaccinations.

While the curve for the proportion of exposed and recovered population began slowly, the curve for the proportion of recovered population started rising after 30 days, indicating that those most susceptible would be infected with influenza epidemics in the coming years. These SEIR models were examined using the RK4 order method and the Euler method.

3.2 Approximations Method

This section illustrated the problem of the system of first order nonlinear ordinary differential equations (ODEs) that had been considered. Figure 2(a) until 2(d) described the Influenza Infection model using ode45 in MATLAB compared with approximate methods which were RK4 order and Euler method. The problem was tested with three different step sizes, h to determine good approximation to the behaviour of each of the SEIR variable for each method. The step sizes that were considered are when $h = 0\ .1$, 0.25 to ensure approximation to the behaviour of each of the SEIR variable with time interval, days and were compared with a smaller step size, $h = 0.01$.

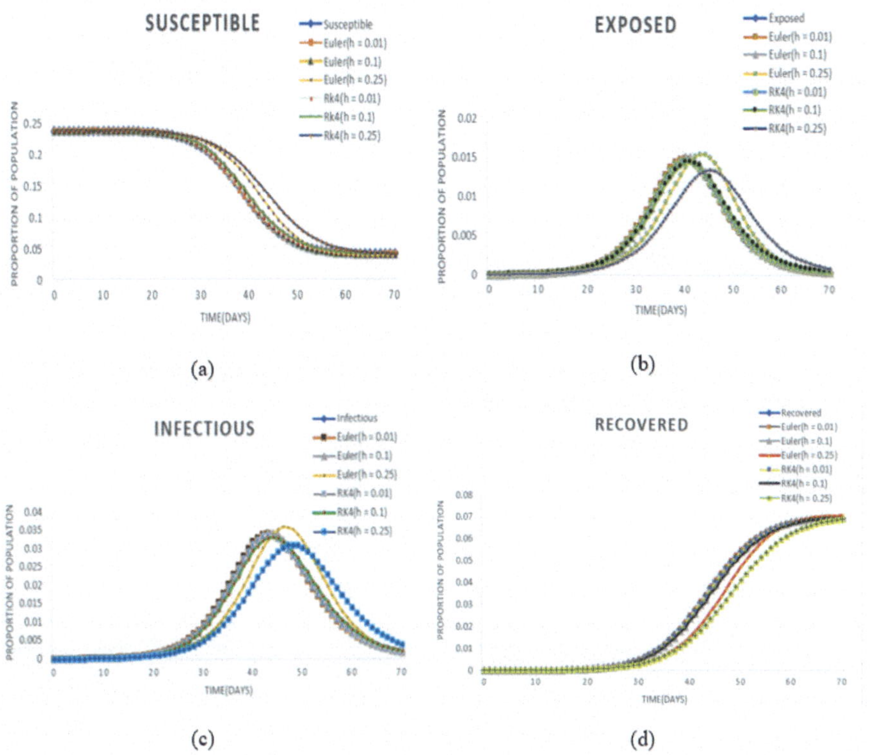

Fig. 2. Proportion of SEIR Model of population between approximations method with different step sizes (a) Susceptible; (b) Exposed; (c) Infected; (d) Recovered.

3.3 Error Analysis

The numerical error is the difference between the SEIR model using ode45 in MATLAB and the approximate method. In this study, the SEIR model compared with approximate methods, which were RK4 and Euler method with different step sizes. The approximation to the behaviour of each of the SEIR variable of the solution depended on the step size, as shown in this study $h = 0.01, 0.1$ and 0.25.

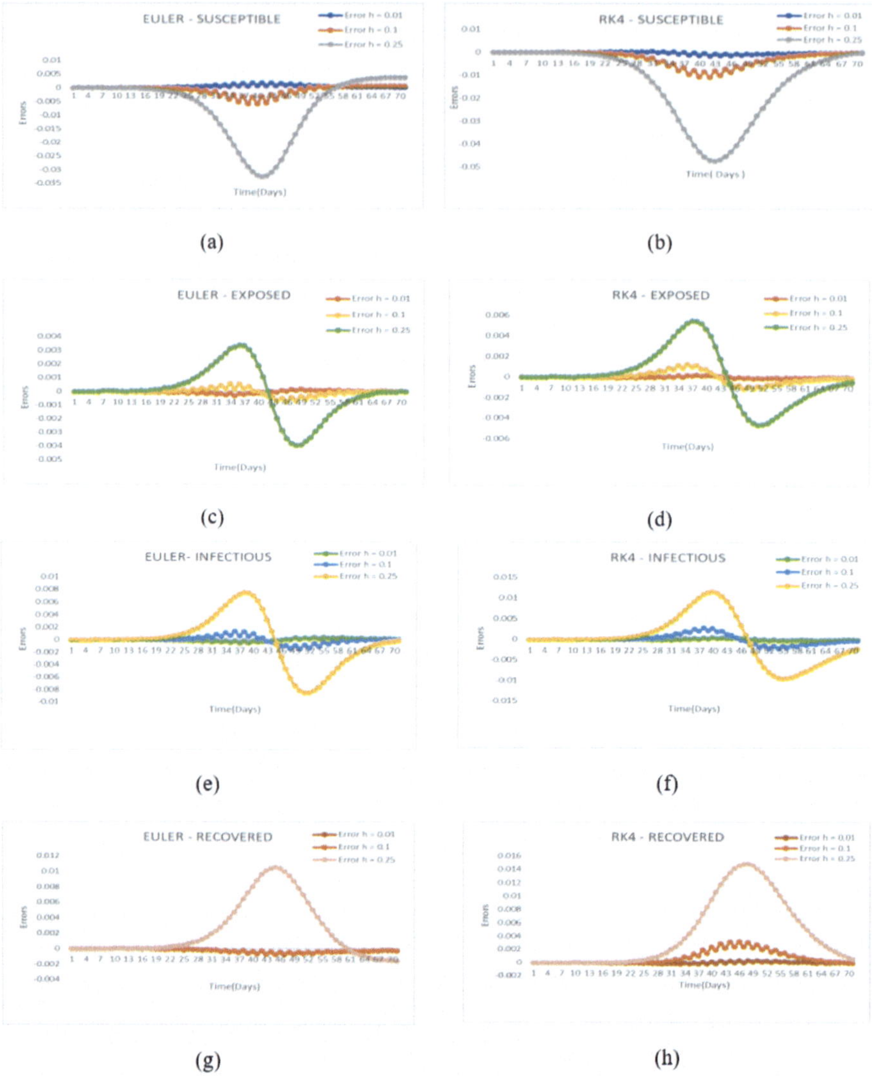

Fig. 3. Errors of various step sizes (a) Euler - Susceptible; (b) RK4 – Susceptible; (c) Euler – Exposed; (d) RK4 - Exposed; (e) Euler - Infected; (e) RK4 – Infected; (f) Euler - Recovered; (g) RK4 – Recovered

3.4 Discussions

The outcomes that were obtained were demonstrated in Fig. 2 for the SEIR model using ode45 in MATLAB. The model was compared with the approximate method which RK4 and Euler method. The numerical approximates solution was compared with step size $h = 0.01, 0.1, 0.25$ as well as errors being evaluated at the step size on the time interval [0,70]. For every method from the above Fig. 3(a) until (h), we stated that an approximate solution is to observe performance of behaviour for each SEIR model using ode45 in MATLAB if step size decreases error in such a way that in limiting case when the step size goes to 0, the error also tends to 0.

We observed that the Euler method with step size 0.1 and 0.25 did not close to SEIR model using ode45 in MATLAB. Nevertheless, for Euler cases in Fig. 3(g) it closer at $h = 0.01$. We observed the comparison between the Runge-Kutta method, it is slightly different because the Runge-Kutta method was a good approximation to the behaviour of each of the SEIR variable. We observed that the RK4 method in Fig. 3(a) until (h) with step size 0.01 nearest from the start to SEIR model using ode45 in MATLAB. The smallest step size $h = 0.01$ for RK4 method fewer errors compared to the Euler method. Therefore, the curve of RK4 method was nearest to SEIR model using ode45 in MATLAB. It is proved that the RK4 method was a good approximation to the behaviour of SEIR variable compared to the Euler method. Finally, we observed that the RK4 method is more closely approximated to the SEIR model than the Euler method, making it a good approximation approach for solving initial value problems associated with ordinary nonlinear differential equations.

4 Conclusion

This research emphasized solving the influenza infection model using ode45 in MATLAB comparing it with the RK4 and Euler methods. The outcomes were then compared to the SEIR model using ode45 in MATLAB and an approximation method with different step sizes. The results were calculated using MATLAB software and revealed that a smaller step size provided a better approximation to the behaviour of SEIR variable. In contrast, a more significant step size provided a less accurate approximation, evident in both the behaviour of the SEIR variables and the associated errors. We observed that the approximations method for the RK4 order method is most close to the SEIR model using ode45 in MATLAB and proved that these methods are one of the most good approximation approach of solving the initial value problem.

Acknowledgement. The authors are very grateful to Universiti Tun Hussein Onn Malaysia, specifically the Department of Mathematics and Statistics in the Faculty of Applied Sciences and Technology, for providing the research facilities and resources.

References

1. Aakash, M., Gunasundari, C., Al-Mdallal, Q.M.: Mathematical modeling and simulation of SEIR model for COVID-19 outbreak: a case study of Trivandrum. Front. Appl. Math. Stat. **9**, 1124897 (2023)
2. Ashgi, R., Pratama, M.F., Purwani, S.: Comparison of numerical simulation of epidemiological model between Euler Method with 4th Order Runge Kutta Method. Int. J. Glob. Oper. Res. **2**(1), 37–44 (2021)
3. Brian, D.S.: Numerical Methods for Ordinary Differential Equations (2006). https://www2.atmos.umd.edu/~rjs/class/fall2014/readings/storey_ode_matlab.pdf. Accessed 4 Mar 2025

4. Centers for Disease Control and Prevention (CDC). https://www.cdc.gov/mmwr/preview/mmwrhtml/mm5816a7.htm. Accessed 17 Sept 2024
5. Dobrushkin, V.: Applied Differential Equations with Boundary Value Problems, 1st edn. Chapman and Hall/CRC, Boca Raton (2017)
6. Fatehah, S.: Simulation Analysis and Forecasting of Covid-19 spreading in Malaysia. University Tun Hussein Onn Malaysia (2023)
7. Goswami, N., Shanmukha, B.: A mathematical model of influenza: stability and treatment. In: Proceedings of the International Conference on Mathematical Modeling and Simulation. BHU, Varanasi (2016)
8. Ibraheem, K., I.: Optimization algorithm based on the Euler method for solving fuzzy nonlinear equations. East.-Eur. J. Enterp. Technol. **4**(115), 13–19 (2022)
9. Kanyiri, C.W., Mark, K., Luboobi, L.: Mathematical analysis of influenza a dynamics in the emergence of drug resistance. Comput. Math. Methods Med. (2018)
10. National Museum of Australia. https://www.nma.gov.au/defining-moments/resources/influenza-pandemic. Accessed 28 Sept 2022
11. Nurujjaman, M.: Enhanced Euler's method to solve first order ordinary differential equations with better accuracy. J. Eng. Math. Stat. **1**(4), 1–13 (2020)
12. Osmani, S.A.B.: Study of numerical solution of ordinary differential equation by Taylor, Euler and Runge-Kutta methods. Int. J. Math. Phys. Sci. Res. **7**(2) (2019)
13. Rabiei, F., Ismail, F., Suleiman, M.: Improved Runge-Kutta methods for solving ordinary differential equations. Sains Malaysiana **42**(11), 1679–1687 (2013)
14. Samsuzzoha, M., Singh, M., Lucy, D.: Parameter estimation of influenza epidemic model. Appl. Math. Comput. **220**, 616–629 (2013)
15. Tay, K.G., Kek, S.L., Han, C.T., Abdul-Kahar, R., Lee, M.F.: The fourth Order Runge-Kutta Spreadsheet calculator using VBA programing for ordinary differential equations. Procedia. Soc. Behav. Sci. **204**, 231–239 (2015)
16. Zarebski, A.E., Dawson, P., McCaw, J.M., Moss, R.A.: Model selection for seasonal influenza forecasting. Infect. Dis. Model. **2**(1), 56–70 (2017)

Modeling and Analysis of Incommensurate Fractional-Order SIR Model with Caputo Derivative

Chang Phang[1(✉)], Jun Hao Tan[1], and Abdulnasir Isah[2]

[1] Universiti Tun Hussein Onn Malaysia, Pagoh Campus, 84600 Pagoh, Johor, Malaysia
pchang@uthm.edu.my

[2] Universiti Teknologi Malaysia, 81310 Skudai, Johor, Malaysia

Abstract. Many mathematical models are formulated by involving fractional differential equations with Caputo derivative. It is widely known that the analytical solutions are complicated to obtain or there may no exact solution in solving the fractional differential equations with Caputo derivative, especially when involving incommensurate fractional-order model. Hence, numerical methods are needed. In this paper, we are interested in solving the incommensurate fractional-order SIR model that involves Caputo derivative. Fractional Runge-Kutta method was used to solve the fractional-order SIR model with the aid of MAPLE software. Subsequently, the solution of different fractional-order SIR models was obtained by changing the values of Caputo fractional parameters and then compared to the real data of COVID-19 cases. It has been shown that the modified incommensurate fractional-order model achieves better results for fitting the real data compared to the other models. The application of the Caputo derivative to the mathematical problem is reliable and can be utilized for other practical problems.

Keywords: Caputo Derivative · Incommensurate Fractional-order SIR Model · Fractional Runge-Kutta Method · MAPLE Software

1 Introduction

Fractional calculus cannot be regarded as a new topic because it has a history of more than 300 years and is similar to the familiar classical calculus. Fractional calculus can be defined as differentiation and integration with arbitrary order, either complex or real [1]. The Riemann-Liouville fractional integral usually represents the fractional integration while there are various definitions of fractional derivatives, for instance, Riemann-Liouville fractional derivative and Caputo fractional derivative. When modelling real-world problems, the Caputo definition is more preferred by the researchers as the derivative of any constant is zero and the initial value condition of the fractional differential equation is similar to that of the integer differential equation [2]. In contrast, the Riemann-Liouville derivative of a constant is not zero and it requires fractional initial conditions.

P. Chang et al. (Eds.): SCIEMATHIC 2024, SPPHY 427, pp. 13–23, 2025.
https://doi.org/10.1007/978-981-96-7679-8_2

In the past few decades, many researchers have applied fractional calculus to many mathematical models to solve real problems in the world such as economic growth modelling, fluid mechanics, financial economics and epidemiology [3–6]. In certain phenomena, the mathematical model with the fractional-order operators performed better than the integer-order operators as the derivative is not limited to the integer-order derivative only. Mathematical models are tools formulated based on the problems that arise in science as well as engineering and are used to solve the problems in real life. As an example, in the field of epidemiology, recently, the SIR (Susceptible-Infectious-Recovered) model had been widely used by researchers or authorities around the world to have a deeper understanding of the transmission of COVID-19 disease. Moreover, they also predict the number of infections and deaths, gained significant insight from the SIR model, thereby formulating effective strategies to control the COVID-19 pandemic. However, some researchers pointed out that the performance of the SIR model can be improved by generalizing the integer-order into fractional-order.

It is known that for the problems that are related to the real world, there may not be a precise solution or the process of solving the problems is complicated, especially when involving fractional derivatives. Therefore, a numerical method such as the fractional Runge-Kutta method plays an important role to solve the fractional differential equations numerically. The fractional Runge-Kutta method is a single-step method derived from the classical Runge-Kutta method and is especially used for solving fractional differential equations [7]. When the single-step method is used, only a previous point and derivative will be referenced to determine the next iteration's value. Before proceeding to the next step, the previous information will be discarded. On the other hand, the multi-step method refers to several previous points will be used to determine the next iteration's value [8]. In other words, the multi-step method will retain and use the information from previous steps instead of discarding it. However, the multi-step methods have a more expensive computational cost.

Hence, this study was constructed to solve the fractional SIR model with Caputo derivative by the fractional Runge-Kutta method. Subsequently, the time series plot for the fractional SIR model with various types has been plotted using the MAPLE software. Then, the SIR prediction models with different types were compared to the real data of COVID-19 cases for Malaysia and China.

2 Methodology

2.1 Fractional Integrals and Fractional Derivatives

In a case of function f which defined on an interval $G = [a, b](-\infty < a < b < \infty)$ and $0 < \alpha < 1$, the Riemann-Liouville left-sided fractional integral, $I_{a+}^{\alpha} f$ and right-sided fractional integral, $I_{b-}^{\alpha} f$ of order $\alpha \in \mathbb{C}(\Re(\alpha) > 0)$ are defined respectively by

$$\left(I_{a+}^{\alpha} f\right)(y) = \frac{1}{\Gamma(\alpha)} \int_{a}^{y} \frac{f(t)dt}{(y-t)^{1-\alpha}} \quad (1)$$

and

$$\left(I_{b-}^{\alpha} f\right)(y) = \frac{1}{\Gamma(\alpha)} \int_{y}^{b} \frac{f(t)dt}{(t-y)^{1-\alpha}} \quad (2)$$

Here, $\Gamma(\alpha)$ is the widely known Gamma function. If α is an arbitrary order, the Riemann-Liouville left-sided fractional derivative, $D^{\alpha}_{a+}f$ and right-sided fractional derivative, $D^{\alpha}_{b-}f$ of order $\alpha \in \mathbb{C}(\Re(\alpha) > 0)$ are defined by

$$(D^{\alpha}_{a+}f)(y) = \left(\frac{d}{dy}\right)^n (I^{n-\alpha}_{a+}f)(y) \tag{3}$$

$$(D^{\alpha}_{b-}f)(y) = \left(-\frac{d}{dy}\right)^n (I^{n-\alpha}_{b-}f)(y) \tag{4}$$

respectively, where $n = [\Re(\alpha)] + 1$ and $[\Re(\alpha)]$ is the greatest integer that less than or equal to $\Re(\alpha)$. While the left-sided and right-sided Caputo derivatives are, respectively, defined by

$$\left({}^C D^{\alpha}_{a+}f\right)(y) = \left(I^{n-\alpha}_{a+}f^{(n)}\right)(y) = \frac{1}{\Gamma(n-\alpha)}\int_a^y \frac{f^{(n)}(t)dt}{(y-t)^{1-n+\alpha}} \tag{5}$$

and

$$\left({}^C D^{\alpha}_{b-}f\right)(y) = (-1)^n\left(I^{n-\alpha}_{b-}f^{(n)}\right)(y) = \frac{(-1)^n}{\Gamma(n-\alpha)}\int_y^b \frac{f^{(n)}(t)dt}{(t-y)^{1-n+\alpha}}. \tag{6}$$

In this paper, only the left-sided fractional derivatives and fractional integrals were utilized since the function of left-sided operators are to describe the process past state while the process future state is described by the right sided-operators [9]. When the current condition of the process does not depend on the outcomes of the future development, the right-sided operators are normally ignored in applications.

2.2 Fractional Runge-Kutta Method

Once in a while, certain ordinary differential equations cannot be solved exactly. Therefore, the Runge-Kutta method is a numerical method or also known as a single-step method that is used to obtain the numerical solutions for the ordinary differential equations. The advantages of using the Runge-Kutta method are very stable and simple to carry out. However, the calculation for the high order Runge-Kutta method will take a longer time than that of the multi-step method. The fractional Runge-Kutta method had been introduced in [10] to solve the fractional differential equation of type:

$$D^{\alpha}y(t) = f(t, y(t)), \quad 0 < \alpha \leq 1. \tag{7}$$

where the initial condition is $y(t_0) = y_0$ and D^{α} represents the Caputo derivative operator. The formula of the fourth-order fractional Runge-Kutta is as follows:

$$y_{n+1} = y_n + (K_1 + 2K_2 + 2K_3 + K_4), \tag{8}$$

with:

$$\begin{aligned} K_1(h) &= S_1 f(t_n, y_n), \\ K_2(h) &= S_1 f(t_n + S_1, y_n + K_1), \\ K_3(h) &= S_1 f(t_n + S_1, y_n + K_2), \\ K_4(h) &= S_1 f(t_n + S_1, y_n + K_3) \end{aligned} \tag{9}$$

where $S_1 = \frac{h^{\alpha}}{\Gamma(\alpha+1)}$ and h is the stepsize.

2.3 Fractional SIR Model

The SIR model is a mathematical model that is used to have a greater understanding of how the diseases spread among the population in the community [11]. It is not only can be used to anticipate the number of future infections and deaths, but it can also provide insights for the competent authorities to take action to control infectious diseases. The performance of the classical SIR model can be improved by generalizing the integer-order differential equations to fractional-order [12]. The fractional-order derivative is typically defined in several ways such as Riemann-Liouville or Caputo, and it is given in the form of D^{α}, where $\alpha \in \mathbb{R}$. The fractional SIR model [13] is presented in the form

$$D^{\alpha_1}S(t) = \frac{-\beta S(t) I(t)}{N},$$

$$D^{\alpha_2}I(t) = \frac{\beta S(t) I(t)}{N} - \gamma I(t),$$

$$D^{\alpha_3}R(t) = \gamma I(t), \tag{10}$$

where the initial conditions are

$$S(0) = S_0,\ I(0) = I_0, R(0) = R_0. \tag{11}$$

Here, $S(t)$, $I(t)$ and $R(t)$ represents the number of susceptible individuals, active infectious individuals and recovered individuals at time t respectively. N represents the total population size while the β and γ are real positive constant, which are known as the infection rate and recovery rate. The Caputo fractional-order differential operators with order α_1, α_2 and α_3 are denoted respectively by D^{α_1}, D^{α_2} and D^{α_3}, while $0 < \alpha_i \leq 1$ for $i = 1, 2, 3$.

In this study, four types of SIR prediction models were simulated by considering different conditions. The types of SIR prediction models were categorized as integer-order, commensurate fractional-order, incommensurate fractional-order and modified incommensurate fractional-order. For the integer-order model, the values of the Caputo fractional parameters were set to an integer value of one. The simulation result of the commensurate fractional-order model was obtained by changing the Caputo fractional parameters' values into a fractional value with the assumption such that $\alpha_1 = \alpha_2 = \alpha_3$ [14]. Otherwise, the model is considered as incommensurate fractional-order when at least one of the Caputo fractional parameters does not have an identical value to the other. We had modified the incommensurate fractional-order model by splitting the time interval into two phases. For the active infectious cases, we split the time interval by assuming the maximum point of the real data plotted in the graph as the cut-point. While for the recovered cases, the middle point of the steepest part from the curve would be considered as the cut-point. Then we find the suitable values of the Caputo fractional parameters that can best fit the real data for each phase.

2.4 Plotting of Time Series Plot via MAPLE

The aforementioned fractional Runge-Kutta method was applied to solve the fractional SIR model numerically. Then, the time series plot has been plotted using the solution obtained by the fractional Runge-Kutta method with the aid of MAPLE software.

MAPLE is a computer algebra system and it is widely used by mathematicians, educators and students [15]. MAPLE consists of various useful features and commands in doing the mathematical task. It has the ability to do exact integer computation as well as numerical computation to mathematical problems. Furthermore, users can also use MAPLE for two-dimensional or three-dimensional plotting. MAPLE can be considered as a simple programming language as the user can easily master the skill of using the MAPLE.

The fractional SIR model prediction was simulated using the real data of COVID-19 cases for Malaysia and China. In this study, the size of interval was set as $h = 0.05$ to find the approximate solution of the initial value problem. Through the fractional Runge-Kutta method, we can determine the next value (y_{n+1}) by adding the current value (y_n) and the weighted average of four increments, where each increment is the product of the estimated slope and the size of the interval, h. Then, we use the approximation as the starting point and repeat the process to generate a solution over time. Last but not least, the time series plot was plotted using the solution obtained by the fractional Runge-Kutta method via MAPLE software. The time series plot for various types of the SIR prediction models such as integer-order, commensurate fractional-order, incommensurate fractional-order and modified incommensurate fractional-order were plotted by changing the value of parameters α_1, α_2 and α_3.

2.5 Data Used

The real data used for the simulation was obtained from the Worldometer website [16]. The data collected for Malaysia was started from January to May, 2020 while the data collected for China was started from January to June, 2020. The data are tabulated in the form of time series where the rows are recorded in time, and the three columns are the total number of infected cases, number of active infectious cases and number of death cases. Consequently, the number of recovered cases can be estimated from the data by subtracting the number of active infectious cases and death cases from the total number of infected cases.

3 Results and Discussion

The values of the parameters N, β and γ that have been obtained via the Nelder-Mead method found in [17] is utilized because the real data used in this study was the same as what they used. Then, we stimulate the prediction model using the SIR model and then find the suitable value of the Caputo fractional parameters that best fit the real data of COVID-19 cases for Malaysia and China by trial-and-error approach. This trial-and-error approach has been applied in [11], where the researchers adjusted the parameters manually to fit the real data as best as possible. Lastly, we compared the model fitting to the real data. In this study, we only analyze the model fitting of active infectious population and recovered population to the real data of COVID-19 cases as it is more useful in providing significant insights on containing the pandemic.

Table 1. Caputo fractional parameters for prediction model of active infectious case for Malaysia

Types of models	α_1	α_2	α_3
Integer-order	1	1	1
Commensurate fractional-order	0.95	0.95	0.95
Incommensurate fractional-order	1	0.98	1
Modified incommensurate fractional-order (Before day 70)	1	0.98	1
Modified incommensurate fractional-order (After day 70)	1	1	1

Table 2. Caputo fractional parameters for prediction model of recovered case for Malaysia

Types of models	α_1	α_2	α_3
Integer-order	1	1	1
Commensurate fractional-order	0.95	0.95	0.95
Incommensurate fractional-order	1	1	0.95
Modified incommensurate fractional-order (Before day 75)	1	1	1
Modified incommensurate fractional-order (After day 75)	1	1	0.88

3.1 Results

The values of the Caputo fractional parameters used for the prediction model of the active infectious case and recovered case for Malaysia are presented in Table 1 and Table 2 respectively. The simulation result was obtained using these parameters values: $N = 5700$, $\beta = 0.1488$, $\gamma = 0.0354$ [17]. The simulation result for the fractional-order SIR model is presented in Fig. 1 below.

Based on Fig. 1, for the active infectious case, the performance of the modified incommensurate fractional-order was almost identical to that of incommensurate fractional order. The modified incommensurate fractional model prediction was simulated by dividing the time interval into two ranges by assuming day 70 as the cut-point. Before day 70, the model was able to reach the highest number of cases of the real data due to the decrease in value of the parameter α_2. After day 70, the values of parameters α_1, α_2 and α_3 was set to an integer value of one as the changing of the parameters will cause the curve to deviate more from the actual line. Hence, both incommensurate and modified incommensurate fractional-order models are better in predicting the peak situation of COVID-19 than the integer-order model in this case.

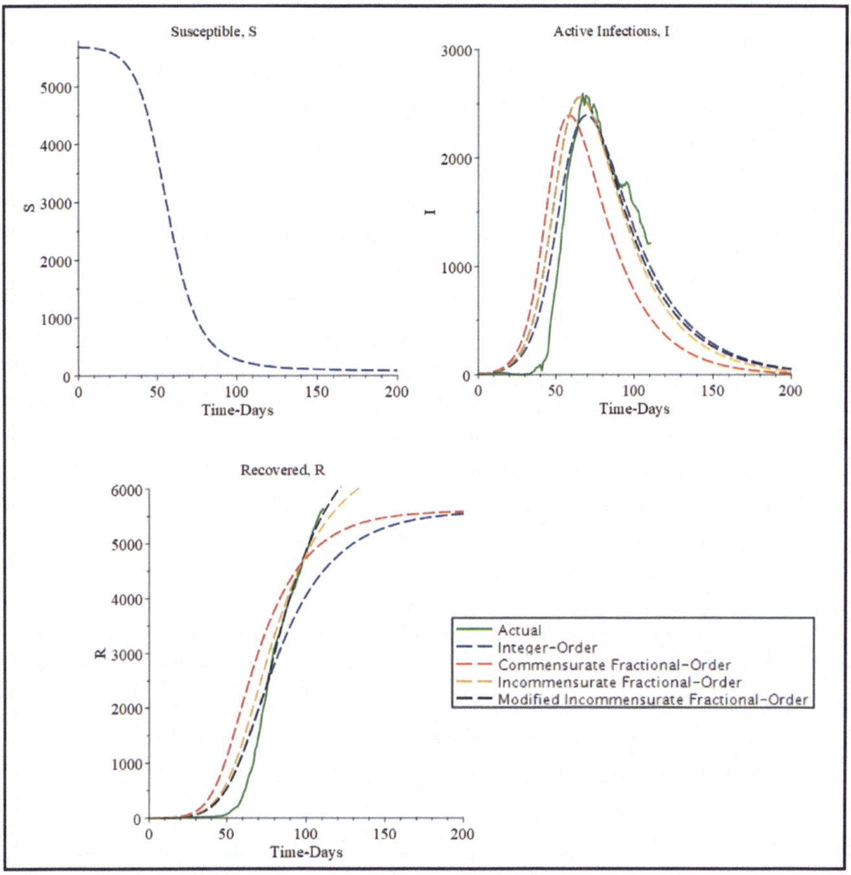

Fig. 1. Malaysia: Model prediction for the period from 30 January to 16 August, 2020 with data from January to May, 2020.

Meanwhile, for the recovered case, the modified incommensurate fractional-order model also shows the best fitting to the real data among the four models. The time interval was divided by a cut-point which is day 75. The values of parameters α_1, α_2 and α_3 was the same which was equal to one before day 75 and the parameter value of α_3 has been decreased to 0.88 after that so that the fitting of the model gets close and close to the real data. This indicates that the fractional-order is preferable rather than the integer-order in this case. Then followed by the incommensurate fractional-order model, it also provides a good fitting compared to the models of commensurate fractional-order and integer-order by decreasing the value of the parameter α_3 only.

On the other hand, the simulation of the SIR prediction model for China was also conducted using these parameters values: $N = 86000$, $\beta = 0.350$, $\gamma = 0.035$ [17]. Figure 2 depicts the model prediction up to 200 days.

The simulation of the prediction model of the active infectious case of China was carried out using the Caputo fractional parameters as shown in Table 3 while Table 4

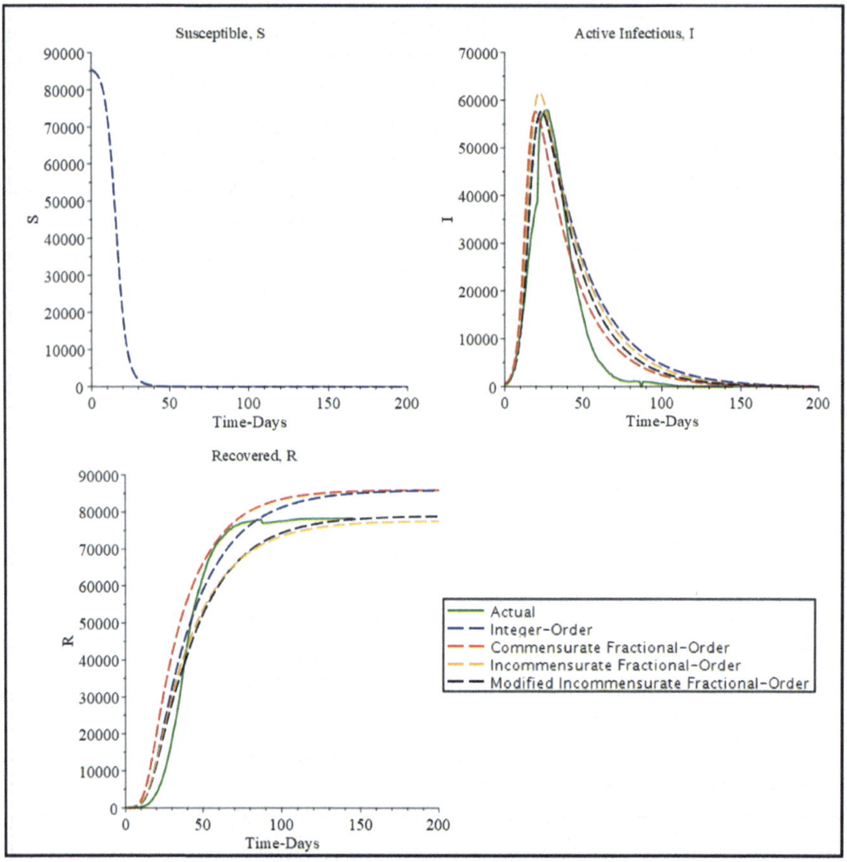

Fig. 2. China: Model prediction for the period from 22 January to 8 August, 2020 with data from January to June, 2020.

presents the values of the Caputo fractional parameters that provided a better fitting to the real data for recovered cases in China.

Table 3. Caputo fractional parameters for prediction model of active infectious case for China

Types of models	α_1	α_2	α_3
Integer-order	1	1	1
Commensurate fractional-order	0.95	0.95	0.95
Incommensurate fractional-order	1	0.98	1
Modified incommensurate fractional-order (Before day 25)	1	1	1
Modified incommensurate fractional-order (After day 25)	1	0.95	1

Table 4. Caputo fractional parameters for prediction model of recovered case for China

Types of models	α_1	α_2	α_3
Integer-order	1	1	1
Commensurate fractional-order	0.95	0.95	0.95
Incommensurate fractional-order	0.95	1	0.98
Modified incommensurate fractional-order (Before day 38)	0.95	1	1
Modified incommensurate fractional-order (After day 38)	1	1	0.95

As can be seen from Fig. 2, for China, the performance of the modified incommensurate fractional-order model in fitting the data of active infectious seemed better compared to other models as it gets closer to the real data. Before day 25, the values of parameters α_1, α_2 and α_3 have been fixed into an integer value of one as it found that the changing of the parameter's value from an integer value to a fractional value will lead to a higher deviation from the actual data. After day 25, the values of parameters α_1, α_2 and α_3 have been changed to 1, 0.95 and 1 respectively so that it shows a good fitting. This indicates that the utilization of a fractional value for the parameter α_2 could reduce the deviation from the actual data.

In addition, the modified incommensurate fractional-order model was the best fitting model for the real data of the recovered cases for China as well. The simulation of the modified incommensurate fractional-order model for the recovered case was started by setting the Caputo fractional parameters as $\alpha_1 = 0.95$, $\alpha_2 = 1$, and $\alpha_3 = 1$ before day 38. It shows that the decrease in the value of the parameter α_1 enhanced the model performance in fitting the real data. After that, the values of the Caputo fractional parameters were changed to $\alpha_1 = 1, \alpha_2 = 1$, and $\alpha_3 = 0.95$ in order to get a better fitting to the actual line. Besides that, the model of incommensurate fractional-order also demonstrates some good fitting for the data as it was not much different to the modified incommensurate fractional-order.

3.2 Discussions

The results show that the condition of the COVID-19 pandemic in Malaysia and China was similar in the early period of this pandemic. There was a drastic rise in the number of infected individuals once the COVID-19 disease outbreak. After that, the number of infected individuals declined steadily as the government or relevant authorities aware of the situation and take action to control the pandemic by implementing the policies such as social distancing. While the number of recovered individuals increase over time due to the recovery of infected individuals. However, the results also show that China reaches their major outbreak earlier than Malaysia which is on day 25.

According to the simulation results of the SIR prediction model, it can be concluded that the modified incommensurate fractional-order model is the best model among the four models as it best fits the real data of COVID-19 cases for both countries. The advantage of using the modified incommensurate fractional-order model is the greater flexibility in determining the values for the Caputo fractional parameters as it can include both integer-order and fractional-order cases. As an example, for the active infectious case of China, the values of the Caputo fractional parameters α_1, α_2, and α_3 were equal to one as it fits the real data better before day 25. After that, the decreasing of the value of α_2 to 0.95 results in a better model fitting.

Unlike other models, the time range for the simulation of the modified incommensurate fractional-order model was split into two phases and then the most suitable parameters' values in each phase were determined by trial-and-error approach. In short, the Caputo fractional parameters were dependent on the trend of the actual data in each phase. Therefore, in order to improve the effectiveness of this model, further studies are required to determine the values of the Caputo fractional parameters that can be used for the future prediction of the COVID-19 cases.

In general, a model with a better model fitting to the real data would have a higher prediction accuracy. Thus, the application of Caputo fractional derivatives to the mathematical model has a significant improvement in the performance of solving real-world problems.

4 Conclusion

In conclusion, the epidemic models such as SIR model have been demonstrated to be an effective tool for predicting the number of COVID-19 cases and it is vital for the planning of preventive and control measures to contain the COVID-19 pandemic. However, the performance of the classical SIR model would be enhanced by applying fractional calculus to the model.

In this paper, the fractional SIR model was solved using the fractional Runge-Kutta method and the different types of SIR prediction models were compared to the real data of COVID-19 cases. This research had proved that the Caputo fractional parameters would boost the performance of the prediction model in fitting the real data in a particular situation, especially for the modified incommensurate fractional-order case proposed in this study. It can be said that if the model fits the actual data better, the accuracy of future forecasting will increase as well. Nevertheless, the prediction of the epidemic trends might not be useful if the citizens do not comply with the standard operating procedure (SOP) established by the government.

It is recommended for future researchers to study another epidemic model such as the SEIR (Susceptible-Exposed-Infectious-Recovered) model instead of the SIR model as the SEIR model has included the exposed group to the model, and then it would be more appropriate as a model for disease such as COVID-19. The exposed group includes individuals who have been exposed to the infection but are not themselves infectious yet. In addition, future researchers can also utilize a longer time frame of data so that the simulation results obtained are more reliable. Apart from that, it is worth to study other incommensurate model [18] or other fractional operators [19, 20] in SIR model to better predict the dynamic of infectious diseases.

References

1. Li, C., Qian, D., Chen, Y.: On Riemann-Liouville and caputo derivatives. In: Discrete Dynamics in Nature and Society, 2011 (2011)
2. Baleanu, D., Agrawal, O.P.: Fractional Hamilton formalism within Caputo's derivative. Czech J. Phys. **56**(10), 1087–1092 (2006)
3. Cheow, Y.H., Ng, K.H., Phang, C., Ng, K.H.: The application of fractional calculus in economic growth modelling: An approach based on regression analysis. Heliyon **10**(15) (2024)
4. Kulish, V.V., Lage, J.L.: Application of fractional calculus to fluid mechanics. J. Fluids Eng. **124**(3), 803–806 (2002)
5. Fallahgoul, H., Focardi, S., Fabozzi, F.: Fractional Calculus and Fractional Processes with Applications to Financial Economics: Theory and Application. Academic Press (2016)
6. Atangana, A.: Application of fractional calculus to epidemiology. In: Fractional Dynamics, pp. 174–190. De Gruyter Open Poland (2016)
7. Milici, C., Machado, J.T., Drăgănescu, G.: Application of the Euler and Runge-Kutta generalized methods for FDE and symbolic packages in the analysis of some fractional attractors. Int. J. Nonlinear Sci. Numer. Simul. **21**(2), 159–170 (2020)
8. Mei, Y., Mo, R., Sun, H., He, B., Wan, N.: Stability prediction in milling based on linear multistep method. Int. J. Adv. Manuf. Technol. **105**(5), 2677–2688 (2019)
9. Caputo, M.C., Torres, D.F.: Duality for the left and right fractional derivatives. Signal Process. **107**, 265–271 (2015)
10. Milici, C., Drăgănescu, G., Machado, J.T.: Introduction to Fractional Differential Equations, vol. 25. Springer, Heidelberg (2018)
11. Cooper, I., Mondal, A., Antonopoulos, C.G.: A SIR model assumption for the spread of COVID-19 in different communities. Chaos Solitons Fractals **139**, 110057 (2020)
12. Alshomrani, A.S., Ullah, M.Z., Baleanu, D.: Caputo SIR model for COVID-19 under optimized fractional order. Adv. Differ. Equ. **2021**(1), 1–17 (2021)
13. Kozioł, K., Stanisławski, R., Bialic, G.: Fractional-order SIR epidemic model for transmission prediction of COVID-19 disease. Appl. Sci. **10**(23), 8316 (2020)
14. Almuneef, A. A. S. (2021). *Fractional derivative models for the spread of diseases.* University of Strathclyde: Ph.D. Thesis
15. Garvan, F.: The Maple Book. CRC Press, Boca Raton (2001)
16. Coronavirus Cases: Coronavirus Worldometer Website. https://www.worldometers.info/coronavirus
17. Ishak, N.Z.: Simulation and prediction for COVID-19's number of cases with SIR model. Unpublished master's thesis, Universiti Tun Hussein Onn Malaysia, Johor, Malaysia (2021)
18. Ng, Y.X., Phang, C., Loh, J.R., Isah, A.: Analytical solutions of incommensurate fractional differential equation systems with fractional order $1< \alpha, \beta< 2$ via bivariate Mittag-Leffler functions. AIMS Math. **7**(2), 2281–2317 (2022)
19. Toh, Y.T., Phang, C., Ng, Y.X.: Temporal discretization for Caputo-Hadamard fractional derivative with incomplete Gamma function via Whittaker function. Comput. Appl. Math. **40**(8), 1–19 (2021)
20. Md Nasrudin, F.S., Phang, C., Kanwal, A.: Fractal-fractional advection–diffusion–reaction equations by Ritz approximation approach. Open Phys. **21**(1), 20220221 (2023)

Curve Fitting Technique on First-Order Linear Ordinary Differential Equation for Dynamic System Modelling

Sie Long Kek[1](✉), Fong Peng Lim[2], and Chong Mun Ho[3]

[1] Universiti Tun Hussein Onn Malaysia, Pagoh Campus, 84600 Parit Raja, Johor, Malaysia
slkek@uthm.edu.my

[2] Universiti Putra Malaysia, 43400 UPM Serdang, Selangor, Malaysia

[3] Universiti Malaysia Sabah, 88999 Kota Kinabalu, Sabah, Malaysia

Abstract. This paper discusses a curve-fitting technique based on the first-order linear ordinary differential equation model for predicting the solution of dynamic systems. First, a least squares optimization problem is defined to minimize the differences between the system and the model. It is assumed that the system's actual solution exists. Second, the model parameters are updated iteratively by applying the gradient method, in turn, to minimize the mean square errors and to determine the model's solution optimally. When convergence is achieved, the optimal parameters employed in the linear model give a best-fit solution to the system solution with a minimum mean square error value. Third, a chemical reaction model consisting of three nonlinear ordinary differential equations is studied for illustration. The simulation results show the accuracy of the curve-fitting solution to dynamic systems with a small mean square error value. Hence, the technique's efficiency is demonstrated, and the best fitting of the nonlinear dynamic system using the linear model is satisfied.

Keywords: Curve Fitting · Linear Ordinary Differential Equation · Mean Square Error · Dynamic System · Prediction

1 Introduction

Dynamic systems are often applied to deal with real-world problems in various fields of engineering and sciences. Since the behavior changes over time, dynamic systems can express the situations occurring in the following period for these problems. Thus, dynamic system modelling is vital to system design, optimization, and control [1]. However, we might not know the actual mathematical model for a dynamic system [2]. From this point of view, collecting data from an experiment or a process, even performing a data-driven analysis of a dynamic system, is necessary to understand and predict the solution of a dynamic system [3].

In dynamic modelling, the challenge is recognizing a system's nonlinear structure and identifying a stochastic system disturbed by random disturbances [4]. Many dynamic

P. Chang et al. (Eds.): SCIEMATHIC 2024, SPPHY 427, pp. 24–31, 2025.
https://doi.org/10.1007/978-981-96-7679-8_3

systems with nonlinear and random behavior exist in real-world problems. Hence, computational methods have been developed in the literature to handle these systems [5–7]. Nevertheless, solving a linear dynamic system expressed in a first-order linear ordinary differential equation (ODE) is relatively simple and easy. This fact provides a potential investigation of using a linear dynamic system for nonlinear dynamic system modelling [8, 9].

Assuming we only know the solution curve of a nonlinear dynamic system or the trend of collected data, where the actual mathematical model is unknown. We suggest a mathematical model to fit the solution curve by estimating the model parameters. We refer to this way as curve-fitting modelling [10]. In the literature, the curve-fitting technique is a statistical approach used to select the best model for a dataset [11]. Recently, curve fitting has been widely covered in neural networks, machine learning, and artificial intelligence applications [12, 13]. This approach enables the approximation of complex relationships and dynamic behaviors in systems, including handling high-dimensional data, capturing intricate system dynamics, and providing uncertainty quantification [14–16], where the exact mathematical form is not easily discernible.

Moreover, data-driven modelling methods such as physics-informed neural networks (PINNs) and symbolic regression have shown great potential in identifying governing equations and parameterizing models based on the observed data trends [17]. The ability to integrate domain knowledge with data-driven techniques has further enhanced the accuracy and interpretability of the models [18]. These developments underline the growing relevance of curve-fitting and data-driven modelling approaches in modern engineering and scientific applications. Therefore, the novel and improved techniques shall enhance the accuracy and efficiency of the fitting solutions, especially for complex, nonlinear and uncertain dynamic system modelling.

In this paper, we propose a computational algorithm based on the first-order linear ODE model for curve fitting in dynamic system modelling [19–21]. For this aim, we introduce a least square optimization problem, where the differences between the system and the model are measured, and the necessary conditions are derived. Using the gradient method, we estimate the value of the model parameters and minimize the differences between the system and the model [11]. After achieving convergence, we apply these optimal parameter estimates in the linear model to predict the system's solution curve. For illustration, we study a chemical reaction model. The simulation results show the accuracy and efficiency of the proposed method.

The paper is organized as follows. Section 2 describes the curve-fitting problem on the first-order linear differential equation model. Section 3 discusses the methodology used in the paper. The computational algorithm is summarized for managing curve fitting task. Section 4 illustrates an example of a chemical reaction model. Finally, a concluding remark is made.

2 Problem Description

Consider a general class of a dynamic system [19, 21],

$$\dot{x} = f(x(t), t), \tag{1}$$

where x is an n-vector of variables and f is a vector-valued function, denoted as $f : \Re^n \times \Re \to \Re^n$. While, $\dot{x}$ is the rate of change of the system over time t. Here, (1) can be written as

$$dx = f(x(t), t)dt, \tag{2}$$

and its solution is given by

$$x(t) = x(\tau) + \int_{\tau}^{t} f(x(t), t)dt, \tag{3}$$

for $t > \tau$. The initial condition is $x(0) = x_0$. We assume that the solution (3) exists, and the data-driven set points of the solution can be generated by using a built-in ODE solver.

To predict the solution (3), we consider a first-order linear ODE model [3, 20], given by

$$\dot{x} = Ax + b, \tag{4}$$

where A is an $n \times n$ coefficient matrix and b is an $n \times 1$ coefficient vector. A least square optimization problem [11] is defined as follows:

Problem (P): To minimize the objective function,

$$J(A, b) = \frac{1}{m}(x - \hat{x})^{\mathrm{T}}(x - \hat{x}) \tag{5}$$

subject to the linear model (4), where m is the number of set points and $^{\top}$ is the transposition operator. Here, x is the data-driven observed curve from the system (1) and $\hat{x}$ is the solution curve obtained from the model (4).

Referring to this optimization model, we aim to estimate the model parameters A and b to minimize the objective function J. With the minimum value of the objective function (5), the linear model (4) can approximate closely the dynamic function in the system (1). With this, the linear model solution will approximate satisfactorily the system solution. Therefore, this problem is a curve-fitting problem with a first-order linear ODE model for dynamic systems. We remark that only solving the linear model (4) will not give the true solution of the system (1) due to the different structures. Instead, by solving the least squares optimization problem (5), the linear model solution provides the approximate solution to the dynamic system despite the model-reality differences.

3 Curve Fitting Technique

Now, we find the gradient of the objective function (5) with respect to the matrix A and the vector b as follows,

$$\frac{\partial J}{\partial A} = \frac{2}{m}(x - \hat{x})(-\hat{x})^{\mathrm{T}}, \tag{6}$$

$$\frac{\partial J}{\partial b} = \frac{2}{m}(x - \hat{x})(-1)^{\mathrm{T}}, \tag{7}$$

where $1 = (1, 1, \cdots, 1)^{\mathrm{T}}$ is a vector with all elements of 1s. Refer to the first-order necessary conditions [11, 19], that is, $\partial J/\partial A = 0$ and $\partial J/\partial b = 0$, we derive the updating rules,

$$A^{i+1} = A^i - \alpha \times \left(\frac{\partial J}{\partial A}\right)^i, \tag{8}$$

$$b^{i+1} = b^i - \beta \times \left(\frac{\partial J}{\partial b}\right)^i, \tag{9}$$

with A_0 and b_0 are the initial conditions, where α and β are the step sizes. Once convergence is achieved, the following stopping criteria

$||A^{i+1} - A^i|| < \varepsilon_1$ and $||b^{i+1} - b^i|| < \varepsilon_2$,

are satisfied, where ε_1 and ε_2 are small positive values. Hence, the optimal estimates for the matrix A and the vector b are determined and the minimum value of the objective function J is calculated.

3.1 Analytical Solution

Denote $\hat{A}$ and $\hat{b}$ as the optimal estimates obtained from (8) and (9). The linear ODE model (4) is expressed by

$$\dot{x} = \hat{A}x + \hat{b}. \tag{10}$$

Rewrite the model (10) and multiply $e^{-\hat{A}t}$ to both sides of (10),

$$e^{-\hat{A}t}\dot{x} - e^{-\hat{A}t}\hat{A}x = e^{-\hat{A}t}\hat{b}. \tag{11}$$

Notice that the left-hand side of (11) can be simplified to be

$$\frac{d}{dt}\left(e^{-\hat{A}t}x\right) = e^{-\hat{A}t}\hat{b}. \tag{12}$$

Integrate both sides of (12) with respect to t to give

$$e^{-\hat{A}t}x(t) - e^{-\hat{A}\tau}x(\tau) = \int_{\tau}^{t} e^{-\hat{A}t}\hat{b}dt \tag{13}$$

for $t > \tau$. Thus, we obtain the analytical solution [19] to (10),

$$\hat{x}(t) = e^{\hat{A}(t-\tau)}x(\tau) - \hat{A}^{-1}(\hat{b} - e^{\hat{A}(t-\tau)}\hat{b}), \tag{14}$$

where the inverse matrix $\hat{A}$ exist. Here, the solution (14) is the predicted solution to the system (1), and the prediction error is represented by

$$x_e(t) = x(t) - \hat{x}(t). \tag{15}$$

3.2 Analytical Solution

From the discussion above, we summarize the calculation procedure as an iterative algorithm, as given below.

Data A, b, x_0, m, t, f.
Step 1 Solve the linear model (4) with the initial value x_0. Set $i = 0$.
Step 2 Evaluate the objective function J^i from (5).
Step 3 Calculate the gradients $(\partial J/\partial A)^i$ and$(\partial J/\partial b)^i$ given by (6) and (7), respectively.
Step 4 pdatethe matrix A^i and the vector b^i from (8) and (9), respectively.
Step 5 Obtain the predicted solution $\hat{x}^i$ from (14) and the prediction error x_e^i from (15).
Step 6 Check the convergence. If $A^{i+1} \approx A^i$ and $b^{i+1} \approx b^i$, stop. Otherwise, set $i = i + 1$ and repeat from Step 2

4 Illustrative Example

Consider a chemical reaction model [22, 23],

$$\dot{x}_1 = a(x_2 - x_1x_2 + x_1 - bx_1^2), \tag{16}$$

$$\dot{x}_2 = a^{-1}(x_3 - x_1x_2 - x_2), \tag{17}$$

$$\dot{x}_3 = c(x_1 - x_3), \tag{18}$$

for $0 < t < 400$ min, with a, b and c are constant reaction rates, where x_1 is the concentration of bromous acid (HBrO2), x_2 is the concentration of bromide ions (Br-), and x_3 is the concentration of cerium oxides (Ce(IV)). This model is also known as the Belousov-Zhabotinsky (BZ) reaction and the Oregonator model. We set the initial conditions $x_1(0) = 1.2$, $x_2(0) = 5.0$ and $x_3(0) = 2.4$, while the values of kinetic constants are $a = 77.27$, $b = 0.12$ and $c = 0.03$, respectively.

We simplify the reaction model given in (16)–(18) as a linear model,

$$\dot{x} = Ax + b, \tag{19}$$

before implementing the computational algorithm with the initial matrix and the initial vector given by

$$A_0 = \begin{bmatrix} 77.27 & 77.27 & 0 \\ 0 & -0.0129 & 0.0129 \\ 0.03 & 0 & -0.03 \end{bmatrix} \text{ and } b_0 = \begin{bmatrix} 400 \\ -3 \\ 0.1 \end{bmatrix},$$

respectively. During the calculation procedure, the entries in the matrix A and the vector b are updated. Once the convergence is achieved, the following linear model,

$$\begin{bmatrix} \dot{x}_1 \\ \dot{x}_2 \\ \dot{x}_3 \end{bmatrix} = \begin{bmatrix} 77.5250 & 77.5440 & 0.19517 \\ -0.39970 & -0.82386 & -0.49525 \\ 0.33992 & -0.0091087 & 0.16187 \end{bmatrix} \begin{bmatrix} x_1 \\ x_2 \\ x_3 \end{bmatrix} + \begin{bmatrix} 400.1241 \\ -3.2473 \\ 0.1676 \end{bmatrix} \tag{20}$$

is applied to predict the solution of the reaction model.

Table 1 shows the simulation results after running the computational algorithm. It takes 378 iteration numbers to converge within a tolerance 1×10^{-5} in 26 s. The accuracy of the predicted solution to the solution of the reaction model is satisfactorily accepted since the mean square error of 1.0185×10^{-2} is small.

Table 1. Simulation result.

Iteration number	Mean square error	Tolerance	Elapsed time (s)
378	1.0185×10^{-2}	1×10^{-5}	26.8601

Figure 1 shows the state trajectories of the reaction model. The concentration of HBrO2, denoted by x_1, is initially increased to 4.5 units at 100 s, then reduced to 3.8 units and stayed at this value after 200 s. The concentration of Br-, denoted by x_2, expresses a dramatic drop from the initial value to the value of 0.4 units and a bit increase to 0.6 units and stays along this value after 140 s. The concentration of Ce(IV)), denoted by x_3, has a slight reduction, but it increases gradually after 45 s and stays at 3.7 units after 200 s. Therefore, the concentrations in the reaction model reached their respective equilibrium 200 s before the termination time at 400 min.

Figure 2 shows the simplified model we extracted from the reaction model, as given in (19). The concentrations seem to stay at the initial value; only the con centration of HBrO2 presents a slow increase under two units. Thus, the linear model (19) is not a practical representation for predicting the solution curve of the reaction model. However, applying the proposed computational algorithm, the linear model (20) with the optimal parameter estimates $\hat{A}$ and $\hat{b}$ can success fully predict the trajectories of the reaction model. The lines without asterisk (*) shown in Fig. 3 give a satisfactory prediction solution for the solution curve of the reaction model. Figure 4 shows the prediction errors of the linear model we used to predict the solution of the reaction model. These prediction errors reach zero and stay at this value for 200 s. Hence, the small value of the prediction error supports the accuracy of the predicted solution.

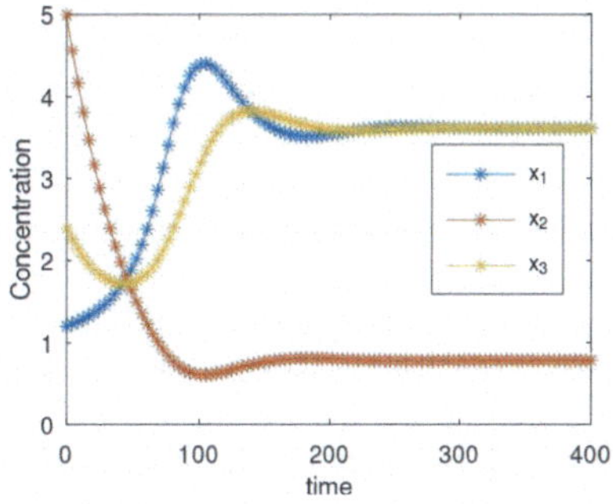

Fig. 1. State trajectories of chemical reaction model.

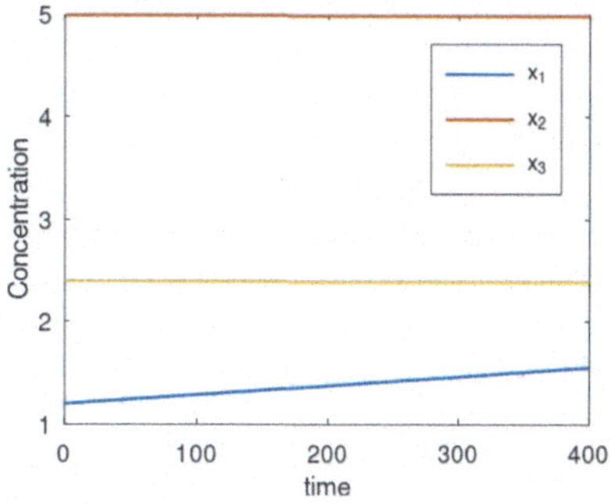

Fig. 2. State trajectories of linear model.

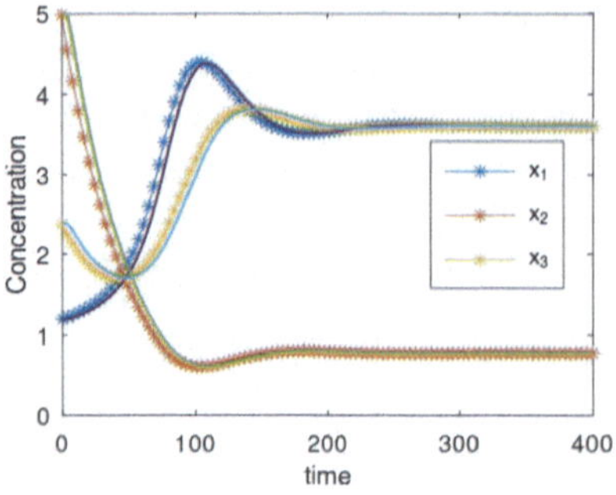

Fig. 3. State trajectories of predictive model.

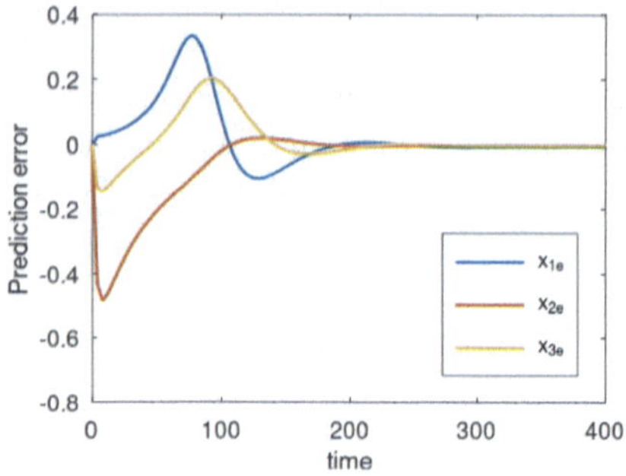

Fig. 4. Trajectories of prediction error between models.

5 Concluding Remarks

This paper discussed a curve-fitting technique for managing a dynamic system. In dynamic system modelling, using a linear dynamic model to fit the solution curve of a dynamic system provides new insight into curve-fitting problems. The advantage is the existing solution of a linear dynamic model, and the solution procedure of a linear dynamic model is the standard procedure. Hence, we only need to estimate the model parameters in the dynamic model from the observed nonlinear system or process. The proposed computational algorithm conducts the fitting task using a linear first-order ordinary differential equation model. For illustration, we studied the dynamic modelling of a chemical reaction model, and the simulation results have shown the accuracy of the curve-fitting solution. Therefore, the curve-fitting technique in this paper is efficient for dynamic modelling. In future research, we recommend applying a linear dynamic model to approximate the trajectories of more complex dynamic systems, including stochastic and uncertain systems.

Acknowledgments. This research was supported by Ministry of Higher Education (MOHE) Malaysia through Fundamental Research Grant Scheme (FRGS/1/2023/STG06/UTHM/02/2). The authors would like to thank the financial support from Faculty of Applied Sciences and Technology (FAST), Universiti Tun Hussein Onn Malaysia.

References

1. Astrom, K.J., Murray, R.M.: Feedback Systems: An Introduction for Scientists and Engineers. Princeton University Press, Princeton (2008)
2. Dorfler, F., Coulson, J., Markovsky, I.: Bridging direct and indirect data-driven control formulations via regularizations and relaxations. IEEE Trans. Autom. Control **68**(2), 883–897 (2023). https://doi.org/10.1109/TAC.2022.3148374
3. Brunton, S.L., Kutz, J.N.: Data-Driven Science and Engineering: Machine Learning, Dynamical Systems, and Control, 2nd edn. Cambridge University Press, Cambridge (2022)
4. Pillonetto, G., Chen, T., Chiuso, A., Nicolao, G. D., Ljung, L.: Regularized System Identification: Learning Dynamic Models from Data. Springer Cham (2022)

5. Yu, R., Wang, R.: Learning dynamical systems from data: an introduction to physics-guided deep learning. Proc. Natl. Acad. Sci. **121**(27), e2311808121 (2024). https://doi.org/10.1073/pnas.2311808121
6. Kutz, J.N., Brunton, S.L., Brunton, B.W., Proctor, J.L.: Dynamic Mode Decomposition: Data-Driven Modeling of Complex Systems. SIAM, Philadelphia (2016)
7. You, F., Vossen, T.: An approximate dynamic programming approach to dynamic stochastic matching. INFORMS J. Comput. (2024). https://doi.org/10.1287/ijoc.2021.0203
8. Gao, T.T., Barzel, B., Yan, G.: Learning interpretable dynamics of stochastic complex systems from experimental data. Nat. Commun. **15**, 6029 (2024). https://doi.org/10.1038/s41467-024-50378-x
9. Shobana, R., Kumar, R., Jaint, B.: Nonlinear dynamical system approximation and adaptive control based on hybrid-feed-forward recurrent neural network: Simulation and stability analysis. Expert. Syst. **41**(9), e13619 (2024). https://doi.org/10.1111/exsy.13619
10. Carey, M., Ramsay, J.O.: Fast stable parameter estimation for linear dynamical systems. Comput. Stat. Data Anal. **156**, 107124 (2021). https://doi.org/10.1016/j.csda.2020.107124
11. Seber, G.A.F., Wild, C.J.: Nonlinear Regression. Wiley, Hoboken (2003)
12. Bishop, C. M.: Pattern Recognition and Machine Learning. Springer, Heidelberg (2016)
13. Bataineh, A. A., Kaur, D.: A comparative study of different curve fitting algorithms in artificial neural network using housing dataset. In NAECON 2018- IEEE National Aerospace and Electronics Conference, Dayton, OH, USA, 2018, pp. 174–178 (2018). https://doi.org/10.1109/NAECON.2018.8556738
14. Lejarza, F., Baldea, M.: Data-driven discovery of the governing equations of dynamical systems via moving horizon optimization. Sci. Rep. **12**, 11836 (2022). https://doi.org/10.1038/s41598-022-13644-w
15. Schulz, E., Speekenbrink, M., Krause, A.: A tutorial on Gaussian process regression: Modelling, exploring, and exploiting functions. J. Math. Psychol. **85**, 1–16 (2018). https://doi.org/10.1016/j.jmp.2018.03.001
16. Li, M., Li, L.D.: A novel method of curve fitting based on optimized extreme learning machine. Appl. Artif. Intell. **34**(12), 849–865 (2020). https://doi.org/10.1080/08839514.2020.1787677
17. North, J.S., Wikle, C.K., Schliep, E.M.: A review of data-driven discovery for dynamic systems. Int. Stat. Rev. **91**(3), 464–492 (2023). https://doi.org/10.1111/insr.12554
18. Zhang, W., Gu, X., Hong, L., Han, L., Wang, L.: Comprehensive review of machine learning in geotechnical reliability analysis: algorithms, applications and further challenges. Appl. Soft Comput. **136**, 110066 (2023). https://doi.org/10.1016/j.asoc.2023.110066
19. Layek, G.C.: An Introduction to Dynamical Systems and Chaos. Springer, Singapore (2024). https://doi.org/10.1007/978-981-99-7695-9
20. Ahmad, S., Ambrosetti, A.: A textbook on Ordinary Differential Equations. Springer, Cham (2015). https://doi.org/10.1007/978-3-319-16408-3
21. Butcher, J. C.: Numerical Methods for Ordinary Differential Equations. John Wiley & Sons, Ltd., Hoboken (2016). https://doi.org/10.1002/9781119121534
22. Leonardo, S.D.: Basic concepts of self-organized phenomena on chemical systems. Quim. Nova **44**(5), 646–654 (2021). https://doi.org/10.21577/0100-4042.20170684
23. Gupta, M.K., Sahu, A., Yadav, C.K., Goswami, A., Swarup, C.: KCC theory of the oregonator model for Belousov-Zhabotinsky reaction. Axioms **12**, 1133 (2023). https://doi.org/10.3390/axioms12121133

Exploring Goal Scoring in the English Premier League Using Poisson and Probit Regression Models

Norziha Che Him[1,2](✉), Noor Azliza Abd. Latif[1], Nur Izzah Jamil[3], Mohd Saifullah Rusiman[1], and Yusliandy Yusof[4]

[1] Department of Mathematics and Statistics, Faculty of Applied Sciences and Technology, Universiti Tun Hussein Onn Malaysia (Pagoh Campus), 84600 Pagoh, Muar, Johor, Malaysia
{norziha,norziha,azliza,saifulah}@uthm.edu.my

[2] Computational Intelligence and Data Science Focus Group (CIDaS), Faculty of Applied Sciences and Technology, Universiti Tun Hussein Onn Malaysia (Pagoh Campus), 84600 Pagoh, Muar, Johor, Malaysia

[3] Faculty of Computer and Mathematical Sciences, Universiti Teknologi MARA, Negeri Sembilan Branch Rembau Campus, Jalan Kampung Pilin, 71300 Rembau, Negeri Sembilan, Malaysia
nurizzah@uitm.edu.my

[4] Kolej Kemahiran Tinggi MARA Sri Gading, Batu 11, Jalan Kluang, 86400 Parit Raja, Batu Pahat, Johor, Malaysia

Abstract. English Premier League (EPL) is the top-tier competition in English football and become the most-watched league internationally. This study aims to examine the correlation between an explanatory variable and goal scoring in EPL; to analyse the consistency metrics between goal scoring and the explanatory variable in EPL by using Poisson and Probit Regression Models and to propose an appropriate Poisson and Probit Regression Models that fit to the dataset. The methods that considered in this study are Poisson and Probit Regression Model. The dataset is based on the match final scores for the 2021/2022 season with 380 matches. Therefore, the dataset consists of 380 observations, each representing a match and 22 attributes. The results revealed that the Probit Regression Model is a better fit for the dataset with Akaike Information Criteria (AIC) and Bayesian Information Criteria (BIC) values of 407.38 and 420.8 compared to the Poisson Regression Model with the values of 1169.2 and 1139.7. In conclusion, Manchester City and Liverpool had the highest average goal count among clubs. Both clubs are identified as having high probabilities of securing the top ranks in the season. This is substantiated by the fact that both teams indeed finished at the top in the end as first and second place. Furthermore, both models demonstrate significance in fulfilling the Chi-Square test and overdispersion for the Poisson Regression Model, when it comes to model selection, the Probit Regression Model outperforms the Poisson Regression Model in terms of AIC and BIC values.

Keywords: Goal Scoring · Football · English Premier League · Poisson Regression Model · Probit Regression Model

P. Chang et al. (Eds.): SCIEMATHIC 2024, SPPHY 427, pp. 32–43, 2025.
https://doi.org/10.1007/978-981-96-7679-8_4

1 Introduction

Football, a sport that holds the title for being the most popular on a global scale. Football is frequently referred to as the king of sports due to its flexibility of playing. It can be played practically everywhere, whether indoors or outdoors. Football's widespread appeal is due to its simplicity, universal appeal, and rich history and tradition (Nguyen 2021), with international competitions such as the FIFA World Cup and continental tournaments such as the UEFA Champions League and EPL.

The EPL is the most viewed league on earth drawing viewership from one billion households across 188 countries and well-known for being one of the hardest leagues to predict. The EPL consists of 20 clubs competing for the top division of England's football pyramid. The latest concluded season was the 2021/2022 season, featuring notable events such as the first season post-pandemic. Additionally, Manchester City secured their 6th championship during this season. Furthermore, the return of football icon Cristiano Ronaldo to Manchester United marked a significant highlight, with Ronaldo also emerging as a top scorer in the season. Moreover, several other players, including Romelu Lukaku, made comebacks and rejoining the Chelsea club (Ryan 2022).

Given the unpredictable nature of sports, predicting match outcomes becomes a subject of great interest for fans, sports media, experts, stakeholders, and even the betting industry. The anticipation and speculation leading up to a match are part of the excitement that makes sports so captivating. Meanwhile, the predictions can be challenging due to the many variables at play in each game (Saraiva et al. 2016). Furthermore, within the sports industry, the results of a match are inherently unpredictable (Yang 2021).

A frequently employed statistical modelling technique for predicting football match outcomes is Poisson Regression Model. This method is especially well-suited for scenarios in which the dependent variable pertains to count or frequency, such as the number of goals scored in a football match (Fredrizzi *et al.* 2022). In addition, Poisson Regression Model provides a statistical approach that specifically addresses scenarios involving discrete, count-based data (Bruinsma 2020). Its application in football analytics allows researchers and analysts to estimate match outcomes focusing on the frequencies of goals scored. Another statistical model used is Probit Regression Model. The assumption of interdependence between the goals scored by the two teams holds a main effect (Goddard 2005; Skogsvik and Skogsvik 2013). The selection between the Probit and Poisson Regression Models, determining which is the superior choice, relies on the specific nature of the prediction task. Probit Regression Model is well-suited for binary outcomes, such as predicting whether a team will win or lose a game. In contrast, Poisson Regression Model is preferable for count-based scenarios, particularly when the focus is solely on goal scoring. This study aims to examine the correlation between an explanatory variable and goal scoring in EPL; to analyse the consistency metrics between goal scoring and the explanatory variable in EPL by using Poisson and Probit Regression Models and to propose an appropriate Poisson and Probit Regression Models that fit to the dataset.

2 Material and Method

2.1 Data Description

The dataset of this study consists of match final scores of the 2021/2022 season EPL games with a complete season of the EPL with a total of 380 matches by all the teams combined. Therefore, the dataset consists of 380 observations, each represent a match and 22 attributes. The whole dataset has been retrieved from webpage data world (Table 1).

Table 1. Data Description.

Variables	Description
Home Team	The team that plays home
Away Team	The team that plays away home
FTHG	Full Time Home Team Goals
FTAG	Full Time Away Team Goals
FTR	Full Time Result (H = Home Win, D = Draw, A = Away Win)
HTHG	Half Time Home Team Goals
HTAG	Half Time Away Team Goals
HTR	Half Time Result (H = Home Win, D = Draw, A = Away Win)
HS	Shots from Home Team
AS	Shots from Away Team
HST	Shots on Target from Home Team
AST	Shots on Target from Away Team
HF	Corners from Home Team
AF	Corners from Away Team
HC	Fouls Committed from Home Team
AC	Fouls Committed from Away Team
HY	Yellow Cards from Home Team
AY	Yellow Cards from Away Team
HR	Red Cards from Home Team
AR	Red Cards from Away Team

2.2 Spearman's Rank Correlation Coefficient

The Spearman's Rank Correlation Coefficient assesses the linear association between two variables and provides a dimensionless measure with the value ranges from -1, indicating a simple and linear negative relationship, to $+1$, indicating a simple and linear positive relationship (Ali 2022). In addition, Mukaka (2020) revealed that a correlation coefficient of zero shows the absence of a linear relationship between two continuous variables. The formula of Spearman's Rank Correlation Coefficient is represented by Eq. (1).

$$r_s = \frac{1 - 6\sum d^2}{n(n^2 - 1)} \tag{1}$$

2.3 Generalised Linear Model

The Generalised Linear Model (GLM) is an extended concept of linear regression to handle non-normal response variables. In GLM, the relationship between the explanatory variables and the response variable is modelled using a link function and a linear predictor. According to GLM theory, it is possible to represent the canonical link function as a linear combination of the explanatory variables (Roback and Legler 2021).

There are three components in the GLM; first component is a random component that identifies the response variable (Y) and assume or determine its probability distribution, second component is a linear predictor and it represents the explanatory variables. The linear predictor predicts the linear relationship between explanatory variables (predictors) and the response variable. The model equation of linear predictor is represented in Eq. (2) with $\beta_0, \beta_1, \beta_2, \ldots, \beta_p$, are the coefficients associated with predictor variables.

$$\beta_0 + \beta_1 x_1 + \beta_2 x_2 + \ldots + \beta_p x_p \tag{2}$$

The third component known as a link function shows the relationship between the expected value of the response variable $E(y)$ and the linear predictor. The expected value $\mu = E(y)$ of a probability distribution of y, , changes based on the explanatory variable. The link function defines a transformation of the expected value that allows it to be related to the linear predictor (Abdul Kabir *et al.* 2015). Meanwhile, the link function specifies a function g that relates μ to the linear predictor as shown in Eq. (3).

$$g(\mu) = \beta_0 + \beta_1 x_1 + \beta_2 x_2 + \ldots + \beta_p x_p \tag{3}$$

2.4 Poisson Regression Model

Poisson Regression Model is a type of GLM used to model count data and particularly suitable when the response variable follows a Poisson distribution (Yang and Berdine 2015). In Poisson Regression Model, the aim is to model the expected value of the response variable, λ as a function of the explanatory variables and β as the coefficient of the predictor variable. The relationship between λ and the explanatory variables is typically assumed to be exponential. The assumption of Poisson distribution is the mean and variance are assumed to be equal. The probability for Poisson regression is shown in Eq. (4) with the random variable, $x = 1, 2, 3, \ldots$ and μ is the mean also equal to the variance. However, in the linear function, the variance is assumed to be constant throughout the whole explanatory variables. To avoid these situations, the model equation of Poisson Regression Model is denoted in the log function (Van den Berg 2022) and the model could be represents as in Eq. (5).

$$f(x; \mu) = \frac{\mu^x e^{-\mu}}{x!} \tag{4}$$

$$\log(\lambda) = \beta_0 + \beta_1 x_1 + \beta_2 x_2 + \ldots + \beta_p x_p \tag{5}$$

2.5 Probit Regression Model

Probit Regression Model is a statistical technique used to analyse binary or dichotomous outcome variables (Nisbet et al., 2018). The probability of the binary outcome is transformed using the inverse standard normal distribution where this transformed probability could be represented as a linear combination of predictor variables in the model. Moreover, Probit Regression Model to estimate the conditional probability of default for a given observation. These models offer more flexibility and do not require as restrictive assumptions (DePamphilis 2022).

Binary response models directly model the probabilities, denoted as $P(y_i = 1)$ of the dependent variable y_i. The model considers a sample of N independently and identically distributed observations, labelled as $i = 1, 2, 3, \ldots$ where y_i represents the dependent binary variable. Additionally, x_i predictor variables, is used to determine the probability that the dependent variable (Schmidheiny 2015). The model can be seen in Eq. (6) and with note in Eq. (7).

$$P(Y = 1|x_1, x_2, \ldots x_k) = \phi(\beta_0 + \beta_1 x_1 + \beta_2 x_2 + \ldots + \beta_k x_k) \tag{6}$$

$$\Phi(z) = P(Z \leq z), Z \sim N(0, 1) \tag{7}$$

2.6 Model Selection

To assess the goodness-of-fit for a Poisson and Probit Regression Models, the most common test used is the Chi-Square test. The Chi-Square test looked if association between two categorical variables exist and the formula as shown in Eq. (8) with O_i is an observed value and e_i is an expected value.

$$X^2 = \sum r_i^2 = \sum \frac{(O_i - e_i)^2}{e_i} \tag{8}$$

Selecting a model with less parameters could lead to the incorrect assumptions, bias and poor prediction. Somehow, if choosing a model with too many parameters could lead to overfitting model. To avoid both issues, the Akaike Information Criterion (AIC) and Bayesian Information Criterion (BIC) are used. The AIC and BIC could be defined as in Eq. (9) and Eq. (10) respectively, where $\ln L$ is the maximized log-likelihood of the model, N is the sample size and k is the number of parameters estimated.

$$AIC = -2\ln L + 2k \tag{9}$$

$$BIC = -2\ln L + k\ln N \tag{10}$$

3 Results and Discussion

3.1 Spearman's Rank Correlation Coefficient

Table 2 shows the first analysis known as Spearman's Rank Correlation Coefficient and the result present the high positive correlation with full-time home goal (FTHG) scored with the value of 0.686 and high negative correlation with full-time away goal (FTAG) at −0.705. Besides, the result shows the low positive associated with half-time goal scored and number of shots on target with 0.410 (Home), −0.480 (Away) and 0.468 (Home), −0.507 (Away) respectively. The relationship between the final result and number of shots, foul committed, corner, red card and yellow card are negligible as the values are less than 0.40. Hence, only FTAG and FTHG are taken for used.

Table 2. Consistency Matrix of Spearman Rank Correlation Coefficient.

	FTHG	FTAG	FTR	HTHG	HTAG	HTR	HS	AS	HST	AST
FTHG	1.000	−0.177	**0.686**	0.665	−0.113	0.515	0.391	−0.250	0.619	−0.203
FTAG	−0.177	1.000	**−0.705**	−0.065	0.703	−0.487	−0.269	0.417	−0.183	0.647
FTR	0.686	−0.705	1.000	0.410	−0.480	0.596	0.356	−0.369	0.468	−0.507
HTHG	0.665	−0.065	**0.410**	1.000	−0.060	0.710	0.167	−0.072	0.389	−0.062
HTAG	−0.113	0.703	**−0.480**	−0.060	1.000	−0.675	−0.149	0.274	−0.102	0.471
HTR	0.515	−0.487	0.596	0.710	−0.675	1.000	0.202	−0.224	0.327	−0.332
HS	0.391	−0.269	0.356	0.167	−0.149	0.202	1.000	−0.525	0.649	−0.385
AS	−0.250	0.417	−0.369	-0.072	0.274	−0.224	−0.525	1.000	−0.375	0.736
HST	0.619	−0.183	**0.468**	0.389	−0.102	0.327	0.649	−0.375	1.000	−0.254
AST	−0.203	0.647	**−0.507**	−0.062	0.471	−0.332	−0.385	0.736	−0.254	1.000

3.2 Consistency Metrics Between Goal Scorings

The probability of scoring number of goal is calculated by using Poisson probability formula. Based on Table 3, the Poisson probability of all clubs scoring two to three goals is very close, at 0.2. However, for scoring five goals and above, four clubs such as: Leeds, Leicester, Liverpool, and Manchester City manage to exhibit high probabilities, exceeding 0.2, compared to other clubs with only 0.1. The Poisson probability indicates that the average number of goals scored per club is two.

The probability of scoring number of goal is calculated by using Probit probability formula as shown in Table 4. It indicates the probability of all clubs scoring three goals is very close, at 0.2. Probit probability shows the mean of all clubs is three goals. However, for scoring of five goals and above, also four clubs such as: Leeds, Leicester, Liverpool, and Manchester City manage to exhibit high probabilities, exceeding 0.3.

Table 3. Consistency Matrix of Poisson Regression Model.

Goals/Probability	0	1	2	3	4	5 +
All Clubs	0.0597	0.1683	**0.2371**	**0.2228**	0.157	0.1552
Arsenal	0.0568	0.1629	0.2336	0.2234	0.1602	0.1631
Aston Villa	0.0615	0.1714	0.2391	0.2223	0.155	0.1507
Brentford	0.0648	0.1773	0.2426	0.2213	0.1514	0.1426
Brighton	0.104	0.2354	0.2664	0.201	0.1137	0.0795
Burnley	0.1013	0.232	0.2655	0.2027	0.116	0.0825
Chelsea	0.0568	0.1629	0.2336	0.2234	0.1602	0.1631
Crystal Palace	0.0987	0.2285	0.2646	0.2043	0.1183	0.0856
Everton	0.0568	0.1629	0.2336	0.2234	0.1602	0.1631
Leeds	0.0414	0.1319	0.2099	0.2228	0.1774	**0.2166**
Leicester	0.0414	0.1319	0.2099	0.2228	0.1774	**0.2166**
Liverpool	0.0425	0.1343	0.212	0.2231	0.1762	**0.2119**
Man City	0.0373	0.1226	0.2017	0.2211	0.1818	**0.2355**
Man United	0.0498	0.1494	0.224	0.224	0.168	0.1847
Newcastle	0.0615	0.1714	0.2391	0.2223	0.155	0.1507
Norwich	0.0599	0.1685	0.2373	0.2227	0.1568	0.1548
Southampton	0.0553	0.1601	0.2318	0.2236	0.1618	0.1674
Tottenham	0.0568	0.1629	0.2336	0.2234	0.1602	0.1631
Watford	0.0539	0.1574	0.2299	0.2238	0.1634	0.1717
West Ham	0.0539	0.1574	0.2299	0.2238	0.1634	0.1717
Wolves	0.1186	0.2529	0.2696	0.1915	0.1021	0.0653

Table 4. Consistency Matrix of Probit Regression Model.

Goals/Probability	0	1	2	3	4	5 +
All Clubs	0.0466	0.0928	0.1736	**0.2301**	0.2162	0.2398
Arsenal	0.0452	0.0898	0.1691	0.2269	0.2170	0.2508
Aston Villa	0.0474	0.0945	0.1762	0.2319	0.2156	0.2334
Brentford	0.0490	0.0979	0.1811	0.2352	0.2142	0.2218
Brighton	0.0662	0.1343	0.2300	0.2573	0.1880	0.1241

(*continued*)

Table 4. (*continued*)

Goals/Probability	0	1	2	3	4	5 +
Burnley	0.0651	0.1319	0.2271	0.2565	0.1902	0.1291
Chelsea	0.0452	0.0898	0.1691	0.2269	0.2170	0.2508
Crystal Palace	0.0640	0.1296	0.2242	0.2557	0.1923	0.1341
Everton	0.0452	0.0898	0.1691	0.2269	0.2170	0.2508
Leeds	0.0372	0.0733	0.1430	0.2054	0.2173	**0.3203**
Leicester	0.0372	0.0733	0.1430	0.2054	0.2173	**0.3203**
Liverpool	0.0378	0.0745	0.1450	0.2073	0.2176	**0.3146**
Man City	0.0349	0.0686	0.1351	0.1980	0.2158	**0.3429**
Man United	0.0416	0.0825	0.1577	0.2181	0.2181	0.2799
Newcastle	0.0474	0.0945	0.1762	0.2319	0.2156	0.2334
Norwich	0.0467	0.0929	0.1738	0.2303	0.2161	0.2392
Southampton	0.0444	0.0883	0.1668	0.2252	0.2174	0.2566
Tottenham	0.0452	0.0898	0.1691	0.2269	0.2170	0.2508
Watford	0.0437	0.0868	0.1645	0.2234	0.2177	0.2624
West Ham	0.0437	0.0868	0.1645	0.2234	0.2177	0.2624
Wolves	0.0721	0.1470	0.2449	0.2599	0.1757	0.1003

3.3 Poisson Regression Model and Probit Regression Model

FTHG (Full Time Home Goals), FTAG (Full Time Away Goals) both are significant from the Spearman's Rank Correlation Coefficient analysis and the models for both Poisson and Probit Regression Models then be developed based on both variables. There are 20 clubs in EPL, FTHG and FTAG of respective clubs are filtered and assign to respective FTHG. For instance, the Full-Time Home Goals (FTHG) for Arsenal are designated as FTHG1, the second club Aston Villa is designed as FTHG2 and so on. All coefficient of model is kept to predict the results of all 20 clubs.

3.4 Poisson Regression Model

Table 5 shows a summary model of Poisson Regression Model (Home and Away) with the *p*-value is set as significant at 0.05. Therefore, there are only eight clubs that are significant at Home Model including Arsenal, Burnley, Chealsea, Leeds, Manchester City, Norwich, Watford, and Wolves. Next is there are only two clubs that are significant known as Liverpool and Norwich in Away Model.

Table 5. Summary for Poisson Regression Model (Home and Away).

Variables	Estimate	Pr(>|z|)	Variables	Estimate	Pr(>|z|)
FTHG1	0.6109	<0.001	FTAG1	0.31365	>0.1
FTHG2	−0.1880	>0.1	FTAG2	−0.1226	>0.1
FTHG3	−0.4643	>0.1	FTAG3	−7.9E−10	>0.1
FTHG4	−0.6109	<0.1	FTAG4	−0.1226	>0.1
FTHG5	−0.6649	<0.1	FTAG5	−0.4855	>0.1
FTHG6	0.0555	>0.1	FTAG6	0.40546	>0.1
FTHG7	−0.2595	>0.1	FTAG7	−0.1226	>0.1
FTHG8	−0.2595	>0.1	FTAG8	−0.4855	>0.1
FTHG9	−0.6109	<0.1	FTAG9	−0.1226	>0.1
FTHG10	−0.0289	>0.1	FTAG10	0.07410	>0.1
FTHG11	0.3364	>0.1	FTAG11	0.54856	<0.1
FTHG12	0.5050	<0.1	FTAG12	0.45547	>0.1
FTHG13	−0.0896	>0.1	FTAG13	−0.0392	>0.1
FTHG14	−0.2972	>0.1	FTAG14	−0.3677	>0.1
FTHG15	−1.0704	<0.01	FTAG15	−0.8602	<0.1
FTHG16	−0.4198	>0.1	FTAG16	−0.2623	>0.1
FTHG17	0.0822	>0.1	FTAG17	0.17589	>0.1
FTHG18	−0.7221	<0.1	FTAG18	−0.4248	>0.1
FTHG19	−0.0588	>0.1	FTAG19	0.0377	>0.1
FTHG20	−0.5596	<0.1	FTAG20	−0.3677	>0.1

3.5 Probit Regression Model

Meanwhile, Table 6 displays a summary model of Probit Regression Model (Home and Away) with the *p*-value is set as significant at 0.05. However, there are only three clubs significant known as Arsenal, Norwich and Wolves in Home Model. There are obviously only one club is significant at 0.05 known as Leicester in Away Model.

Table 6. Summary for Probit Regression Model (Home and Away).

Variables	Estimate	Pr(>|z|)	Variables	Estimate	Pr(>|z|)
FTHG1	1.0031	<0.001	FTAG1	0.4795	>0.1
FTHG2	3.3E-15	>0.1	FTAG2	−2.5E-15	>0.1
FTHG3	−0.5236	>0.1	FTAG3	0.7726	>0.1

(*continued*)

Table 6. *(continued)*

Variables	Estimate	Pr(>|z|)	Variables	Estimate	Pr(>|z|)
FTHG4	−0.9371	<0.1	FTAG4	0.1541	>0.1
FTHG5	−0.8039	>0.1	FTAG5	0.5236	>0.1
FTHG6	0.6167	>0.1	FTAG6	−0.5455	>0.1
FTHG7	−0.1985	>0.1	FTAG7	0.5236	>0.1
FTHG8	−0.3695	>0.1	FTAG8	0.7726	>0.1
FTHG9	−0.5236	>0.1	FTAG9	0.5236	>0.1
FTHG10	−0.1985	>0.1	FTAG10	1.1403	>0.1
FTHG11	4.5727	>0.1	FTAG11	−0.4134	>0.1
FTHG12	0.2489	>0.1	FTAG12	−0.5455	>0.1
FTHG13	−0.3695	>0.1	FTAG13	0.5236	>0.1
FTHG14	4.92E-15	>0.1	FTAG14	0.5236	>0.1
FTHG15	−1.0691	<0.1	FTAG15	0.5236	>0.1
FTHG16	4.03E-15	>0.1	FTAG16	0.7726	>0.1
FTHG17	−0.1985	>0.1	FTAG17	−0.2803	>0.1
FTHG18	−0.8039	>0.1	FTAG18	0.5236	>0.1
FTHG19	4.5727	>0.1	FTAG19	0.3250	>0.1
FTHG20	−0.9371	<0.1	FTAG20	−0.1434	>0.1

3.6 Comparison for Poisson Regression Model and Probit Regression Model

Table 7 shows the summary comparison for Poisson and Probit Regression. The null deviance represents the goodness-of-fit for a model with no predictor variables. The lowest deviance indicates a better fit model to the dataset. The residual deviance measures the goodness-of-fit for the model with predictor variables. Also, the lowest residual deviance suggests a better fit model to the dataset. Hence, this study concludes that both Probit Regression Models is better fit model than Poisson Regression Model. Besides, AIC and BIC also considered a lowest value as better fit model. The Probit Regression Model presents lowest AIC and BIC values than the Poisson Regression Model. Hence, the Probit Regression Model shows a better fit model. In summary, the Probit Regression Model have a better fit model than Poisson Regression Model. In addition, the Chi-Square test has been considered at significance level of 0.05 to identify the independency of models.

Table 7. Summary Comparison for Poisson and Probit Regression Model.

	Home Poisson Model	Home Probit Model	Away Poisson Model	Away Probit Model
Null deviance	492.61	422.92	516.85	422.92
Degrees of freedom	379	379	379	379
Residual deviance	413.19	367.38	460.47	380.80
Degrees of freedom	360	360	360	360
AIC	1169.2	407.38	1139.7	420.8
BIC	1247.97	11.88	1218.53	431.32
Chi-Square Test (*p*-value)	<0.001	<0.01	<0.001	<0.001

4 Conclusion

Overall, the potential future improvements for this similar study such as by including additional predictor such as player statistics. Manchester City and Liverpool are found to have higher mean in number of goals than other clubs. Manchester City and Liverpool had the highest predicted probabilities of finishing in the top positions. This is substantiated by the fact that both teams indeed finished at the top in the end as first and second place. Furthermore, both models demonstrate significance in fulfilling the Chi-Square test and overdispersion for the Poisson Regression Model. However, for the model selection, the Probit Regression Model outperforms Poisson Regression Model in terms of AIC and BIC values.

Acknowledgement. The authors would like to express gratefully heartfelt thanks to the Universiti Tun Hussein Onn Malaysia (UTHM) and Research Management Centre (RMC), Univrsiti Tun Hussein Onn Malaysia (UTHM) for the research grant awarded (Q553).

References

Abdulkabir, M., Edem, U., Tunde, R., Kemi, B.: An empirical study of generalized linear model for count data. J. Appl. Comput. Math. **4**, 253 (2015)

Ali Abd Al-Hameed, K.: Spearman's correlation coefficient in statistical analysis. Int. J. Nonlinear Anal. Appl. **13**(1), 3249–3255 (2022)

Bruinsma, R.S.: Using Poisson regression to model football scores and exploit inaccuracies in the online betting market (Doctoral dissertation) (2020)

DePamphilis, D.M.: Alternative exit and restructuring strategies: bankruptcy, reorganization, and liquidation. In: Mergers, Acquisitions, and Other Restructuring Activities (Eleventh Edition), pp. 485–505 (2022)

Fedrizzi, G., Canal, L., Micciolo, R.: UEFA EURO 2020: an exciting match between football and probability. Teach. Stat. **44**(3), 119–125 (2022)

Goddard, J.: Regression models for forecasting goals and match results in association football. Int. J. Forecast. **21**(2), 331–340 (2005)

Mukaka, M.M.: Statistics corner: a guide to appropriate use of correlation coefficient in medical research, vol. 24 (2012)

Nguyen, Q.: Poisson modeling and predicting english premier league goal scoring. In: Graduate Research Symposium (2021)

Nisbet, R., Elder, J., Miner, G.D.: Handbook of Statistical Analysis and Data Mining Applications, 2nd edn. Academic press (2018)

Roback, P., Legler, J.: Beyond Multiple Linear Regression: Applied Generalized Linear Models and Multilevel Models in R. CRC Press (2021)

Ryan, K.: Premier League top scorers 2021–22: Salah, Son, Ronaldo & the race for the Golden Boot | Goal.com Malaysia (2022)

Saraiva, E.F., Suzuki, A.K., Louzada, F.: Predicting football scores via Poisson regression model: applications to the National Football League. Commun. Stat. Appl. Methods **23**(4), 297–319 (2016)

Schmidheiny, K.: Short Guides to Microeconometrics Fall 2022 (2015). Accessed 8 July 2023

Skogsvik, K., Skogsvik, S.: On the choice-based sample bias in probabilistic bankruptcy prediction. Invest. Manag. Finan. Innov. **10**(1), 29–37 (2013)

Van den Berg, S.M.: Analysing data using linear models, 6th edn. (2022)

Yang, S., Berdine, G.: Poisson regression. Southwest Resp. Crit. Care Chron. **3**(9), 61–64 (2015)

Yang, W.: Predict soccer match outcome based on player performance. Front. Sport Res. **3**(3) (2021)

Solving Two Species Lotka-Volterra Biology Model Using Differential Transform Method

Noor Azliza Abd Latif(✉) and Norziha Che Him

Department of Mathematics and Statistics, Faculty of Applied Sciences and Technology, Universiti Tun Hussein Onn Malaysia (Pagoh Campus), 84600 Pagoh, Muar, Johor, Malaysia
{azliza,norziha}@uthm.edu.com.my

Abstract. In this research, the Lotka-Volterra model, commonly recognized as the predator-prey model, is addressed using the differential transform method (DTM), a semi-numerical-analytic approach. This study encompasses two Lotka-Volterra model which are the classic Lotka-Volterra model and the modified Lotka-Volterra model. The computation of the solution is acquired using Maple 2020 software. The utilization of Maple 2020 software is also aiding in the comparison of outcomes obtained through the DTM with results from other numerical methods such as Variational iteration method (VIM), Adomian decomposition method (ADM), and fourth-order Runge-Kutta (RK4) accomplished by previous researchers. By the end of this study, it was demonstrated that DTM proficiently solved the Lotka-Volterra model and it is a high accuracy method when compared to 3-term ADM and 2-iteration VIM.

Keywords: Lotka-Volterra · Differential transform method · nonlinear · ordinary differential equations

1 Introduction

The Lotka-Volterra model, commonly referred to as a predator-prey model, is a mathematical model that illustrates the interactions between two species in an ecosystem. The population dynamics of the predator and prey are thought to be interconnected in the model. The prey population serves as a food source for the predator population, which in turn controls the prey population by consuming them. The Lotka-Volterra model depicts this relationship through a set of nonlinear first order ordinary differential equations (ODEs). It is usual practice to use the Lotka-Volterra model to illustrate the shifts in population size of two groups that are directly competing with one another within an ecosystem. It was conceived to investigate how these species interact with one another as well as the growth patterns that they exhibit [12]. This model had widespread application in modelling a variety of biological and socioeconomic processes. For instance, [17] studied the impact of applying biological control to *Plutella xylostella* and [9] used the model to make forecasts for sales for two retail formats that compete with one another. In addition, [19] investigated the development of marine clusters by applying the Lotka-Volterra model to their data. However, there occur difficulties in finding the exact or

P. Chang et al. (Eds.): SCIEMATHIC 2024, SPPHY 427, pp. 44–54, 2025.
https://doi.org/10.1007/978-981-96-7679-8_5

analytical solutions caused by the nonlinear part. [4] have proposed a study which fulfilled their objective of obtaining a series of solutions for multispecies Lotka-Volterra model and [6] have done research regarding solving multispecies Lotka-Volterra equations. Therefore, the differential transform method (DTM) is employed in this study in solving the Lotka-Volterra model.

The DTM is recognized as a convenient and robust approach for solving diverse types of differential equations. [20] is credited with pioneering the DTM concept which has been successfully utilized in solving both linear and nonlinear initial value problems within the realm of electric circuit analysis. Providing a novel formalization of the Taylor series, the DTM is frequently characterized as a semi-numerical-analytic approach. It presents a new way of applying the Taylor series to mathematical analysis. The provided differential equation and its related beginning conditions are turned into a recurrence equation using the DTM [10]. It dispenses with the requirement of employing linearization or perturbation methods, and it enables the avoidance of extensive computational efforts and round-off errors [8]. DTM distinguishes itself from the typical high-order Taylor series method, as it does not require symbolic computation of the necessary derivatives of the given functions [14]. The Taylor series method can be computationally intensive, especially for higher orders, leading to increased processing time. On the other hand, the DTM reduces the computational domain and offers ease of application to a diverse array of problems [2]. Mathematical models have found extensive application in understanding and predicting the dynamics of ecological populations affected by different infectious diseases. These diseases play a crucial role in influencing the size of species populations. Differential equations are commonly employed to describe these models, with the prey-predator model holding particular significance for both ecologists and mathematicians [18]. [13] have proposed research about solving an array of non-linear differential equations representing the mathematical model of dengue transmission by employing DTM, [7] explored the use of DTM on initial value problems, [11] has discovered numerical solutions for linear and nonlinear systems of ODE by applying the DTM while [1] applied DTM so that they could arrive at semi-analytical solutions for the SIS and SI epidemic models. Thus, this study aims to solve the Lotka-Volterra model which commonly known as prey-predator model that constituted by nonlinear first order ordinary differential equations using DTM.

2 Materials and Method

2.1 Lotka-Volterra Model

This study encompasses two Lotka-Volterra models which are classic Lotka-Volterra model and modified Lotka-Volterra model [3]. The computation of the solution is acquired using Maple 2020 software.

2.1.1 Classic Lotka-Volterra Model

The classic Lotka-Volterra model is presented as follows [15]:

$$\frac{dx}{dt} = x(t)(a - by(t)), \tag{1}$$

$$\frac{dy}{dt} = y(t)(dx(t) - c). \tag{2}$$

where
$x =$ prey population
$y =$ predator population
$a, b, c, d =$ positive real constants
$t =$ time

2.1.2 Modified Lotka-Volterra Model

The modified Lotka-Volterra model is outlined as below from the study of [5]:

$$\frac{dN_1}{dt} = N_1(b_1 + a_{11}N_1 + a_{12}N_2), \tag{3}$$

$$\frac{dN_2}{dt} = N_2(b_2 + a_{21}N_1 + a_{22}N_2). \tag{4}$$

where
$N_1 =$ prey population
$N_2 =$ predator population
$a_{11}, a_{12}, a_{21}, a_{22}, b_1, b_2 =$ positive real constants
$t =$ time

The competition between two species that inhabit the same ecological niche is described by Eq. (3) and Eq. (4).

2.2 Theorems and Algorithms of Differential Transform Method

The differential transform method (DTM) is a semi-analytic numerical technique developed on the Taylor expansion. This method generates an analytical solution in polynomial form. The subsequent essential explanations and characteristics are presented below, as stated in the study conducted by [16].

The following is the definition of the one-dimensional differential transform of function $c(x)$:

$$C(k) = \frac{1}{k!}\left[\frac{\partial^k}{\partial x^k}c(x)\right]_{x=x_0}, \tag{5}$$

and the differential inverse transform of C(k) is defined as:

$$c(x) = \sum_{k=0}^{\infty} C(k)(x - x_0)^k. \tag{6}$$

In the condition where $x_0 = 0$, the Eq. (5), the one-dimensional differential transform of function $c(x)$ can be written as:

$$C(k) = \frac{1}{k!}\left[\frac{\partial^k}{\partial x^k}c(x)\right]_{x=0}. \tag{7}$$

In this context, $c(x)$ represents the initial function while $C(k)$ denotes the altered function. On the other hand, the differential inverse transform of $C(k)$ is defined as:

$$c(x) = \sum_{k=0}^{\infty} C(k)x^k. \tag{8}$$

By substituting the Eq. (7) into Eq. (8), we obtain

$$c(x) = \sum_{k=0}^{\infty} \frac{1}{k!}\left[\frac{\partial^k}{\partial x^k}c(x)\right]_{x=0} x^k. \tag{9}$$

In actual application, the function $c(x)$by a finite series of Eq. (8) can be written as:

$$c(x) = \sum_{k=0}^{n} C(k)x^k, \tag{10}$$

and Eq. (7) indicates that

$$c(x) = \sum_{k=n+1}^{\infty} C(k)x^k. \tag{11}$$

is ignored because of its small size. Typically, the value of x is determined by a convergency of the series coefficients. By examining the definitions provided in the Eq. (7) and Eq. (8), it can be easily demonstrated that the transformed functions adhere to fundamental mathematical operations. The fundamental operations that are executed throughout the differential transformation process that applied in the study are presented in Table 1.

Table 1. Fundamental operations by differential transformation used in the study.

Original function	Transformed function
$c(t) = u(t) \pm v(t)$	$C(k) = U(k) \pm V(k)$
$c(t) = \alpha u(t)$	$C(k) = \alpha U(k)$, where α is constant
$c(t) = \frac{\partial}{\partial t}u(t)$	$C(k) = (k+1)U(k+1)$
$c(t) = u(t)v(t)$	$C(k) = \sum_{r=0}^{k} U(r)V(k-r)$

2.3 Application of Differential Transform Method in Classic Lotka-Volterra Model

Referring to Eq. (1) and Eq. (2) and performing the elementary steps of DTM which are outlined in Table 1, we obtained the equations as shown below:

$$X(k+1) = \frac{1}{k+1}\left[aX(k) - b\sum_{r=0}^{k} X(r)Y(k-r)\right], \tag{12}$$

$$Y(k+1) = \frac{1}{k+1}\left[d\sum_{r=0}^{k} X(r)Y(k-r) - cY(k)\right]. \tag{13}$$

2.4 Application of Differential Transform Method in Classic Lotka-Volterra Model

By consulting Eq. (3) and Eq. (4) and applying the fundamental procedures of DTM detailed in Table 1, we derived the following equations as:

$$N_1(k+1) = \frac{1}{k+1}\left[b_1 N_1(k) + a_{11}\sum_{r=0}^{k} N_1(r)N_1(k-r) + a_{12}\sum_{r=0}^{k} N_1(r)N_2(k-r)\right], \tag{14}$$

$$N_2(k+1) = \frac{1}{k+1}\left[b_2 N_2(k) + a_{21}\sum_{r=0}^{k} N_1(r)N_2(k-r) + a_{22}\sum_{r=0}^{k} N_2(r)N_2(k-r)\right]. \tag{15}$$

3 Results and Discussion

3.1 Classic Lotka-Volterra Model Case 1

By referring to the model in Eq. (12) and Eq. (13), consider the specific values illustrated below as $a = 0.1$, $b = 1$, $c = 1$, $d = 1$ with the initial conditions $x(0) = 14$ and $y(0) = 18$ [15]. The approximate solution when $n = 3$ is obtained and written as:

$$x(t) = 14 - 250.6t + 604.8700000t^2 + 19364.94233t^3, \tag{16}$$

$$y(t) = 18 + 234t - 734.4000000t^2 - 19099.98000t^3. \tag{17}$$

From the approximation derived through the implementation of DTM and the approximation obtained using VIM [15], we can attain the numerical solution by setting t from 0 to 0.10 with intervals of 0.01. The computation is carried out using Maple 2020 software, and the results are presented in Table 2. Additionally, a graph is plotted, as illustrated in Fig. 1, to depict the trend of the equations.

3.1.1 Numerical Result of Case 1

Table 2. Numerical solution of DTM and VIM [15] in Case 1.

t	DTM		VIM	
	X	Y	X	Y
0.0	14.00000000	18.00000000	14.00000000	18.00000000
0.01	11.57385194	20.24746002	11.57385194	20.24746002
0.02	9.384867539	22.23344016	9.38486752	22.23344016
0.03	7.549236443	23.84334054	7.54923638	23.84334054
0.04	6.183148309	24.96256128	6.18314816	24.96256128
0.05	5.402792791	25.47650250	5.40279250	25.47650250
0.06	5.324359543	25.27056432	5.32435904	25.27056432
0.07	6.064038219	24.23014686	6.06403742	24.23014686
0.08	7.738018473	22.24065024	7.73801728	22.24065024
0.09	10.46248996	19.18747458	10.46248826	19.18747458
0.10	14.35364233	14.95602000	14.35364000	14.95602000

Based on Table 2 and Fig. 1, it can be clearly seen that the results have minimal disparity which can be said that the implementation of DTM compared to VIM in solving Lotka-Volterra has a similar accuracy. The approximate solution of DTM involved four terms which are same with VIM. From Fig. 1, it showed that when Y increases, X declines. This condition indicates that when the number of predators grows, the number of prey drops.

3.2 Classic Lotka-Volterra Model Case 2

By referring to the model in Eq. (12) and Eq. (13), consider the specific values illustrated below as $a = 1, b = 1, c = 0.1, d = 1$ with the initial conditions $x(0) = 16$ and $y(0) = 10$ [15]. The approximate solution when $n = 3$ is obtained and written as:

$$x(t) = 16 - 144t - 624.0000000t^2 + 6602.400000t^3, \tag{18}$$

$$y(t) = 10 + 159.0t + 544.0500000t^2 - 6828.535000t^3. \tag{19}$$

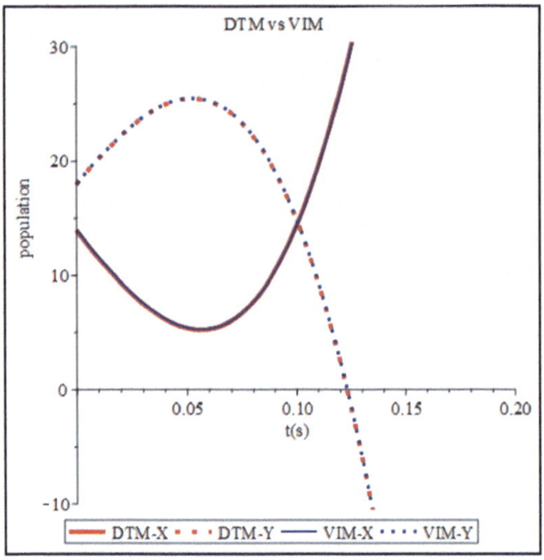

Fig. 1. The comparison of the results by DTM and VIM [15] in Case 1.

From the approximation derived through the implementation of DTM and the approximation obtained using VIM [15], we can attain the numerical solution by setting t from 0 to 0.10 with intervals of 0.01. The computation is carried out using Maple 2020 software, and the results are presented in Table 3. Additionally, a graph is plotted, as illustrated in Fig. 2, to depict the trend of the equations.

3.2.1 Numerical Result of Case 2

Table 3. Numerical solution of DTM and VIM [15] in Case 2.

t	DTM		VIM	
	X	Y	X	Y
0.0	16.00000000	10.00000000	16.00000000	10.00000000
0.01	14.50420240	11.63757646	14.50420241	11.63757648
0.02	12.92321920	13.34299172	12.92321925	13.34299182
0.03	11.29666480	15.07527456	11.29666496	15.07527488
0.04	9.664153600	16.79345376	9.664153984	16.79345453
0.05	8.065300000	18.45655812	8.065300750	18.45655962
0.06	6.539718400	20.02361644	6.539719696	20.02361903
0.07	5.127023200	21.45365750	5.127025258	21.45366161
0.08	3.866828800	22.70571008	3.866831872	22.70571622
0.09	2.798749600	23.73880298	2.798753974	23.73881173
0.01	1.962400000	24.51196500	1.962406000	24.51197700

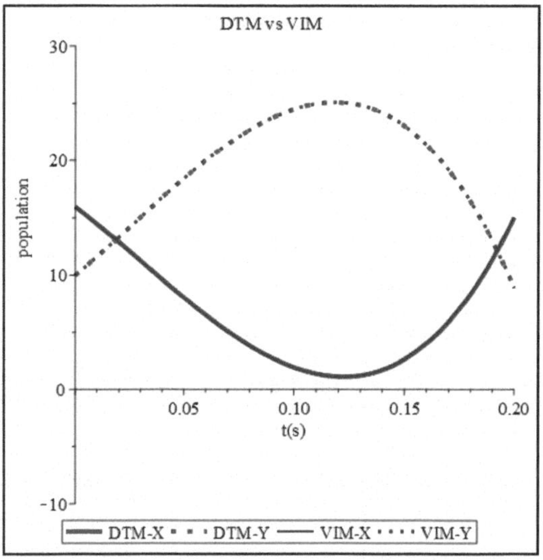

Fig. 2. The comparison of the results by DTM and VIM [15] in Case 2.

Upon examining Table 3 and Fig. 2, it is evident that the results exhibit minimal differences, suggesting a similar accuracy in the implementation of DTM and VIM for solving the Lotka-Volterra model. It is noteworthy that both DTM and VIM utilized a four-term approximate solution. As depicted in Fig. 2, an increase in Y results in a decline in X. This pattern signifies that a rise in the predator population correlates with a reduction in the prey population.

3.3 Modified Lotka-Volterra Model

This section discussed the results for modified Lotka-Volterra model. By referring to Eq. (14) and Eq. (15), the model is solved. The parameters are $a_{11} = -0.0014$, $a_{12} = -0.0012$, $a_{21} = -0.0009$, $a_{22} = -0.001$, $b_1 = 0.1$, $b_2 = 0.08$ with the initial conditions $N_1(0) = 4, N_2(0) = 10$ [5]. The approximate solution when $n = 3$ is obtained and written as:

$$N_1(t) = 4 + 0.3296t + 0.01106304000t^2 + 0.0001173886293t^3, \tag{20}$$

$$N_2(t) = 10 + 0.6640t + 0.01724160000t^2 + 0.0000783313067t^3. \tag{21}$$

By the application of DTM, we derived the approximate solution as shown in Eq. (20) and Eq. (21). The result is compared with the outcomes of several numerical methods from the findings of [5] by setting t from 0 to 1 with intervals of 0.10. The numerical solution of DTM is presented in Table 4 while the numerical solution of 3-term Adomian decomposition method (ADM), 2-iteration Variational iteration method (VIM) and fourth order Runge-Kutta method (RK4) from [5] is showed in Table 5 (Fig. 3).

3.3.1 Numerical Result for Modified Lotka-Volterra Model

Table 4. Numerical solution of DTM.

t	DTM	
	N_1	N_2
0.0	4.000000000	10.00000000
0.1	4.033070747	10.06657250
0.2	4.066363461	10.13349029
0.3	4.099878843	10.20075385
0.4	4.133617599	10.26836367
0.5	4.167580434	10.33632019
0.6	4.201768050	10.40462390
0.7	4.236181154	10.47327525
0.8	4.270820449	10.54227473
0.9	4.305686638	10.61162280
1.0	4.340780429	10.68131993

Table 5. Numerical solution of 3-term ADM [5], 2-iteration VIM [5] and RK4 [5].

t	ADM, ϕ_3		2-iteration VIM		RK4, $h = 0.001$	
	N_1	N_2	$N1$	$N2$	N_1	N_2
0.0	4.000000000	10.00000000	4.000000000	10.00000000	4.000000000	10.00000000
0.1	4.033070630	10.06657242	4.033070759	10.06657251	4.033070747	10.06657249
0.2	4.066362522	10.13348966	4.066363553	10.13349043	4.066363454	10.13349028
0.3	4.099875674	10.20075174	4.099879151	10.20075432	4.099878808	10.20075380
0.4	4.133610086	10.26835866	4.133618314	10.26836476	4.13361749	10.26836347
0.5	4.167565760	10.33631040	4.167581808	10.33632225	4.167580159	10.33631970
0.6	4.201742694	10.40460698	4.201770394	10.40462736	4.201767478	10.40462287
0.7	4.236140890	10.47324838	4.236184811	10.47328059	4.236180090	10.47327334
0.8	4.270760346	10.54223462	4.270825823	10.54228247	4.270818626	10.54227146
0.9	4.305601062	10.61156570	4.305694169	10.61163354	4.305683707	10.61161755
1.0	4.340663040	10.68124160	4.340790588	10.68133427	4.340775941	10.68131190

In this study, RK4 is used as the control. The numerical solution of DTM, 3-term ADM and 2-iteration VIM are compared with the numerical solution of RK4 to determine the accuracy of the methods. From Table 6, it can be seen that the absolute error between RK4 and DTM is the smallest compared to 3-term ADM and 2-iteration VIM. Hence, DTM is considered more accurate the 3-term ADM and 2-iteration VIM.

Table 6. Absolute error of RK4 [5] with DTM, 3-term ADM [5] and 2-iteration VIM [5].

t	Absolute error					
	RK4 – DTM		RK4 – ADM, ϕ_3		RK4 – 2-iteration VIM	
	$N1$	$N2$	N_1	N_2	N_1	N_2
0.0	0.000000000	0.000000000	0.000000000	0.000000000	0.000000000	0.000000000
0.1	0.000000000	0.000000010	0.000000117	0.000000070	0.000000012	0.000000020
0.2	0.000000007	0.000000010	0.000000932	0.000000620	0.000000099	0.000000150
0.3	0.000000035	0.000000050	0.000003134	0.000002060	0.000000343	0.000000520
0.4	0.000000109	0.000000200	0.000007404	0.000004810	0.000000824	0.000001290
0.5	0.000000275	0.000000490	0.000014399	0.000009300	0.000001649	0.000002550
0.6	0.000000572	0.000001030	0.000024784	0.000015890	0.000002916	0.000004490
0.7	0.000001064	0.000001910	0.000039200	0.000024960	0.000004721	0.000007250
0.8	0.000001823	0.000003270	0.000058280	0.000036840	0.000007197	0.000011010
0.9	0.000002931	0.000005250	0.000082645	0.000051850	0.000010462	0.000015990
1.0	0.000004488	0.000008030	0.000112901	0.000070300	0.000014647	0.000022370

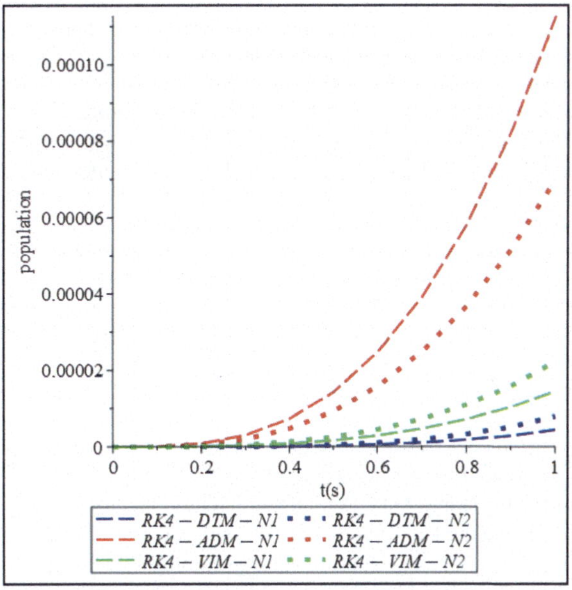

Fig. 3. Comparison of absolute error between RK4 [5] with DTM, 3-term ADM [5] and 2-iteration VIM [5].

4 Conclusion

This study has successfully proved the capability of DTM in solving Lotka-Volterra model. Throughout the study, the application of DTM in solving the Lotka-Volterra is accomplished by applying the theorem and computations are conducted using Maple

2020 software. In addition, the approximate solutions of the classic Lotka-Volterra model and modified Lotka-Volterra model are obtained via DTM. As a result, it can be deduced that DTM is capable of both solving the equation and obtaining approximations. Lastly, it showed that DTM has higher accuracy than 3-term ADM and 2-iteration VIM in solving Lotka-Volterra model when the results of RK4 is used as a control. Therefore, DTM can be considered as a semi-numerical-analytic approach with high accuracy.

References

1. Ahmad, M.Z., Alsarayreh, D., Alsarayreh, A., Qaralleh, I.: Differential transformation method (DTM) for solving sis and SI epidemic models. Sains Malaysiana **46**(10), 2007–2017 (2017)
2. Al-Sawalha, M.M., Noorani, M.S.M.: On solving the Lorenz system by differential transformation method. Chin. Phys. Lett. **25**(4), 1217–1219 (2008)
3. Anisiu, M.C.: Lotka Volterra and their model. In ResearchGate (2016)
4. Bataineh, A.S., Noorani, M., Hashim, I: Series solution of the multispecies Lotka-Volterra equations by means of the homotopy analysis method. Int. J. Differ. Eq., 1–14 (2008)
5. Batiha, B., Noorani, M.S.M., Hashim, I.: Variational iteration method for solving multispecies Lotka-Volterra equations. Comput. Math. Appl. **54**(7–8), 903–909 (2007)
6. Chowdhury, M.S.H., Rahman, M.M.: An accurate solution to the Lotka-Volterra equations by modified homotopy perturbation method. Int. J. Mod. Phys. Conf. Ser. **09**, 326–333 (2012)
7. Devi Sarita, J.M.: Applications of differential transform method applications of differential transform method. Int. Res. J. Eng. Technol., 544–549 (2021)
8. El-Shahed, M.: Application of differential transform method to non-linear oscillatory systems. Commun. Nonlinear Sci. Numer. Simul. **13**(8), 1714–1720 (2008)
9. Hung, H.C., Chiu, Y.C., Huang, H.C., Wu, M.C.: An enhanced application of Lotka-Volterra model to forecast the sales of two competing retail formats. Comput. Ind. Eng. **109**, 325–334 (2017)
10. Khatib, A.: Differential Transform Method for Differential Equations (2016)
11. Mirzaee, F.: Differential transform method for solving linear and nonlinear systems of ordinary differential equations. Appl. Math. Sci. **5**(70), 3465–3472 (2011)
12. Morris, S.A., Pratt, D.: Analysis of the Lotka-Volterra competition equations as a technological substitution model. Technol. Forecast. Soc. Chang. **70**(2), 103–133 (2003)
13. Ndii, M.Z., Anggriani, N., Supriatna, A.K.: Application of differential transformation method for solving dengue transmission mathematical model. In: AIP Conference Proceedings, vol. 1937 (2018)
14. Odibat, Z.M.: Differential transform method for solving Volterra integral equation with separable kernels. Math. Comput. Model. **48**(7–8), 1144–1149 (2008)
15. Rafei, M., Daniali, H., Ganji, D.D.: Variational iteration method for solving the epidemic model and the prey and predator problem. Appl. Math. Comput. **186**(2), 1701–1709 (2007)
16. Thongmoon, M., Pusjuso, S.: The numerical solutions of differential transform method and the Laplace transform method for a system of differential equations. Nonlinear Anal. Hybrid Syst **4**(3), 425–431 (2010)
17. Tonnang, H.E.Z., Nedorezov, L.V., Ochanda, H., Owino, J., Löhr, B.: Assessing the impact of biological control of Plutella xylostella through the application of Lotka-Volterra model. Ecol. Model. **220**(1), 60–70 (2009)
18. Madhusudanan, V., Tapas Bapu, B.R., Nagaraju, V., Pradeep, S.: Application of backstep control in diseased prey-predator system. Appl. Math. Inf. Sci. **11**(3), 789–798 (2017)
19. Zhang, W., Lam, J.S.L.: Maritime cluster evolution based on symbiosis theory and Lotka-Volterra model. Marit. Policy Manag. **40**(2), 161–176 (2013)
20. Zhou, J.K.: Differential Transformation and its Applications for Electrical Circuits. Huazhong University Press, Wuhan (1986)

Regression and Correlation Analysis of Air Pollutants and Air Quality Index in Johor

Lau Yu Fong and Srimazzura Basri(✉)

UTHM Kampus Cawangan Pagoh, Hub Pendidikan Tinggi Pagoh, KM 1, Jalan Panchor, 84600 Pagoh, Muar, Johor, Malaysia
mazzura@uthm.edu.my

Abstract. Air pollution is a significant challenge for many developed and developing countries. In fact, the issue is even more severe in the economies that are developing the fastest, as well as in the areas that have seen the greatest rises in industrialization. Beyond its impact on climate change, air pollution also affects both public and individual health. This study employs regression and correlation analysis to analyze the linear relationship between the air quality index (AQI) and the annual average concentration of various types of air pollutants at the eight selected stations in Johor from 2017 until 2021. There are five types of air pollutants that were tested with AQI: sulphur dioxide (SO_2), carbon monoxide (CO), ozone (O_3), suspended particulate matter (PM_{10}), and nitrogen dioxide (NO_2). The result shows that the concentration of air pollutants such as SO_2, O_3, and PM_{10} have a linear relationship corresponding to the value of AQI. PM_{10} has the highest correlation coefficient with AQI compared to others, highlighting its strong influence on air quality. In short, there is a significant impact of SO_2, O_3, and PM_{10} to the AQI. Therefore, it is crucial to control emissions of these pollutants, as they play a substantial role in global warming and climate change.

Keywords: Air pollution · Air quality index (AQI) · Regression · Correlation

1 Introduction

The air is a vital resource for human life, just like land and water. It is essential for respiration, enabling living organisms to convert oxygen into energy. According to [1], air emissions include air pollution emissions and greenhouse gas emissions. Air pollutants, such as carbon monoxide (CO), sulphur dioxide (SO_2), nitrogen oxides (NOx), volatile organic compounds (VOCs), ozone (O_3), heavy metals, and respirable particulate matter ($PM_{2.5}$ and PM_{10}), differ in their chemical composition, reaction properties, emission, time of disintegration and ability to diffuse in long or short distances [2].

Air pollution is a critical environmental issue with global consequences. It results from the release of harmful pollutants into the atmosphere from various sources, including industrial facilities, power plants, vehicles, and airplanes. Natural sources, such as volcanic eruptions and wildfires, also contribute to air pollution. This issue cannot be overlooked, as air pollutants play a significant role in altering weather patterns. Greenhouse gas levels are so high primarily because humans have released them into the air by

P. Chang et al. (Eds.): SCIEMATHIC 2024, SPPHY 427, pp. 55–66, 2025.
https://doi.org/10.1007/978-981-96-7679-8_6

burning fossil fuels [3]. These gases absorb solar energy and retain it close to the Earth's surface. Then, the gases in the atmosphere trap heat, contributing to global warming.

Air pollution impacts not only climate change but also poses serious risks to public and individual health. A study by the Centre for Research on Energy and Clean Air (CREA) and Greenpeace Malaysia found that an estimated 32,000 avoidable deaths occur in Malaysia every year because of air pollution [4]. Given its harmful effects on human health, it is crucial to identify and understand the relationships between emission sources, deteriorating air quality, and associated health risks. Long-term exposure to poor air quality can lead to severe health consequences. We should reduce the levels of key air pollutants that contribute to climate change to improve air quality levels. Reducing global greenhouse gas emissions to slow climate change could prevent millions of premature deaths due to air pollution over the next century, according to a study funded by the National Institute of Environment Health Sciences (NIEHS) [5].

Consequently, the aim of this research is to analyse the relationship between the air quality index and the annual average concentration of the different types of air pollutants. Three objectives have been established for this research. Firstly, to determine the correlation between the air quality index and the annual average concentration of the different types of air pollutants. Secondly, to examine the extent of variation in the linear relationship between air pollutant concentrations and the air quality index. Last but not least, to determine the linear relationship between the air quality index and the annual average concentration of the different types of air pollutants.

2 Methodology

2.1 Data Collection

The data collected on the daily air quality index (AQI) and the annual average concentration of the different types of air pollutants for eight selected stations in Johor from 2017 until 2021 was obtained from the historical air quality data platform provided by the Malaysian Department of the Environment [6]. The annual average concentrations of sulphur dioxide (SO_2), carbon monoxide (CO), ozone (O_3), suspended particulate matter (PM_{10}), and nitrogen dioxide (NO_2) and the calculated average value of AQI for eight stations in Johor state from 2017 to 2021 were recorded in Table 1 and Table 2. The air quality index is calculated on average based on the daily air quality index obtained in that year.

2.2 Statistical Tests

Simple linear regression is a straight line that models the relationship between an independent variable, x and a dependent variable, y. A simple linear regression model is defined as $y = \beta_0 + \beta_1 x + \varepsilon$, where the intercept, β_0 and the slope, β_1 are unknown regression coefficients and ε is a random error component. The errors are assumed to have a mean of zero and an unknown variance, σ^2. The parameters β_0 and β_1 play an important role in the total analytical procedure. The slope β_1 is the change in the mean of the distribution of y produced by a unit change in x [7]. In other words, the parameter β_1 is the change in y per unit increase in x [8].

The coefficient of determination, R^2 measures the proportion of variance in y that is explained by x in the regression model,

$$R^2 = \frac{SSR}{SST} = 1 - \frac{SSE}{SST} \tag{1}$$

where SSR is the sum of squares of regression, SST is the sum of squares total and SSE is the sum of squares error. A higher R^2 value indicates a stronger explanatory power of the model.

Correlation analysis measures the strength of a linear relationship that exists between two variables. Correlation also known as the degree to which a straight line fits the scatter graph represents the relationship between two variables. The correlation is given by $r = \pm\sqrt{R^2}$. A plus or minus sign is attached to this measure according to whether the slope of the fitted regression line is positive or negative [9]. The range of r is $-1 \leq r \leq 1$. If the correlation is equal to $+1.00$, it implies that there is a perfect positive linear relationship between x and y. If the points all fall on a straight line, going from bottom right to top left, the correlation will be –1.00, which means that it is a perfect negative linear relationship [8].

2.3 Analysis of Variance (ANOVA)

The probability associated with R^2, or the probability of obtaining an R^2 value as high as it is in the situation that the actual value in the population is zero, defines the significance of the value F [10]. Thus, a further approach for determining the significance of regression is to use the analysis of variance (ANOVA) method. ANOVA is a multiple regression output generally used to examine differences between group means. This statistical method separates observed variance data into different components to use for additional tests.

Thus, the total variability in the response variable is divided into two significant components, which are $SST = SSR + SSE$. Besides, the F value can be obtained from the ANOVA table, where the value of the test statistic (f_{test}), whereas the *Significance F* is the probability of the F value. Consider test statistic F is

$$F = \frac{SSR}{\frac{\sigma^2}{1}} \div \frac{SSE}{\frac{\sigma^2}{n-2}} = \frac{MSR}{MSE} \tag{2}$$

where MSR is the mean squared regression and MSE is the mean squared error, following the F distribution with 1 and (n – 2) degrees of freedom under the null hypothesis $H_0 : \rho = 0$. Correlation analysis should involve the estimation of ρ. It is often useful to test the hypothesis of whether the correlation coefficients are equal to zero or not, that is, $H_0 : \rho = 0$ versus $H_1 : \rho \neq 0$.

Failing to reject H_0 implies that there is no linear relationship between x and y. Alternatively, if H_0 is rejected, this implies that x is of value in explaining the variability in y. The appropriate test statistic for this hypothesis is

$$t = \frac{r\sqrt{n-2}}{\sqrt{1-r^2}} \tag{3}$$

which follows the t-distribution with $(n - 2)$ degrees of freedom. If $|t| > t_{\alpha/2,\, n-2}$, we reject H_0 and conclude that the correlation coefficient $\rho \neq 0$, gives strong statistical evidence of a linear association between two variables.

2.4 Significance Testing of Regression Coefficients

Regression and correlation are closely related, where regression is a more powerful tool in many situations, which is useful in developing quantitative relationships between variables and can be used in prediction. Meanwhile, correlation is a measure of association between two variables. The hypotheses for testing the significance of regression coefficient, such as $H_0 : \beta_1 = 0$ versus $H_1 : \beta_1 \neq 0$.

These hypotheses relate to the significance of regression. Failing to reject H_0 implies that there is no linear relationship because there is no slope between x and y. Alternatively, if H_0 is rejected, this implies that the straight-line model is adequate and there is a linear effect of x. The appropriate test statistic is

$$t = \frac{\beta_1}{\sqrt{\hat{\sigma}^2 / S_{xx}}} \tag{4}$$

follows the t-distribution with $(n - 2)$ degrees of freedom. If $|t| > t_{\alpha/2,\, n-2}$, then we can make a conclusive decision to reject H_0 [7].

3 Result and Discussion

3.1 Annually Average Concentration of the Different Types of Air Pollutants

According to Malaysian Ambient Air Quality Guidelines, the concentration of sulphur dioxide (SO_2) not exceeding 0.035 ppm, carbon monoxide (CO) not exceeding 9 ppm, ozone (O_3) not exceeding 0.1 ppm, particulate matter < 10 μm (PM_{10}) not exceeding 120 μg/m^3, and nitrogen dioxide (NO_2) not exceeding 0.16 ppm [11]. This study focuses on the annual average concentrations of these air pollutants from 2017 to 2021 at eight selected monitoring stations in Johor, measured at specific times of the year, as detailed in Table 1.

Based on Table 1, the annual average concentration of SO_2, CO, O_3, PM_{10}, and NO_2 remained within the limits set by the Malaysian Ambient Air Quality Guidelines. To further analyze air quality trends, statistical tests will be conducted to identify the linear relationship between the annual average concentration of these pollutants and the value of air quality index (AQI). The results of this correlation analysis will be discussed in Subsect. 3.3.

Table 1. Annually average concentration of the different types of air pollutants in the air for eight selected stations [11].

Year	Stations	Pollutant				
		Sulphur Dioxide (SO_2)	Carbon Monoxide (CO)	Ozone (O_3)	Particulate Matter (PM_{10})	Nitrogen Dioxide (NO_2)
2017	Batu Pahat	0.0012	0.576	0.017	24.22	0.0049
	Kluang	0.0000	0.000	0.000	25.20	0.0000
	Kota Tinggi	0.0015	0.589	0.019	21.62	0.0061
	Larkin	0.0021	0.831	0.014	29.59	0.0138
	Pasir Gudang	0.0027	0.796	0.015	29.29	0.0144
	Pengerang	0.0015	0.000	0.000	25.22	0.0072
	Segamat	0.0012	0.761	0.026	24.52	0.0065
	Tangkak	0.0000	0.000	0.000	0.000	0.0000
2018	Batu Pahat	0.0014	0.656	0.017	24.42	0.0050
	Kluang	0.0000	0.000	0.000	22.88	0.0000
	Kota Tinggi	0.0012	0.571	0.019	19.37	0.0047
	Larkin	0.0013	0.775	0.015	29.32	0.0129
	Pasir Gudang	0.0016	0.734	0.015	29.50	0.0130
	Pengerang	0.0013	0.000	0.000	25.50	0.0059
	Segamat	0.0009	0.726	0.024	23.38	0.0057
	Tangkak	0.0008	0.478	0.018	28.69	0.0041
2019	Batu Pahat	0.0015	0.597	0.019	30.28	0.0058
	Kluang	0.0000	0.000	0.000	28.50	0.0000
	Kota Tinggi	0.0011	0.544	0.021	22.53	0.0041
	Larkin	0.0018	0.734	0.017	35.60	0.0135
	Pasir Gudang	0.0015	0.691	0.017	31.47	0.0115
	Pengerang	0.0016	0.000	0.000	24.09	0.0042
	Segamat	0.0017	0.700	0.023	31.83	0.0060
	Tangkak	0.0013	0.571	0.021	36.50	0.0042
2020	Batu Pahat	0.0018	0.522	0.016	16.94	0.0045
	Kluang	0.0000	0.000	0.000	14.05	0.0000
	Kota Tinggi	0.0011	0.459	0.018	15.16	0.0030
	Larkin	0.0018	0.363	0.014	22.98	0.0089
	Pasir Gudang	0.0011	0.579	0.013	20.63	0.0094

(continued)

Table 1. (*continued*)

Year	Stations	Pollutant				
		Sulphur Dioxide (SO_2)	Carbon Monoxide (CO)	Ozone (O_3)	Particulate Matter (PM_{10})	Nitrogen Dioxide (NO_2)
	Pengerang	0.0013	0.000	0.000	14.43	0.0038
	Segamat	0.0017	0.562	0.015	19.14	0.0047
	Tangkak	0.0011	0.481	0.014	20.72	0.0039
2021	Batu Pahat	0.0016	0.521	0.018	18.73	0.0045
	Kluang	0.0000	0.000	0.000	16.67	0.0000
	Kota Tinggi	0.0012	0.386	0.021	15.66	0.0033
	Larkin	0.0019	0.429	0.016	24.22	0.0097
	Pasir Gudang	0.0014	0.571	0.013	21.11	0.0104
	Pengerang	0.0014	0.000	0.000	17.49	0.0035
	Segamat	0.0021	0.605	0.016	18.83	0.0049
	Tangkak	0.0014	0.600	0.016	23.31	0.0045

3.2 Air Quality Index (AQI)

Finding out the current air quality in real-time can be very challenging. Therefore, this study utilizes the annual average air quality index (AQI) to assess air quality trends. The AQI for some stations is inadequate for 365 days due to technical problems [11]. As a result, the AQI values are calculated based on the available data.

Based on Table 2, the AQI value from 2017 to 2018 indicates that the air quality status at the selected eight stations was generally good, meaning pollution levels were low and posed no significant risk to human health. In 2019, air quality at most selected stations declined to a moderate status, except for Kota Tinggi, which showed that the status was good. This situation was primarily attributed to the 2019 Southeast Asian haze.

From 2020 to 2021, air quality improved, with most stations returning to a good status. However, Larkin recorded a moderate status. This improvement was largely due to the 2020 Malaysia Movement Control Order (MCO), which led to a significant reduction in motor vehicle traffic on the road, and industry operations were halted, which reduced the amount of hazardous air pollutants released into the atmosphere [12]. As a result, there was a significant improvement in air quality status from 2019 to 2021. A statistical analysis will be conducted in Subsect. 3.3 to examine the relationship between the annual average concentration of air pollutants and AQI values in greater detail.

Table 2. Air quality index (AQI) for eight selected stations.

Year	Stations	AQI	Air Quality Status	Year	Stations	AQI	Air Quality Status
2017	Batu Pahat	25	Good	2020	Batu Pahat	40	Good
	Kluang	24	Good		Kluang	35	Good
	Kota Tinggi	33	Good		Kota Tinggi	33	Good
	Larkin	36	Good		Larkin	51	Moderate
	Pasir Gudang	35	Good		Pasir Gudang	47	Good
	Pengerang	24	Good		Pengerang	35	Good
	Segamat	29	Good		Segamat	46	Good
	Tangkak	0	Good		Tangkak	46	Good
2018	Batu Pahat	33	Good	2021	Batu Pahat	46	Good
	Kluang	32	Good		Kluang	41	Good
	Kota Tinggi	30	Good		Kota Tinggi	41	Good
	Larkin	38	Good		Larkin	55	Moderate
	Pasir Gudang	37	Good		Pasir Gudang	49	Good
	Pengerang	33	Good		Pengerang	43	Good
	Segamat	35	Good		Segamat	48	Good
	Tangkak	39	Good		Tangkak	50	Good
2019	Batu Pahat	57	Moderate				
	Kluang	54	Moderate				
	Kota Tinggi	49	Good				
	Larkin	62	Moderate				
	Pasir Gudang	59	Moderate				
	Pengerang	51	Moderate				
	Segamat	61	Moderate				
	Tangkak	64	Moderate				

3.3 Statistical Test

Considering the sample data from Table 1 and Table 2, we can generate summary regression statistics using Microsoft Excel. From this output, we can particularly obtain the value of the sample correlation coefficient, r directly by taking the square root of R^2. The correlation coefficient measures the strength and direction of the linear relationship between two variables.

If the correlation coefficient range is $0.5 < r < 1$, it indicates a strong positive linear relationship and $0 < r < 0.5$ means a weak positive linear relationship. There is a strong negative linear relationship when the correlation coefficient range is $-1 < r < -0.5$ and a weak negative linear relationship when it is $-0.5 < r < 0$. Meanwhile, a value of 0 implies that the two variables are not linearly related [13].

Table 3. Summary output regression statistics.

AQI and air pollutants	r	R^2
SO_2	0.371125	0.137734
CO	0.301596	0.090960
O_3	0.328246	0.107745
NO_2	0.273001	0.074530
PM_{10}	0.569248	0.324043

Table 3 presents the summary of regression statistics for air quality index (AQI) and air pollutants such as sulphur dioxide (SO_2), carbon monoxide (CO), ozone (O_3), suspended particulate matter (PM_{10}), and nitrogen dioxide (NO_2). Since the correlation coefficients for SO_2, CO, O_3, and NO_2 are all below 0.5, they indicate a weak positive linear relationship with AQI. However, PM_{10} has a correlation coefficient of 0.569248, which exceeds 0.5, suggesting a strong positive linear relationship with AQI.

Meanwhile, the obtained R^2 of SO_2, CO, O_3, NO_2, and PM_{10} were 0.137734, 0.090960, 0.107745, 0.074530, and 0.324043, respectively. This means that approximately 13.77%, 9.10%, 10.77%, 7.45%, and 32.40% of the variation in AQI is accounted by a linear relationship with SO_2, CO, O_3, NO_2, and PM_{10}, respectively.

The value of the test statistic (f_{test}) can be obtained from the ANOVA table in Table 4 whereas the *Significance F* is the probability of the *F* value.

Table 4. ANOVA table.

AQI and air pollutants	*F*	*Significance F*
SO_2	6.069927	0.018390
CO	3.802351	0.058587
O_3	4.588736	0.038653
NO_2	3.060204	0.088305
PM_{10}	18.21660	0.000126

Next, we test the hypotheses $H_0 : \rho = 0$ versus $H_1 : \rho \neq 0$, where $H_0 : \rho = 0$ means there is a nonlinear relationship between AQI and the air pollutants, meaning we cannot construct an approximate 95% confidence interval on the correlation coefficient. While $H_1 : \rho \neq 0$ means there is a linear relationship between AQI and air pollutants.

In order to make a decision on the significance of regression, we compare the *Significance F* with the chosen significance level α, as we will reject H_0 if *Significance F* $< \alpha$ and fail to reject H_0 if *Significance F* $\geq \alpha$. In this study, we chose $\alpha = 0.05$.

Since the values of *Significance F* for SO_2, O_3, and PM_{10} are lower than 0.05, we reject H_0 and conclude that the data on AQI and these pollutants provide strong evidence for the existence of linearity in the regression model at 0.05 significance level. Conversely, the *Significance F* values for CO and NO_2 are greater than 0.05, meaning we fail to reject H_0. So, we conclude that the data do not provide strong evidence of a linear relationship between AQI and CO or NO_2 at 5% significance level. As a result, we cannot construct a reliable 95% confidence interval on the correlation coefficient.

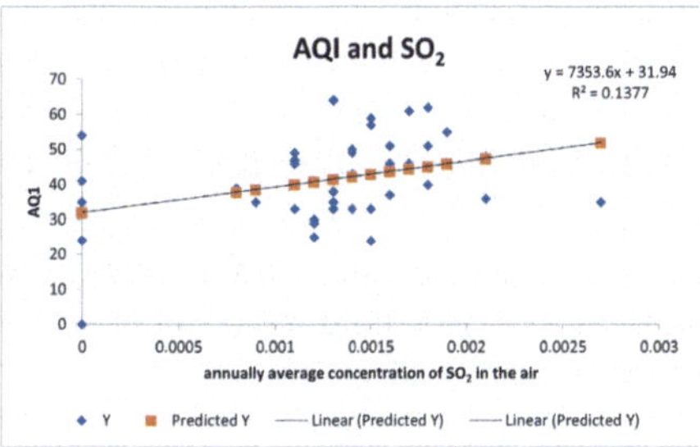

Fig. 1. Fitted regression line of AQI and SO2

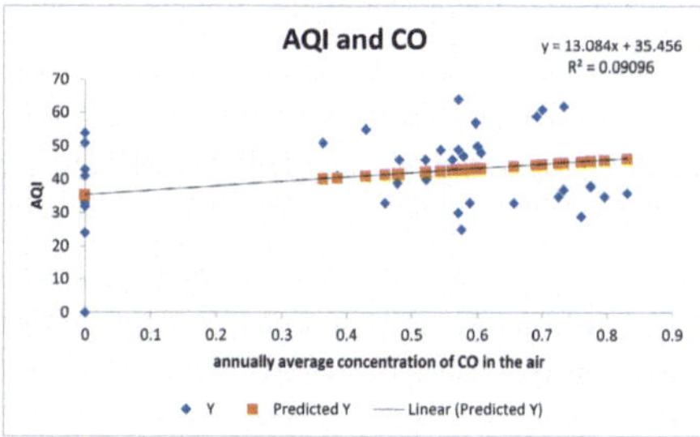

Fig. 2. Fitted regression line of AQI and CO

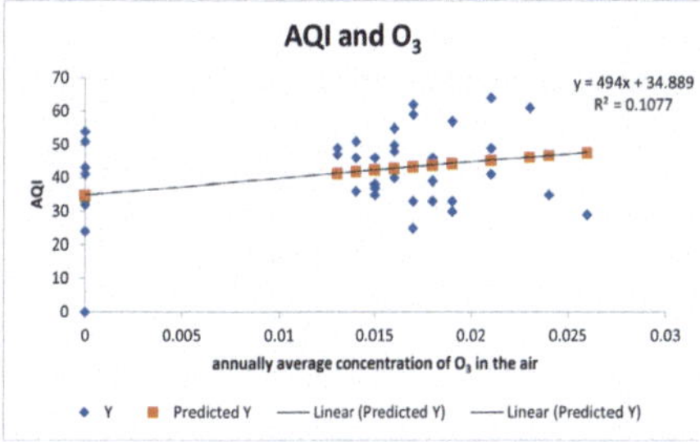

Fig. 3. Fitted regression line of AQI and O_3

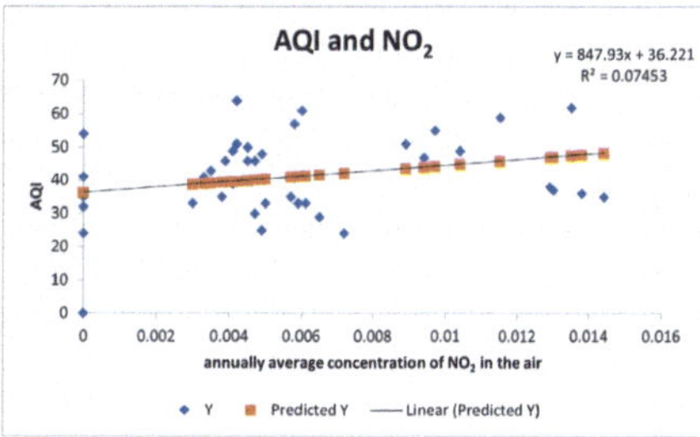

Fig. 4. Fitted regression line of AQI and NO_2

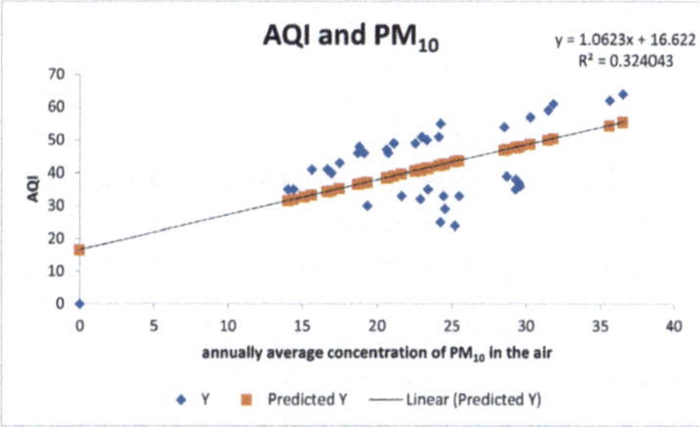

Fig. 5. Fitted regression line of AQI and PM_{10}

Furthermore, a scatter plot was generated to illustrate the fitted values and paired measurements. Figure 1, 2, 3, 4 and 5 depict the regression lines for AQI in relation to SO_2, CO, O_3, NO_2, and PM_{10}. The estimated regression equations for AQI are SO_2: $y = 7353.6x + 31.94$, CO: $y = 13.084x + 35.456$, O_3: $y = 494x + 34.889$, NO_2: $y = 847.93x + 36.221$, and PM_{10}: $y = 1.0623x + 16.622$.

From Fig. 1, 3, and 5, we can see the prediction slope increase with a lot of data points closely clustered around it. Notably, PM_{10} is the highest correlated with AQI, indicating a strong positive linear relationship. This is evident in Fig. 5, where the slope is steeper compared to the others. These findings confirm the existence of a linear relationship between AQI and the annual average concentrations of SO_2, O_3, and PM_{10} in the air.

On the other hand, Fig. 2 and 4 show that the prediction slope slightly increases with data points more scattered and farther from the prediction line. It proved that the correlation between AQI and CO or NO_2 is weakly positive. This is because only 9.10% and 7.45% of the variation in AQI can be attributed to CO and NO_2, respectively.

Additionally, we test whether the slope of the regression line, β_1 is equal to zero, which indicates no linear relationship between AQI and air pollutants. The appropriate hypotheses are $H_0 : \beta_1 = 0$ versus $H_1 : \beta_1 \neq 0$. This hypothesis test determines the significance of regression. Failing to reject H_0 implies that there is no linear relationship between AQI and air pollutants. Alternatively, if H_0 is rejected, this implies that the air pollutant plays a role in explaining the variability in AQI and could indicate that the air pollutant has a linear effect or that the straight-line model is adequate.

To make this decision, we compare the *P*-values with the chosen significance level, $\alpha = 0.05$. Table 5 shows the estimates of regression coefficients, β_1.

Table 5. Estimates of regression coefficients, β_1.

AQI and air pollutants	*t*	*P-value*
SO_2	2.463722	0.018390
CO	1.949962	0.058587
O_3	2.142134	0.038653
NO_2	1.749340	0.088305
PM_{10}	4.268091	0.000126

The *P*-value for β_1 for SO_2, O_3 and PM_{10} were $P(T > 2.463722) = 0.01839$, $P(T > 2.142134) = 0.038653$, $P(T > 4.268091) = 0.000126$, respectively. Since these values are lower than the significance level, 0.05, we reject H_0, confirming that the slopes for SO_2, O_3 and PM_{10} are significantly non-zero at 5% significance level. This indicates that SO_2, O_3, and PM_{10} have a significant impact on AQI, suggesting a linear relationship between these pollutants and AQI. Conversely, the *P*-value for β_1 for CO and NO_2 were $(T > 1.949962) = 0.058587$ and $(T > 1.74934) = 0.088305$, respectively, which are greater than the significance level, 0.05, meaning the slopes for CO and NO_2 are not significantly different from zero. Consequently, this suggests that CO and NO_2 do not have a statistically significant linear relationship with AQI. So, CO and NO_2 do not impact changes in the AQI.

4 Conclusion

This study provides several important conclusions regarding the relationship between air quality index (AQI) and air pollutants. First, the results indicate that AQI is correlated with sulphur dioxide (SO_2), carbon monoxide (CO), ozone (O_3), suspended particulate matter (PM_{10}), and nitrogen dioxide (NO_2), confirming a relationship between AQI and these five pollutants. Among them, PM_{10} exhibits the strongest correlation with AQI, where PM_{10} alone accounts for 32.40% of the total variation in AQI, the highest proportion among all pollutants analyzed. Based on hypothesis testing, the study confirms a linear relationship between AQI and PM_{10}, SO_2, and O_3. Both the correlation coefficient test and the regression slope test support this finding, indicating that changes in AQI are significantly influenced by variations in PM_{10}, SO_2, and O_3. However, for CO and NO_2, the hypothesis tests show that AQI has no significant linear relationship with these pollutants, suggesting that their changes do not have a strong impact on AQI. These findings have important implications for air quality management. Since PM_{10}, SO_2, and O_3 are directly linked to AQI, controlling emissions of these pollutants can help improve air quality. As these pollutants contribute to global warming, reducing their concentrations can also play a role in climate change mitigation.

References

1. Tan, Z.: Air pollution and greenhouse gases: from basic concepts to engineering applications for air emission control. Springer Science & Business Media, Singapore (2014). https://doi.org/10.1007/978-981-287-212-8
2. Kampa, M., Castanas, E.: Environmental pollution. Human Health Eff. Air Pollut. **151**(2), 362–367 (2008). https://www.sciencedirect.com/science/article/abs/pii/S0269749107002849
3. Nunez, C.: Carbon dioxide levels are at a record high (2019). https://www.nationalgeographic.com/environment/article/greenhouse-gases
4. Greenpeace. The state of air quality in Malaysia (2022). https://www.greenpeace.org/malaysia/press/49795/the-state-of-air-quality-in-malaysia/
5. West, J.: Reducing greenhouse gas emissions benefits air quality save lives (2013). https://sph.unc.edu/sph-news/reducing-greenhouse-gases-benefits-air-quality-saves-lives/
6. Institution & University Registration. Air Quality Historical Data Platform (2023). https://aqicn.org/data-platform/register/
7. Montgomery, D.C., Peck, E.A., Vining, G.G.: Introduction to Linear Regression Analysis, 6th edn. India: John Wiley and Sons, Inc. (2021). https://books.google.com.my/books?hl=en&lr=&id=tCIgEAAAQBAJ&oi=fnd&pg=PR13&dq=linear+regression&ots=lfzfTpm0Nm&sig=L1HuYhGawNRzpZR5QRhpMxQ7ME&redir_esc=y#v=onepage&q=linear%20regression&f=false
8. Myers, R.H.: Classical and Modern Regression with Applications, 2nd edn. Duxbury, Canada (1990)
9. Kutner, M.H., Nachtsheim, C.J., Neter, J.: Applied Linear Regression Models, 4th edn. McGraw-Hill Higher Education, New York (2004)
10. Miles, J., Shevlin, M.: Applying Regression & Correlation. SAGE Publications, London (1988)

11. Department of Statistics Malaysia. Environment Statistics Johor 2022. Putrajaya: Department of Statistics Malaysia (2022). https://www.dosm.gov.my/uploads/release-content/file_20221216093823.pdf
12. Abdullah, S., et al.: Air quality status during 2020 Malaysia Movement Control Order (MCO) due to 2019 novel coronavirus (2019-nCoV) pandemic. Sci. Total Environ. **729**, 139022 (2020). https://doi.org/10.1016/j.scitotenv.2020.139022
13. Kahane, L.H.: Regression Basics, 2nd edn. SAGE Publications, London (2008)

Benchmarking Shortest Path Solutions in Mazes: A* vs IDA* Algorithms

Lee Siaw Chong(✉) and Jessie Heng Xue Yun

Department of Mathematics and Statistics, Faculty of Applied Sciences and Technology, UTHM Kampus Cawangan Pagoh, Hab Pendidikan Tinggi Pagoh, KM 1, Jalan Panchor, 84600 Pagoh, Muar, Johor, Malaysia

sclee@uthm.edu.my

Abstract. Mazes are puzzles that require a generous amount of time to locate an optimal route that leads from the start point to the end node. Pathfinding algorithms are often employed to find the optimal solution route. This study aims to compare the performances of the A* and IDA* (Iterative Deepening A*) algorithms in solving two distinct types of mazes: straightforward and complex. Additionally, four different heuristics (1/2 Manhattan, Chebyshev, Euclidean, and Octile distances) were utilised in conjunction with these algorithms to estimate the cost of any node in the maze to the target node. The findings revealed that Octile distance provided the best approximation among the four heuristics tested, followed by Euclidean, Chebyshev, and 1/2 Manhattan. All mazes used in this study were of size 18×18. In cases where the mazes had straightforward paths, the IDA* algorithm outperformed the A* algorithm by a margin of 31.41% to 48.43% when using the Octile distance. On the other hand, for complex mazes, A* algorithm computed the optimal path faster than the IDA* algorithm by at least 99.68% to 99.86%. Furthermore, both A* and IDA* algorithms demonstrated faster computations when the applied heuristic had a higher initial value and a greater increment rate. These results suggest that A* is generally better for complex mazes, while IDA* may be preferable for straightforward mazes under memory-constrained environments.

Keywords: A* algorithm · Iterative Deepening A* (IDA*) algorithm

1 Introduction

A maze can be defined as a path or a collection of paths that connect an entrance to a goal. Mazes are solved by searching for a route, starting from the initial point, that leads to the endpoint while avoiding obstacles along the way. Pathfinding techniques are often used when solving a maze to avoid taking unnecessary or repeated paths so that a user can reach the destination in the shortest time possible.

To implement pathfinding techniques, a maze is first interpreted as a graph. Each unit of the maze map is a node, and its adjacent nodes are those accessible from the current node. Then, graph pathfinding techniques can be utilised to compute the optimal

P. Chang et al. (Eds.): SCIEMATHIC 2024, SPPHY 427, pp. 67–81, 2025.
https://doi.org/10.1007/978-981-96-7679-8_7

route. Pathfinding is the procedure of plotting the route between two points, an origin and a destination, by using a computer application [1]. It has been the core focus area in the development of video games for years. Pathfinding algorithms should achieve two main objectives: first, finding a route from the starting point (origin) to the destination (endpoint), and second, navigating around any obstacles that may obstruct the path.

Various search algorithms can either compute the shortest possible path or calculate the path in the shortest amount of time possible, but rarely both [2]. This is because the algorithms that can calculate the shortest path usually take a lot of computation time. In contrast, algorithms that can find the path in the shortest amount of time frequently fail to come up with the shortest path available.

Search algorithms can be classified into two main categories: uninformed search, also known as blind search, and informed search, also called heuristic search [3, 4]. An uninformed search is a method that does not provide prior knowledge of the solution to the problem. Uninformed search includes algorithms such as Depth-First Search (DFS), Breadth-First Search (BFS), and Dijkstra's algorithm. In contrast to uninformed search algorithms, an informed search is guided by heuristics or additional information about its domain in the search process. This approach prioritizes the expansion of nodes that are likely to lead to a solution, allowing the algorithm to avoid less promising paths. Informed search techniques include the A* algorithm and the Iterative Deepening A* (IDA*) algorithm.

Dijkstra's algorithm, an uninformed search algorithm, calculates the shortest path from a start point to every other graph node [5] and therefore is inferior on large graphs since it requires a lot of storage space to keep all possible paths [6]. The A* algorithm, unlike the uninformed Dijkstra's algorithm, is an extension that incorporates informed search techniques, making it more efficient [7]. A* algorithm uses a heuristic function that provides additional knowledge of how far away the goal node is from the current expanded node. A heuristic function assists the A* algorithm by estimating the minimum cost from any vertex in a graph to the destination node. Commonly used heuristics are Manhattan distance, Chebyshev distance, Euclidean distance, and Octile distance. One potential downside of the A* algorithm is that it records every node explored in the puzzle, resulting in relatively high memory usage to hold all the computed data. The IDA* algorithm can be used as an alternative, which avoids this memory issue by performing a series of depth-first searches with iteratively increasing cost limits [8].

IDA* algorithm is a memory-efficient alternative to the A* algorithm because it uses an iterative deepening search to ensure it will return the shortest-length solution [9]. An iterative deepening search sets a depth limit on the search and increases it iteratively. This algorithm does not keep track of the nodes it has explored but redo all its work from the previous iteration with a new depth limit, otherwise known as threshold. This ensures that the IDA* algorithm does not consume too much memory, but for large mazes that require many detours, this may result in longer computation time.

This research aims to return an optimal path in 18×18 mazes that allow eight-direction movement in the shortest possible time by using the A* and IDA* algorithms. The time used by these two algorithms will be benchmarked. Since a node can be expanded in eight directions, Manhattan distance will overestimate the actual heuristic value, and hence, a modified version of Manhattan distance, 1/2 Manhattan distance, will be used. As a result, in addition to the A* and IDA* algorithms, the time used by 1/2 Manhattan distance, Chebyshev distance, Euclidean distance and Octile distance will also be evaluated.

2 Methodology

In the previous section, we mentioned the four heuristics used in this work. This section will dive deeper into the details of each heuristic and then provide an overview of the general framework for the A* and IDA* search algorithms.

2.1 Admissible Heuristics

A* and IDA* algorithms are distinguished from other uninformed searches by their heuristics. A heuristic function, $h(n)$, for node n, estimates the minimum cost from any node in a graph to the target node. Nonetheless, the knowledge provided will only be useful if the heuristic function returns a value that is always less than or equal to the actual cost of travelling from node n to the target node. In contrast, if the heuristic value exceeds the actual cost, the algorithm is not guaranteed to find the shortest possible path, although the computation might be faster. Thus, an admissible heuristic should always underestimate the cost from node n to the goal node [10, 11]. This study employs four admissible heuristics: 1/2 Manhattan, Chebyshev, Euclidean, and Octile distances (see Fig. 1).

Manhattan distance is a heuristic typically applied to a square grid, allowing movement in four directions, namely North, South, East, and West, as shown in Fig. 1. The Manhattan distance of the node $n(x_n, y_n)$ to the destination node (x_{end}, y_{end}) can be expressed as in Eq. (1).

$$h_M(n) = |x_n - x_{end}| + |y_n - y_{end}| \tag{1}$$

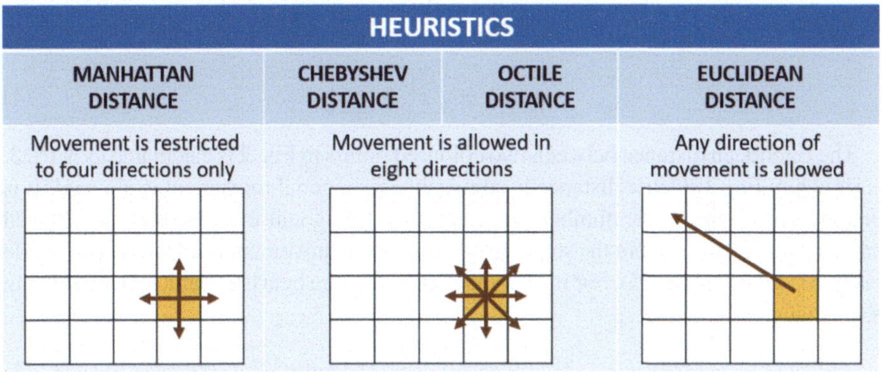

Fig. 1. Manhattan, Chebyshev, Octile and Euclidean heuristics.

Manhattan distance is deemed non-admissible for this research as it will overestimate diagonal routes. Hence, a modified version of Manhattan distance, 1/2 Manhattan, will be used. Equation (2) shows how the 1/2 Manhattan distance is calculated for a node (x_n, y_n) to (x_{end}, y_{end}). Letting $\hat{x} = x_n - x_{end}$ and $\hat{y} = y_n - y_{end}$,

$$h_{\frac{M}{2}}(n) = \frac{|\hat{x}| + |\hat{y}|}{2} \tag{2}$$

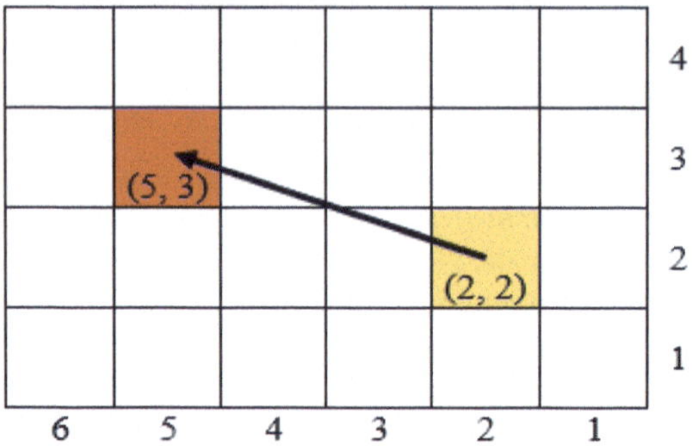

Fig. 2. A simple maze

For the maze shown in Fig. 2, the 1/2 Manhattan distance is calculated as 2. On the other hand, as shown in Fig. 1, Chebyshev's distance can be applied to a maze that permits eight directions of movement. This heuristic measures the distance between two nodes of the graph as the maximum difference over any x-axis or y-axis, as in Eq. (3).

$$h_{c(n)} = \max(|\hat{x}|, |\hat{y}|) \tag{3}$$

Hence, the Chebyshev distance for the maze shown in Fig. 2 is 3. Euclidean distance is the Pythagorean distance between two nodes in a Euclidean space, as in Fig. 1 and Eq. (4).

$$h_E(n) = \sqrt{\hat{x}^2 + \hat{y}^2} \tag{4}$$

The Euclidean distance between two coloured points in Fig. 2 is calculated as 3.1623. As shown in Fig. 1, Octile distance is admissible if diagonal movement is allowed. It is computed by counting the number of steps taken if diagonal movement is not allowed and then subtracting it from the steps saved if diagonal movement is allowed. The Octile distance is given by Eq. (5). For the example in Fig. 2, the heuristic value obtained using Octile distance is 3.4142.

$$h_o(n) = \max(|\hat{x}|, |\hat{y}|) + \left(\sqrt{2} - 1\right)[\min(|\hat{x}|, |\hat{y}|)] \tag{5}$$

2.2 A* Algorithm

The A* algorithm is a search algorithm used to find the least cost path between a start and end point. The A* algorithm, unlike other transversal techniques, has additional knowledge provided by heuristics. A* algorithm calculates the cost function $f(n)$ of the nodes in its path and only expands nodes with the lowest estimated cost. The cost function is computed by Eq. (6),

$$f(n) = g(n) + h(n) \tag{6}$$

where $f(n)$ is the cost function of node n, $g(n)$ is the cost of getting from the start node to n and $h(n)$ is the estimated cost of getting from n to the goal node according to a heuristic function h.

To avoid getting stuck in a loop, the A* algorithm keeps track of its progress by utilising an open list and a closed list. The open list stores nodes that are visited but whose neighbours have not been explored, whereas the closed list consists of nodes that have been visited and their neighbours inspected. The process begins by assigning the node with the lowest cost function in the open list as the current node and moving this node into the closed list. If the eight adjacent neighbours of the current node do not exist in both lists, they are added to the open list. However, if they are already in the open list, they are only considered to be expanded if their distance from the start node is shorter than the known distance. The process is then repeated until the destination is found. The flowchart of A* can be found in Appendix A.

2.3 IDA* Algorithm

The Iterative Deepening A* (IDA*) algorithm is a combination of the Iterative Deepening Depth-First Search and the A* algorithm. It focuses on exploring the most promising nodes until a certain depth of the graph is reached with the help of heuristic functions. Hence, the amount of memory used is much lower than in A* algorithm since it does not record the nodes it went through previously. However, the IDA* algorithm may be slower than the A* algorithm in terms of time consumption because it may explore the same nodes repeatedly.

The algorithm starts by supplying the initial threshold value, otherwise known as the start node's $f(n)$ value as in Eq. (6). For the neighbours of the initial node, if their $f(n)$ values are greater than the threshold, they are pruned. If the neighbour's $f(n)$ value is lower than the threshold, they are added to the stack and further explored. This process continues until all possible branches have been explored. Then, the threshold value is updated with the highest $f(n)$ value of the pruned nodes. This process is repeated until the goal node is returned. See Appendix B for the flowchart of IDA*.

3 Results and Discussion

This research was conducted using a program written in Python and on an Asus Vivobook with an Intel i5-1240P processor and 8GB DDR4 RAM. In the first part of this section, the performance of four heuristics used in A* and IDA* algorithms, namely 1/2 Manhattan distance, Euclidean distance, Chebyshev distance, and Octile distance, were compared using three mazes of size 18 × 18. In the second part, the performances of the A* and IDA* algorithms were benchmarked using two types of mazes: straightforward and complex.

3.1 Admissible Heuristics

This section compares the performance of four heuristics using the A* and IDA* algorithms. The four heuristics are 1/2 Manhattan, Euclidean, Chebyshev, and Octile distances. To compare their efficiency, the actual distance from a specific point in the maze to the endpoint is compared with the heuristic values.

Table 1. Average time (in ms) used by Octile, Euclidean, Chebyshev and 1/2 Manhattan distances

Maze	Octile		Euclidean		Chebyshev		1/2 Manhattan	
	A*	IDA*	A*	IDA*	A*	IDA*	A*	IDA*
M1	8.52	11.54	34.78	167.70	64.69	4.2×10^3	320.57	1.6×10^4
M2	6.82	8.87	35.5	469.63	95.19	1.2×10^4	694.52	6.4×10^4
M3	8.57	15.71	27.15	306.66	66.58	6.8×10^3	602.42	5.3×10^4

Table 1 shows the average time A* and IDA* algorithms used to find the optimal path of the three randomly generated mazes in Fig. 3 using Octile distance, Euclidean distance, Chebyshev distance, and 1/2 Manhattan distance. The difficulty level for these three mazes was similar as the computation time did not differ much from one another. For A* and IDA* algorithms in all three mazes, the calculation time for Octile distance was the shortest, followed by Euclidean distance, Chebyshev distance, and lastly, 1/2 Manhattan distance. This is because the Octile distance approximated the distance from a specific node to the end node better than the other three heuristics.

Fig. 3. Three 18 × 18 mazes with optimal paths in orange

Figure 4 shows the zoomed-in version of the upper left parts of the mazes in Fig. 3. By randomly selecting a node (circled in red) in these three mazes, respectively, the distance between that node and the green goal node was calculated. Table 2 shows the actual, Octile, Euclidean, Chebyshev, and 1/2 Manhattan distances from the red node to the green node. As shown in Table 2, it can be observed that all four heuristics calculated a distance that is less than the actual distance for the mazes shown in Fig. 4, making them admissible heuristics. Octile distance best estimates the actual heuristic cost, followed by Euclidean, Chebyshev, and 1/2 Manhattan distances.

M1 upper left

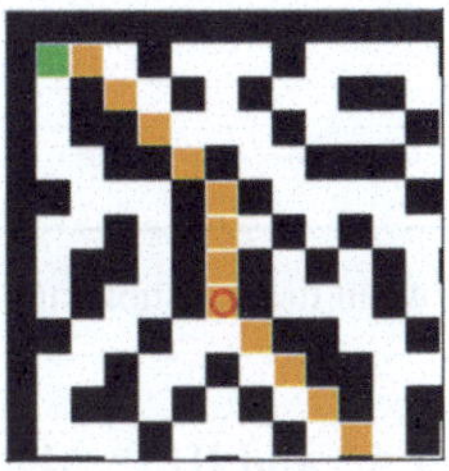
M2 upper left

M3 upper left

Fig. 4. Upper left parts of M1, M2, and M3 with selected node circled in red and path in orange

Table 2. Actual, Octile, Euclidean, Chebyshev and 1/2 Manhattan distances for mazes in Fig. 4

Maze	Actual	Octile	Euclidean	Chebyshev	1/2 Manhattan
M1	14.3137	13.7279	13.4536	10	9.5
M2	9.6569	9.0711	8.6023	7	6
M3	12.8995	12.3137	12.0416	9	8.5

In addition, since the threshold used in calculating the optimal path in the IDA* algorithm depends on the heuristic used, the computation time will be affected by the performance of the four heuristics. In Fig. 5, a graph displaying the characteristics of each heuristic is shown for maze M1. Since the heuristics exhibit the same pattern for all mazes, it is sufficient to analyse only one of the mazes. It can be seen the threshold for both Octile and Euclidean distances has an initial value of 24.0416, which is much greater than the initial value of 17.4142 for Chebyshev and 1/2 Manhattan distances. It is also observed that the threshold for Octile distance increased greatly with each iteration compared to other heuristics. Hence, a heuristic with a higher initial value and a higher threshold increase rate will be able to compute the optimal path faster than other heuristics, as this gives them a good head start and aids in computing the optimal path in the shortest time possible.

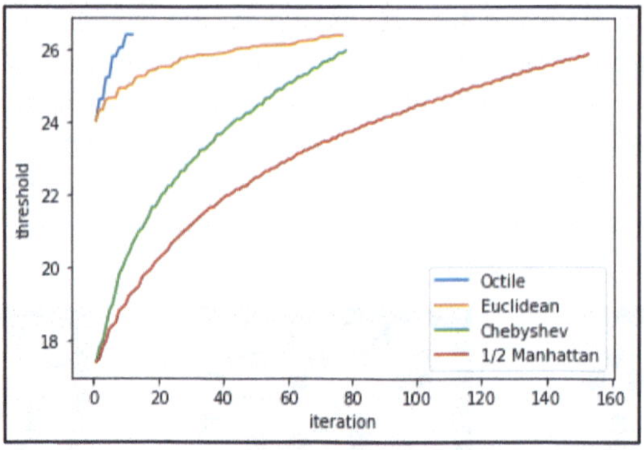

Fig. 5. The relationship between the threshold and iteration number for M1

3.2 Performance Comparison of A* and IDA* Algorithms

The main objective of this section is to benchmark the performance of A* and IDA* in computing the optimal path of mazes. In the previous section, the mazes compared have similar levels of difficulty. In this section, the first part explains situations where A* demonstrates inferior performance than IDA*. The second part will highlight cases where A* surpasses IDA* regarding efficiency. Given that Octile distance has proven to be superior to the other three heuristics, it will be used as the benchmark for assessing the performance of the A* and IDA* algorithms in computing optimal paths.

When IDA* is Faster than A*?

M4 M5 M6

Fig. 6. Three 18 × 18 mazes with paths in orange

The three mazes shown in Fig. 6 all share a common feature: their paths are generally straightforward. This means that only a few nodes are not in line with the main diagonal of the maze. To justify that the mazes in Fig. 6 are mazes with straightforward paths, their threshold values throughout the computation are portrayed in Fig. 7. IDA* outperforms A* in cases like this because these mazes are ideal for IDA*. As shown in Fig. 7, the thresholds of mazes M4, M5 and M6 started with a high value of 24.0416, and IDA* managed to compute the optimal route for these three mazes within five to six iterations. The computation of maze M4 is completed with a threshold value of 24.6274 on the fifth iteration, whereas the optimal paths of mazes M5 and M6 are found with a threshold of 25.2132 on the sixth iteration.

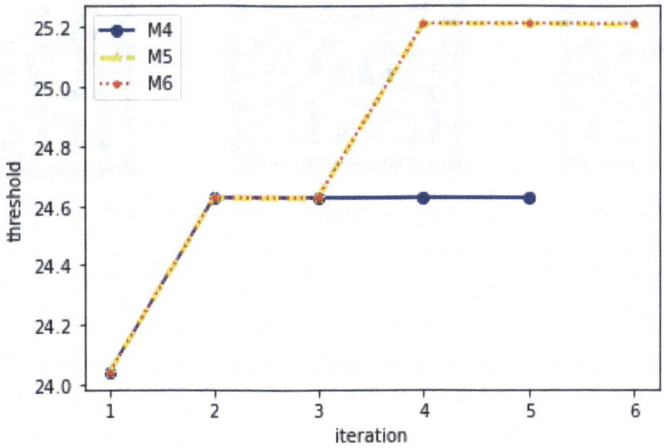

Fig. 7. Increasing threshold of mazes M4, M5 and M6 by iteration

The time used by A* and IDA* to compute the optimal path for these three mazes is displayed in Table 3. IDA* reduced the computation time of the A* by 48.43%, 31.41%, and 36.49% for mazes M4, M5, and M6, respectively. Since the algorithm is programmed to explore nodes in the northwest direction first, the IDA* need not detour much to locate the route with low f values that do not exceed the threshold. In conclusion, IDA* outperforms A* for straightforward mazes like M4, M5 and M6 because these mazes start with high threshold values, and the threshold values do not increase much from the initial value upon completion of the search process.

Table 3. Computation time used by A* and IDA* algorithms for mazes in Fig. 6

Maze	Path length	Computation time (s)		Reduction rate from A* to IDA* (%)
		A*	IDA*	
M4	19	1.0325	0.5325	48.43
M5	20	2.7511	1.8869	31.41
M6	20	2.9711	1.8870	36.49

When A* is Faster than IDA*?

M7 M8 M9

Fig. 8. Three 18 × 18 mazes of increasing difficulty

There are also cases where IDA* will take more time to return to the optimal path than A*. Mazes M7, M8, and M9, as displayed in Fig. 8, all have the same optimal path. However, their difficulty level increases from maze M7 to maze M9 due to the location of the obstacles in their optimal path. If the obstructions are met further along the computation, IDA* takes a longer time to backtrack to the right track. This can be observed in Fig. 9.

Bottom right of M7 Bottom right of M8 Bottom right of M9

Fig. 9. Bottom right of M7, M8 and M9 with dead end in green cross and selected node starred

In Fig. 9, by selecting a node on the optimal path (starred), the number of steps taken for IDA* to backtrack to get back on the right route can be calculated. For maze M7, if the dead end is met during the computation of IDA*, it will need to backtrack at least six steps to get to the optimal path. In a similar situation, for maze M8 and maze M9, the algorithm will need to backtrack at least 10 and 12 steps, respectively. This is worsened by the fact that IDA* does not record the nodes it has expanded before. Thus, there is a possibility that IDA* will backtrack from the same dead end multiple times before reaching the target node, resulting in higher computation time.

From Table 4, it is evident that the computation time of both A* and IDA* algorithms increased from M7 to M9 since the level of difficulty of the mazes is also rising. For the A* algorithm, since it expands node level by level, the time difference for the three mazes is not too big. However, the time taken by the IDA* algorithm to compute the optimal path increased rapidly with the increase in difficulty.

Table 4. Computation time of A* and IDA* for mazes in Fig. 8

Maze	Path length	Computation time (s)		Reduction rate from A* to IDA* (%)
		A*	IDA*	
M7	26	5.10	1.8×10^3	99.72
M8	26	13.95	4.4×10^3	99.68
M9	26	20.34	1.5×10^4	99.86

Table 5 shows the number of nodes expanded by A* and IDA* for mazes M7, M8 and M9. When A* and IDA* algorithms require more nodes for expansion to return the optimal solution, the computation time also increases. It is also apparent that the nodes expanded by A* when looking for the optimal path are much less than the nodes expanded by IDA*, with a percentage reduction of 99.895%, 99.985%, and 99.995% for mazes M7, M8 and M9, respectively. This is because IDA* explored the visited nodes repeatedly until it reached a depth to stop, while A* kept track of its visited nodes in lists to avoid visiting them again.

Table 5. Nodes expanded by A* and IDA* for mazes in Fig. 8

Maze	Path length	Nodes expanded		Reduction rate from A* to IDA* (%)
		A*	IDA*	
M7	26	97	9.2×10^5	99.895
M8	26	102	6.7×10^5	99.985
M9	26	118	2.6×10^6	99.995

In comparison to Fig. 7, it can be seen from Fig. 10 that the number of iterations required and the threshold value to find the optimal path has increased drastically. Mazes M7, M8 and M9 all have an initial threshold of 24.0416 (same as M4, M5 and M6), and the mazes return the optimal route with a threshold of 28.7279. For maze M7, it took 39 iterations to find the optimal path, whereas mazes M8 and M9 required 49 iterations. Hence, complex mazes tend to acquire higher threshold values when the optimal path is found.

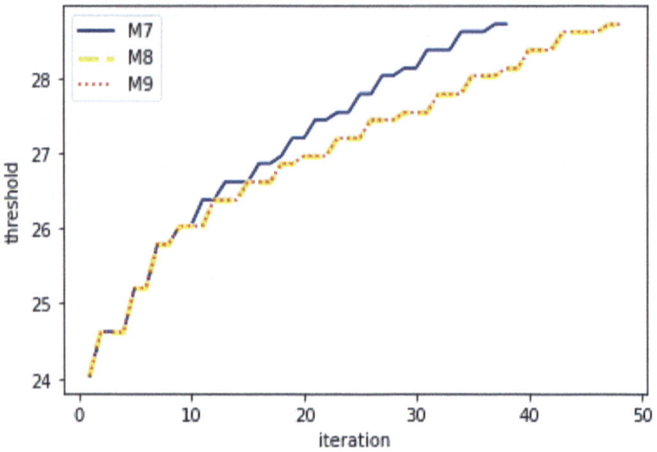

Fig. 10. Increasing threshold of mazes M7, M8 and M9 by iteration

A* and IDA* algorithms have numerous real-life applications across various domains. In robotics, they are commonly used in path planning, a process that navigates robots to their goal by avoiding obstacles. Another application of the algorithms is in puzzle solving, such as the 8-puzzle, where the memory constraint can be challenging for large-sized puzzles. Both algorithms are integral to GPS navigation systems, where A* computes optimal driving routes in real time, while IDA* finds solutions in complex scenarios with limited memory. Collectively, these applications underscore the importance of evaluating solution times based on the complexity of the problems being addressed.

4 Conclusion

In this research, it can be concluded that Octile distance performed the best in calculating the optimal path of a maze, followed by Euclidean distance, Chebyshev distance, and lastly, 1/2 Manhattan distance. As for the search algorithms, the A* algorithm outperformed the IDA* algorithm in most cases except when the maze's path is more straightforward. For mazes with straightforward paths (M4, M5 and M6), IDA* reduced the computation time by 31.41% to 48.43%. However, with the increase of obstacles in the main diagonal of the maze, the difficulty level of finding the maze's optimal path escalated. For situations like this, the A* algorithm can perform better than the IDA* algorithm by at least 99.68% to 99.86% (M7, M8 and M9). Generally, straightforward mazes are not that common in real-life applications. Given its efficiency in complex scenarios, A* is generally preferred for most practical applications, while IDA* remains useful for memory-limited environments with simpler paths.

APPENDIX A: Flowchart of the A* Algorithm

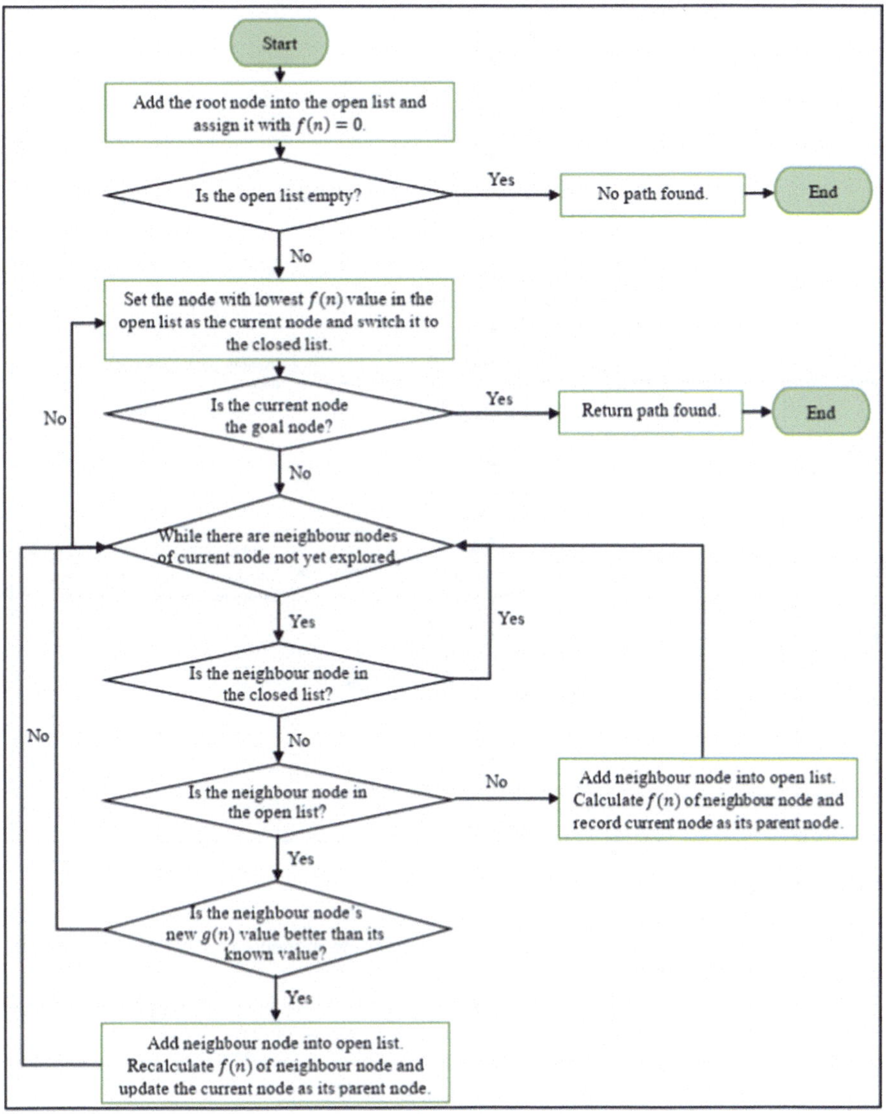

APPENDIX B: Flowchart of the IDA* Algorithm

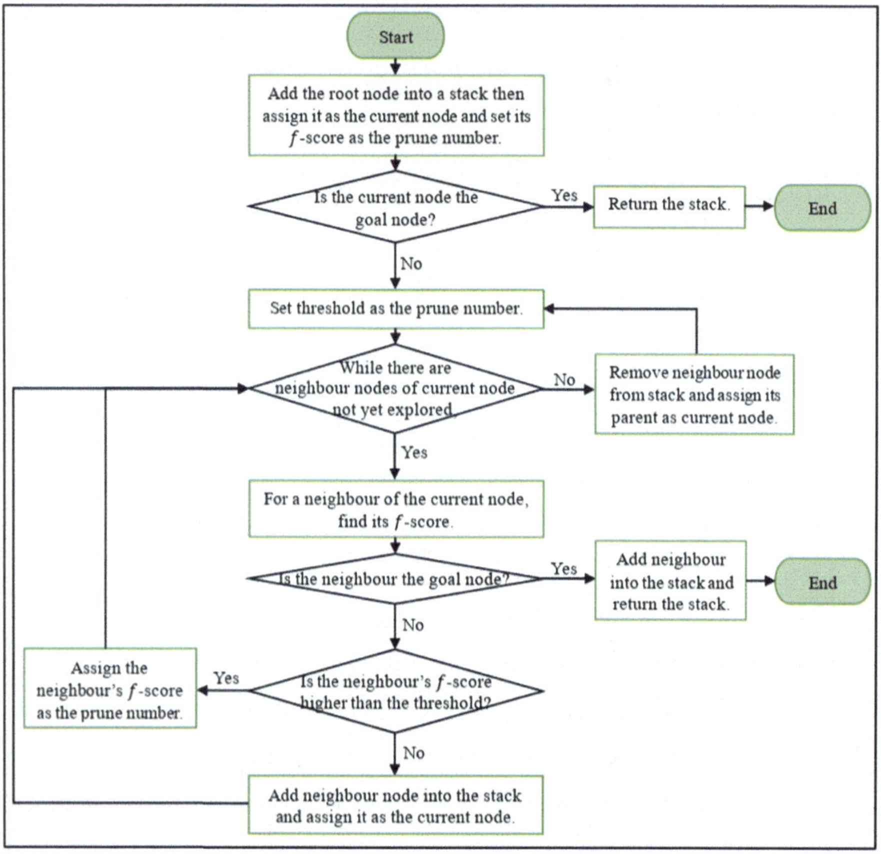

References

1. Mathew, G.E.: Direction based heuristic for pathfinding in video games. Procedia Comput. Sci. **47**, 262–271 (2015)
2. Barnouti, N.H., Al-Dabbagh, S.S.M., Naser, M.A.S.: Pathfinding in strategy games and maze solving using A* search algorithm. J. Comput. Commun. **4**, 11–15 (2016)
3. Algasi, K.A.: Uninformed and informed search techniques in artificial intelligence. Eastern Mediterranean University (EMU): Master's thesis (2017)
4. Noori, A., Moradi, F.: Simulation and comparison of efficiency in pathfinding algorithms in games. Ciência e Natura **37**, 230–238 (2015)
5. Sathyaraj, B.M., Jain, L.C., Finn, A., Drake, S.: Multiple UAVs path planning algorithms: a comparative study. Fuzzy Optim. Decis. Making **7**, 257–267 (2008)
6. Naik, A.S.: User query processing using distance oracle for road networks. Int. J. Eng. Res. Technol. (IJERT) **2**(2) (2013)
7. Andiwijayakusuma, D., Mardhi, A., Savitri, I., Asmoro, T.: A comparative study of the algorithms for path finding to determine the adversary path in physical protection system of nuclear facilities. J. Phys. Conf. Ser. **1198**(9), 092002 (2019)

8. Foead, D., Ghifari, A., Kusuma, M.B., Hanafiah, N., Gunawan, E.: A systematic literature review of A* pathfinding. Procedia Comput. Sci. **179**, 507–514 (2021)
9. Tierney, K., Pacino, D., Voß, S.: Solving the pre-marshalling problem to optimality with A* and IDA. Flex. Serv. Manuf. J. **29**, 223–259 (2017)
10. Hart, P.E., Nilsson, N.J., Raphael, B.: A formal basis for the heuristic determination of minimum cost paths. IEEE Trans. Syst. Sci. Cybern. **4**(2), 100–107 (1968)
11. Nilsson, N.J.: Problem-Solving Methods in Artificial Intelligence. McGraw-Hill, New York (1971)

Physics

Assessment of Sustainability of Triaxial Porcelain with Thermally Treated Palm Oil Fuel Ash (TTPOFA) as a Partial Replacement for Traditional Raw Materials

Maryam Adamu Sunusi[1], Mohamad Zaky Noh[1(✉)], and Sani Garba Durumin-Iya[2]

[1] Physics Material Laboratory, Faculty of Applied Science and Technology, Universiti Tun Hussein Onn Malaysia, 86400 Pagoh, Johor, Malaysia
zaky@uthm.edu.my

[2] Department of Physics, Faculty of Natural and Applied Science, Sule Lamido University Kafin Hausa, Kafin Hausa 048, Jigawa State, Nigeria

Abstract. Industrial development is currently placing considerable pressure on the environment through the extensive use of natural raw materials and waste generation. Palm oil fuel ash (POFA) is an abundant waste material generated by the palm oil industry that is amorphous and has a high silica content. This work aimed to study the possibility of using thermally treated POFA (TTPOFA) in the normal triaxial porcelain composition of kaolin, feldspar, and quartz. The POFA was heat treated at 800 °C in an electric chamber furnace for 90 min. This thermal treatment served to eliminate any excess carbon and other unburned organic materials. The effect of the substitution of quartz with TTPOFA on the physical and mechanical properties of the final product was evaluated. The results showed that the chemical composition of POFA was significantly improved through thermal treatment, with the silica (SiO_2) content increasing from 55.00% to 78.24% after treatment. A maximum compressive strength of 247 MPa and a high bulk density value of 2.44 g/cm^3 were achieved in porcelain with 15 wt. % TTPOFA. X-ray diffraction analysis revealed the presence of quartz and mullite phases, which favor high mechanical strength, while the microstructure showed few small pores, which explains its low water absorption of 0.024%. The thermal treatment of POFA has proven to be an effective method for enhancing its properties, making it a promising waste material for use as a source of silica. Due to these enhanced properties, the addition of TTPOFA showed great potential as a raw material in porcelain applications, which could be regarded as an economic benefit and an important relief on waste disposal issues.

Keywords: Palm oil fuel ash · Porcelain · Thermal Treatment

1 Introduction

Triaxial porcelain is a highly valued ceramic material because of its strength and beauty. It is a ceramic system that has been widely studied for numerous applications, such as stoneware, whiteware, and insulators [1]. Currently, global porcelain production is

P. Chang et al. (Eds.): SCIEMATHIC 2024, SPPHY 427, pp. 85–100, 2025.
https://doi.org/10.1007/978-981-96-7679-8_8

fast growing compared to that of other ceramic types. Global production in 2017 was approximately 13.5 million square meters, which is 5% greater than that in 2012 [2]. It involves three main categories of raw materials, each with a crucial role in determining the quality of the final product [3]. The clay mineral provides plasticity to the body for forming. Feldspar is used as a flux that creates a viscous liquid during sintering and assists in vitrification, and quartz acts as a filler material that reduces shrinkage and deformation of the fired sample [4].

Rapid growth in the world population and industrialization are placing a challenge on the environment through the consumption of natural raw materials and an increase in the rate of waste generated [5]. Considering the extensive utilization of silica in the ceramic industry, it is regarded as the fundamental material for ceramic products. Most manufacturers use quartz, sandstone silica, or quartzite as a source of silica for porcelain production [6]. These are all-natural raw materials and are nonrenewable. The ceramic industry has the potential to utilize several waste materials in body formulations to lower production costs and contribute to solving problems connected to waste generation. Utilizing waste materials such as treated POFA could help lower the burden on natural resources, reduce landfill waste, and promote a circular economy in the ceramic industry. This would not only decrease environmental degradation but also help balance material costs in the industry, supporting long-term sustainability. To make use of this opportunity, waste should be carefully characterized so that its behavior during processing and its effect on the final products can be predicted [3]. Some studies on porcelain production have reported the use of waste materials to partially replace natural raw materials, including rice husk ash [7], fly ash [8], and glass waste [9]. Their results showed an improvement in flexural strength, low water absorption, and reduced firing shrinkage.

Palm oil fuel ash (POFA) is a waste generated by the palm oil industry from the burning of various palm oil biomass materials, including empty fruit bunches, palm kernel shells, fibers, oil palm leaves, fronds, and trunks. These materials are used in palm oil mill boilers to generate electricity for the palm oil extraction process. The ash produced, known as POFA, is dumped in landfills, which has a negative impact on the environment [10]. The rapid expansion of the palm oil industry has led to the generation of substantial amounts of biomass and industrial waste alongside the production of crude palm oil. Globally, the annual production of palm oil residues amounts to approximately 184 million tons, with Malaysia alone producing approximately 53 million tons, and this amount is projected to increase by five percent annually [11]. POFA has been utilized in materials such as concrete as a pozzolanic material [12], in graphene nanosheets [13], and in geopolymers [14]. These materials showed a significant improvement in the mechanical properties after the addition of POFA. Due to its fluxing properties, it has the potential to lower the firing temperature required, contributing to energy savings and reduced greenhouse gas emissions. Few studies have investigated the utilization of POFA in porcelain [15]. Despite numerous studies on POFA, its effective utilization in porcelain production is limited and remains unclear.

The purification of this agricultural waste material from impurities is essential for proper and optimum utilization in both traditional applications, such as ceramics, and modern applications, such as the production of higher value-added components, such as pure silicon [16]. Therefore, to improve efficiency and improve the properties of the final product, pretreatment of the POFA is crucial. The influence of utilizing thermally treated POFA (TTPOFA) at 800 °C on physical and mechanical properties of porcelain has not yet been well established. Hence, TTPOFA could yield significant advantages and beneficial results, which in turn could assist in reducing environmental impacts by minimizing the amount of waste disposed of in landfills.

2 Materials and Methods

2.1 Raw Materials

Kaolin, feldspar, and quartz were Kaolin, feldspar, quartz, and POFA were used as raw materials in this study. Kaolin, feldspar, and quartz were obtained from a local supplier, Maju Saintifik Sdn. Bhd. POFA was collected from a palm oil mill located in Kluang Johor, Malaysia. POFA from different regions may have distinct properties due to variations in soil composition, climate, and palm tree varieties. Other factors that could lead to POFA variation include combustion temperatures, type of palm oil residue used, and processing conditions. These variations can influence the chemical and physical properties of POFA. Differences in chemical composition of POFA, such as silica and alumina content, can affect the properties of the final product. Also, variations in particle size distribution can influence the mechanical and microstructural properties of the final product, leading to differences in porosity, density, and strength. Therefore, through a proper characterization of POFA, its effect on the final product could be predicted.

It was first dried in an oven using a Memmert UF110plus at 110 °C for 24 h to remove any moisture and then dry milled using an RS-1S ball mill with three different sizes of alumina balls as the grinding media with a ball-to-powder (BP) ratio of three to one for 72 h. The POFA was then passed through a 50 μm mesh sieve (ISO3310-1, BS-410-1). Subsequently, about 100 g of the POFA was heat treated at 800 °C in an electric chamber furnace (Protherm PLF 130/15) for 90 min. This thermal treatment served to eliminate any excess carbon and other unburned organic materials present. The TTPOFA was further subjected to further grinding using the same laboratory ball mill to obtain more fine particles due to the agglomeration after combustion.

2.2 Sample Preparation

Six compositions were prepared by progressively replacing quartz with TTPOFA from 0–25 wt. %, while keeping kaolin and feldspar constant, as shown in Table 1. The control sample was denoted P0 (0 wt. %), whereas the samples containing TTPOFA were denoted P5 (5 wt. %), P10 (10 wt. %), P15 (15 wt. %), P20 (20 wt. %), and P25 (25 wt. %). Each composition was measured for five samples and mixed with 1 wt. % polyvinyl alcohol (PVA) as a binder using a planetary centrifugal mixer model SK-350TII (Kakahunter, Japan) to obtain a homogeneous mixture. The process involves strong centrifugal and shearing forces simultaneously mixing materials by rotating and revolving containers holding the material at the same time.

To obtain uniform compact sizes, about 5 g of the mixed powder was dry pressed into circular pellets 20 mm in diameter and 7 mm in thickness with a mold using a

Hydraulic Press Carver 3851 at 91 MPa with a constant clamping force of 3 tonnes for a one-minute holding time. The samples were subsequently sintered at 1150 °C in an electric chamber furnace (Protherm PLF 130/15) for 120 min at constant heating and cooling rates of 1 °C per minute and 5 °C per minute, respectively. The samples were allowed to cool naturally in the furnace and then prepared for physical and mechanical testing.

Table 1. Batch composition (wt. %).

Heading level	Kaolin	Feldspar	Quartz	TTPOFA
P0	50	25	25	0
P1	50	25	20	5
P2	50	25	15	10
P3	50	25	10	15
P4	50	25	5	20
P5	50	25	0	25

2.3 Measurements and Characterization

This study used various tests to investigate the physical, mechanical, and microstructural properties of POFA and porcelain samples. Physical and chemical composition tests were carried out to evaluate the physical and chemical characteristics of POFA before and after thermal treatment. The particle size analysis (PSA) was done using a MICROTRAC X-100 particle size analyzer and the color change due to thermal treatment was visually evaluated and reported. The chemical compositions of the raw materials were analyzed using Energy dispersive X-ray fluorescence (EDXRF) (Thermo Scientific, Model ARL Quant'X). The physical properties of the porcelain samples were based on volume shrinkage, mass loss, bulk density, and water absorption. The percentage volume shrinkage that occurred during firing was calculated using the volume of the sample before and after the sintering process and evaluated using Eq. 1.

$$Volume\ shrinkage = V_o - V_1/V_o \times 100\% \tag{1}$$

where V_o and V_1 are the volumes of the sample before and after sintering respectively. The mass of the samples was measured using a ME204T analytical balance (Mettler Toledo). The mass loss was calculated using Eq. 2.

$$Mass\ loss = M_o - M_1/M_o \times 100\% \tag{2}$$

where M_o and M_1 are the masses of green and sintered samples respectively. The bulk density of each sample was measured by the liquid displacement method. The sample was weighed in air and then immersed in water. The sample weight was measured and recorded both in air and liquid, and the bulk density was calculated using Eq. 3.

$$Bulk\ density = W_1/W_3 - W_2 \tag{3}$$

where W_1 is the dry mass, W_2 is the suspended mass, and W_3 is the saturated mass. The water absorption rate of the samples was determined by the boiling water method according to the ASTMC 373-88 standard for material testing [17]. The water absorption rate was calculated using Eq. 4.

$$Water\ absorption = M_S - M_D/M_D \times 100\% \tag{4}$$

where M_S is the saturated mass and M_D is the dry mass of the sample.

The mechanical property was based on the compressive strength of the sintered samples. The compressive strength of the samples was tested using a 1000 kN universal testing machine UN-7001-LS-100 (GOTEC Taiwan) with a 1000 kN load cell capacity at a crosshead speed of 5 mm/min. Three circular samples were used for each sample batch, and the mean value of three samples was calculated.

The phase analysis of the samples was investigated by X-ray diffraction (XRD) using a Bruker D2 Phaser X-ray diffractometer with Cu-Kα radiation ($\lambda = 1.5406$ Å) operated at 45 kV and 40 mA. The analysis was carried out in the 2θ range between 10 and 80° with a step size of 0.01°. The patterns were analyzed by X' pert High score plus software. The crystalline phases observed in the samples were compared with reference patterns from the International Center Diffraction Data (ICDD) database. The microstructure of the samples was analyzed by scanning electron microscopy (SEM) using a COXEM EM-30AX Plus instrument.

3 Results and Discussion

3.1 Particle Size Analysis of POFA

Figure 1a and Fig. 1b depicted the physical appearance of the raw POFA and TTPOFA. The raw POFA exhibited a black color and changed to light brown after treatment. This can be due to the reduction in carbon content through the thermal treatment process. The color of the raw POFA was also reported to be black and became darker when the quantity of unburned carbon increased [18]. PSA was used to characterize the size distribution of particles in a sample.

The particle size distributions for the raw POFA and TTPOFA samples are illustrated in Fig. 2a and Fig. 2b. The lower diameter (d_{10}) for the raw POFA was 5.376 μm, with a mean diameter (d_{50}) of 21.59 μm. Conversely, TTPOFA exhibit lower and mean diameters of 1.479 μm and 3.354 μm, respectively. Notably, this particle size is less than 13 μm, as reported in previous studies following heat treatment of POFA at 600 °C [19]. This is an indication that the grinding and thermal treatment processes greatly influence the reduction in the particle size of POFA, which can help improve the rate of dissolution. In ceramic materials, fine particles melt more easily than coarse particles, and they bond more strongly after sintering, which enhances the hardness and toughness of the material [2].

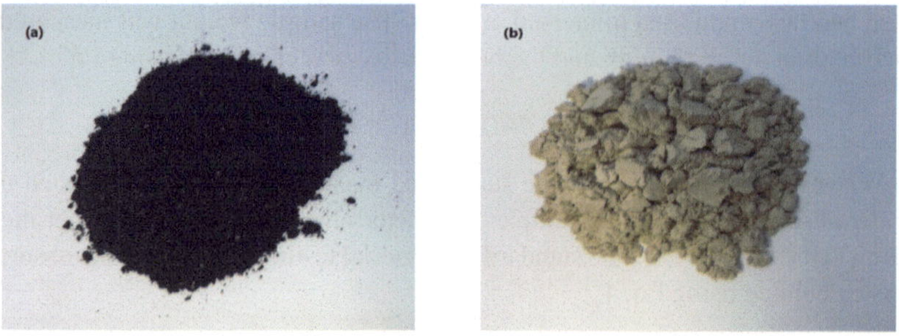

Fig. 1. Physical appearance of (a) raw POFA and (b) TTPOFA.

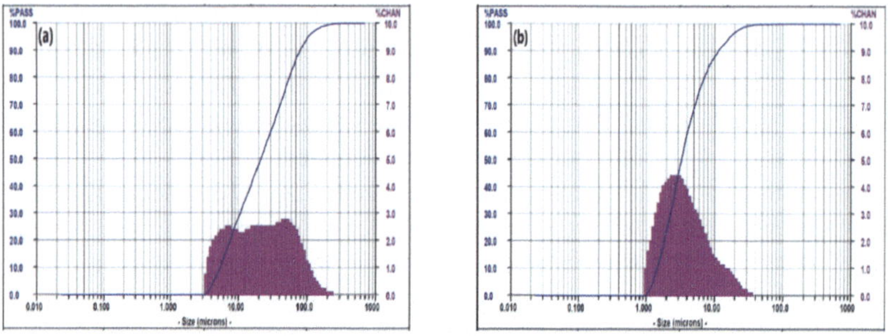

Fig. 2. Particle size distribution of (a) raw POFA and (b) TTPOFA.

3.2 Chemical Composition of Raw Materials

Table 2 presents the results of the chemical composition analysis of the raw materials examined by EDXRF. The results showed that the kaolin, feldspar, and quartz used for this study are within the range used in ceramic material formulations [17]. Kaolin contained mainly SiO_2 (45.99%) and Al_2O_3 (50.27%) with a small amount of K_2O. In feldspar composition, K_2O was the major flux oxide, with the absence of Na_2O. This indicates that the feldspar is characterized as K-feldspar. The quartz has a high purity of silica with 99.40% SiO_2.

The thermal treatment significantly reduced the loss on ignition (LOI) of POFA from 7.18% to 0.19%. It is also worth noting that the contents of chemical constituents other than K_2O especially SiO_2 improved. SiO_2 was the main constituent of the raw POFA and TTPOFA accounting for 55.0% and 78.24%, respectively. The increase in SiO_2 content can be attributed to the decrease in LOI through the thermal treatment process. Shaladi et al. [20] reported that the SiO_2 content in POFA increased after heat treatment at 600 °C from 51.18% to 67.37%, which is directly attributed to the reduction in the LOI. As shown in Table 2, both the raw POFA and TTPOFA contained a significant amount of K_2O, with 8.07% and 7.10%, respectively. This could be due to the nature of the soil where the palm trees were harvested, which will aid in liquid phase formation and help in densification at higher temperatures [21]. The amount of Al_2O_3 in the raw POFA was 1.24%, and that in the TTPOFA increased to 3.91%. Alumina helps to improve the thermal stability and hardness of materials, while the glassy phase generated by the silicate phase has influence on the ceramic properties [22]. The presence of Al_2O_3 in silica-based ceramics also enhances the mechanical properties by acting as a catalyst for the crystallization of fused silica into cristobalite [23].

The presence of MgO assists in the formation of mullite at low temperature. Fe_2O_3 and TiO_2 have common features as coloring oxides, and a small amount is required to prevent changing the sample's natural color [24]. These oxides also play vital roles in vitrification, phase transformation, and mullite growth in porcelain bodies [25]. Calcium oxide (CaO) was also evident (2.90%) in the raw POFA and increased to 3.63%. Other compounds found in both the raw POFA and TTPOFA samples were Na_2O, and P_2O_5. The differences in the chemical constituents of POFA normally vary due to many factors, including burning conditions and the source of materials [26].

Overall, the POFA used for this study could be more economical compared to the traditional silica source (quartz). For example, according to Mordor Intelligence Report 2022, the Malaysian ceramic market is expected to grow from USD 36.20 billion in 2023 to USD 45.48 billion by 2028 at an annual growth rate of 4.67% (Ceramic Industry in Malaysia Size & Share Analysis). This growth will likely increase the environmental impact of mining due to the increased volume of raw materials required to meet production demands. The industry's reliance on mined resources could also generates carbon emissions from processing and transportation, increasing the industry's carbon footprint.

Table 2. Chemical composition of raw materials.

Oxide constituent (wt. %)	Kaolin	Feldspar	Quartz	Raw POFA	TTPOFA
SiO2	45.99	71.87	99.40	55.00	78.24
Al_2O_3	50.27	23.10	0.22	1.24	3.91
Na_2O	0.17	-	-	1.00	1.50
MgO	-	-	-	0.25	1.62
K_2O	1.95	3.48	-	8.07	7.10
CaO	0.003	1.00	-	2.90	3.63
Fe_2O_3	0.73	0.21	-	2.00	2.34
TiO_2	0.77	0.19	-	0.12	0.13
P2O5	-	0.03	-	1.00	1.34
LOI	0.10	0.12	0.10	7.18	0.19

* LOI: Loss on ignition

3.3 Microstructure and Phase Analysis of the POFA

The SEM image of the raw POFA in Fig. 3a shows large irregular and spherical particles. The large particles of POFA can affect the properties of the final product because smaller particles are more reactive than large particles [27]. The particles in the raw POFA had a porous texture. This porous structure also suggested that POFA mostly consists of amorphous materials [28]. In TTPOFA shown in Fig. 3b, the particles had crushed irregular smooth surfaces and are smaller in size and less porous than those of the raw POFA. The reduction in the particle size of TTPOFA was enhanced by Si-O-Si bond breakage after treatment, which caused the agglomerated particles to split into smaller particles [29]. These observed results agree with previous research findings [30].

The XRD patterns of the raw POFA and TTPOFA are shown in Fig. 4a and Fig. 4b. All the patterns are similar, with quartz (SiO_2) (ICDD# 00-046-1045) as the major crystalline phase, while cristobalite (SiO_2) (ICDD# 00-003-0272) and potassium aluminum phosphate ($KAlSiO_4$) (ICDD# 00-012-0134) are the minor crystalline phases. The peak of potassium aluminum silicate identified in both the raw POFA and TTPOFA samples can be attributed to the high amount of K_2O in the POFA, as shown in Table 2. Previous studies have also reported similar findings, noting that silica crystallization into cristobalite occurs at temperatures of 800 °C and above [21]. The presence of the cristobalite crystal phase, however, improves the strength of ceramic materials [17].

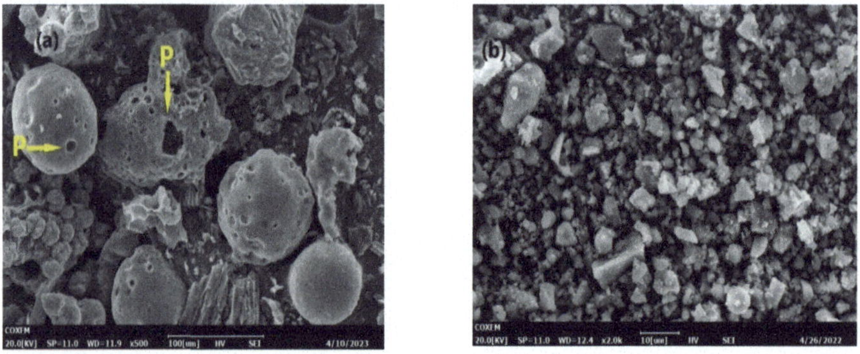

Fig. 3. SEM images of (a) raw POFA and (b) TTPOFA (P: Pore).

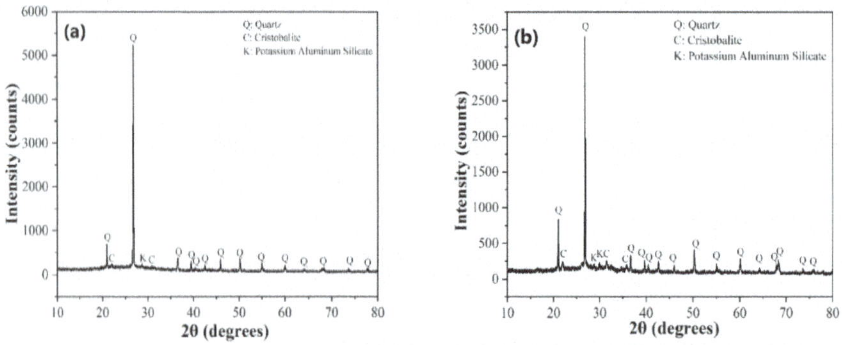

Fig. 4. XRD patterns of (a) raw POFA and (b) TTPOFA (P: Pore).

3.4 Physical and Mechanical Properties of the Porcelain

The physical appearance of the samples is given in Fig. 5. The texture of the sintered samples was smooth, and their color changed from light brown to dark brown as the content of TTPOFA increased. This could be due to the thermal reaction of impurities such as Fe_2O_3, TiO_3, and P_2O_5 present in the TTPOFA, which have a strong influence on the color of the sample, as noted by Njindam et al. [31]. By incorporating whitening agents such as TiO_2, it could help in enhancing the brightness of the porcelain. To maintain a desirable aesthetic appearance, a slight modification in the feldspar composition can influence a glaze application and could help improve the surface gloss and overall appearance.

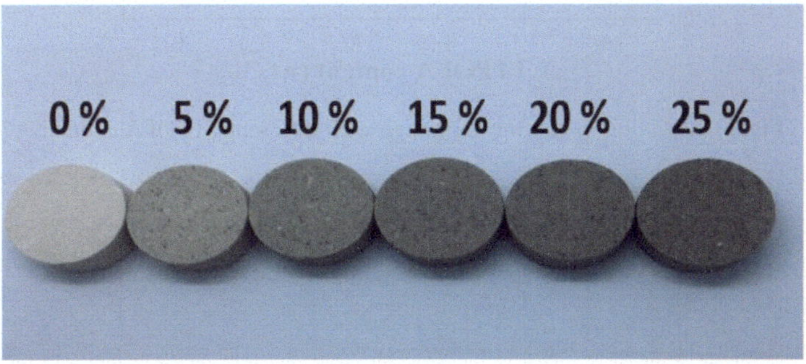

Fig. 5. Physical appearance of porcelain samples with different TTPOFA percentage.

The volume shrinkage of samples with different percentages of TTPOFA is shown in Fig. 6. It increases with TTPOFA addition but decreases sharply at P20 and P25. The shrinkage in the triaxial porcelain can be attributed to the removal of residual water from the clay, decomposition of the binder, liquid formation from the flux and generation of crystalline phases from amorphous materials. The addition of TTPOFA to the body contributed to the production of more silica that aids feldspar, dissolving the particles in the porcelain to enrich the matrix reinforcement [32]. Also, the volume shrinkage of all samples was between 24.01% for P0 and 29.88%, which was the highest value at P15. This is because the total alkali content (Na_2O, K_2O, MgO, CaO) in the TTPOFA, which was used as a substitute for quartz in the composition was high. Increase in the amount of alkaline earth oxides, particularly in ceramic tiles, leads to higher concentration and shrinkage [33].

The change in mass loss with respect to the TTPOFA content is presented in Fig. 7. This has similar variation to volume shrinkage. Although no significant changes in mass loss values were observed, which were all between 6.76 and 7.01%, with P0 having the lowest mass loss of 6.76%, which might be related to the number of closed pores in the sample, while the highest mass loss was 7.01 at P15. The reduction in mass can be attributed to the formation of solid bonds and a decrease in the interparticle space within the sample, leading to a more compacted sample and a decrease in size. Kitouni and Harabi [25] referred to this phenomenon as shrinkage behavior.

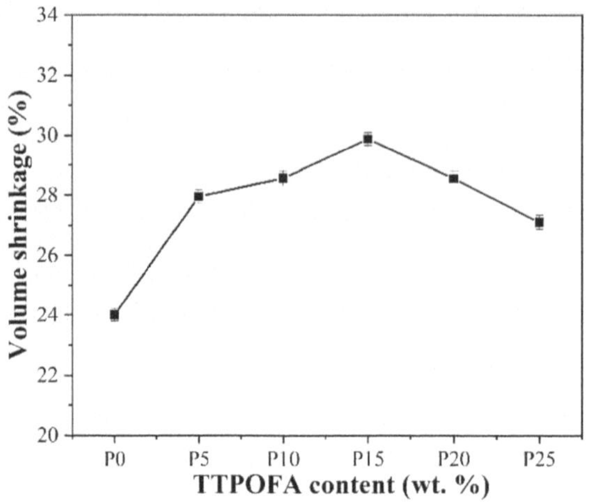

Fig. 6. Variation in volume shrinkage with different TTPOFA content

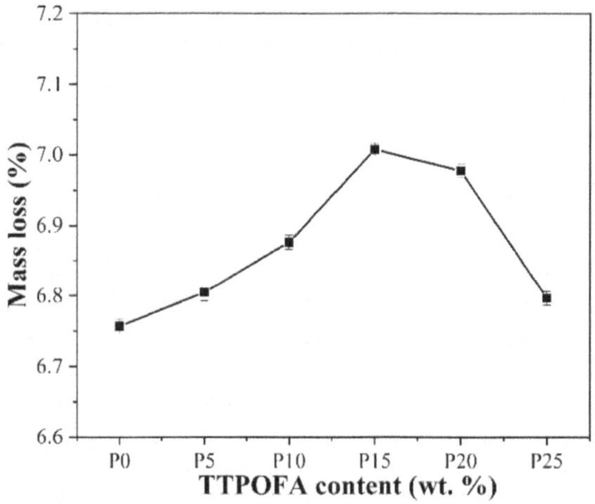

Fig. 7. Variation in mass loss with TTPOFA content

Figure 8 shows the variation in the bulk density of the porcelain samples containing TTPOFA. P0 has a lower bulk density of 2.34 g/cm^3 than the samples with TTPOFA, except for the P25 composition, which has a bulk density of 2.30 g/cm^3. The bulk density increased with an increase in TTPOFA content, reaching 2.44 g/cm^3 at P15 before decreasing. This value is greater than the standard value (~2.40 g/cm^3) for conventional porcelain, as outlined by [34]. This high bulk density value suggests an increase in the compaction and closure of pores by the glassy phase formed during sintering from the silica in the TTPOFA. This could be due to the fine particles of the TTPOFA because fine particles melt easily and bond more readily than coarse particles after sintering, contributing to the enhanced density and strength of the material [17]. Furthermore, increased TTPOFA content promotes liquid phase development and enhances sample bulk density. According to Selli [35], as the amount of glassy phase in a porcelain body decreases, the density increases. However, as the TTPOFA addition exceeds 15 wt. %, the bulk density decreases. This is because a significant amount of silica from the TTPOFA produces additional glassy phases, which lower the sample density.

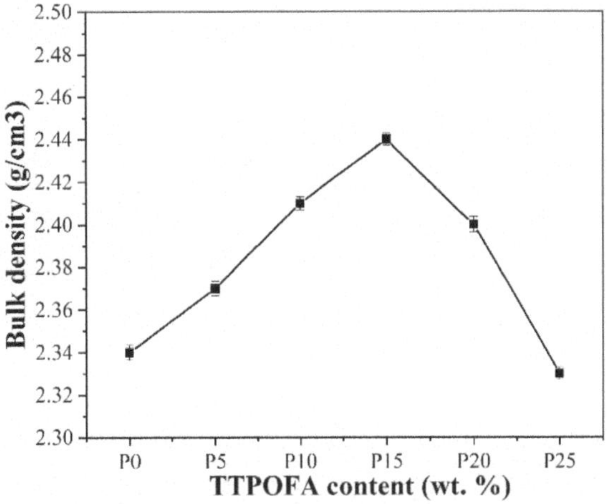

Fig. 8. Variation in bulk density with TTPOFA content

Figure 9 shows a decrease in the water absorption rate as the TTPOFA content increases, and it slightly increases beyond P15. All the sintered samples with 10 wt. % TTPOFA and above showed a relatively low water absorption rate (~0.5). The sample P15 achieved the lowest water absorption of 0.024%, possibly due to the high compaction, which led to the formation of closed pores and caused densification of the structure. The water absorption rate was less than the value of 0.5% required for porcelain according to the ISO 13006 standard [9]. However, P0 shows a relatively high water absorption of 1.894% compared to that of the samples with TTPOFA. A relationship between volume shrinkage and water absorption can be established. The volume shrinkage rate is inversely proportional to the water absorption rate. As the volume shrinkage increases, the water absorption rate decreases, and as the volume shrinkage decreases, the water absorption rate increases.

The compressive strength variation shown in Fig. 10 first increased as the TTPOFA content increased and then decreased. The variation in compressive strength is related to the porcelain bulk density, as both properties have the same trend. The compressive strength increased and reached a maximum value of 247 MPa at P15 and approximately 226 MPa for P0. This could be due to more silica from the TTPOFA, which along with kaolinite produces an aluminum silicate compound phase during the sintering and helps to enhance the mechanical strength of the porcelain.

The decrease in compressive strength for samples above 15 wt. % of TTPOFA can be attributed to the higher content of TTPOFA in the samples. A greater amount of silica can reduce the sintering ability compared to that of clay and feldspar, which increases

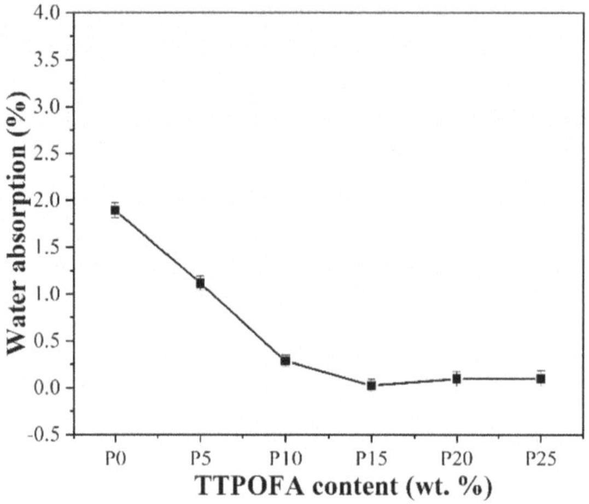

Fig. 9. Variation in water absorption rate with TTPOFA content

the number of pores in the microstructure of the samples, thereby reducing the strength of the body [2]. Therefore, a large amount of silica could affect the strength of porcelain, especially when it is present in greater amounts. Additionally, the maximum compressive strength was obtained when the water absorption rate approached zero. This means that mechanical strength is largely influenced by the microstructure, particularly by defects such as cracks and pores, as reported by Kitouni and Harabi [25].

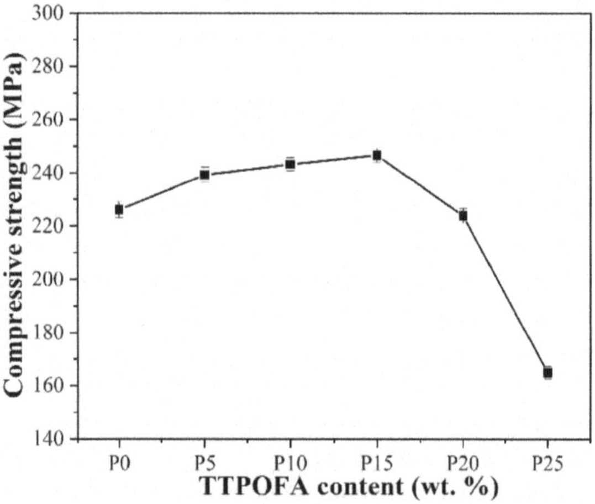

Fig. 10. Variation in compressive strength with TTPOFA content

Figure 11a and Fig. 11b show the XRD patterns of P0 and P15 respectively. The crystalline phases in the vitreous phase were identified as quartz (ICDD# 00-007-9634) and mullite (ICDD# 00-006-6452). Quartz was considered a residual phase formed from the raw materials, while mullite was formed during firing. Mullites are formed from the decomposition of clay material and play a substantial role in the mechanical features of porcelain stoneware bodies [36]. An earlier study by Carty and Udayan [37] indicated that in a porcelain body with excess quartz, the cristobalite phase began to form when the rate of transition of quartz to cristobalite exceeded the rate of quartz dissolution. Despite P15 sample having the highest bulk density and lowest water absorption, some diffraction peaks of the anorthite phase (ICDD# 00-041-1486) were observed in the sample, apparently due to the high amount of CaO in the TTPOFA or due to complete glass formation.

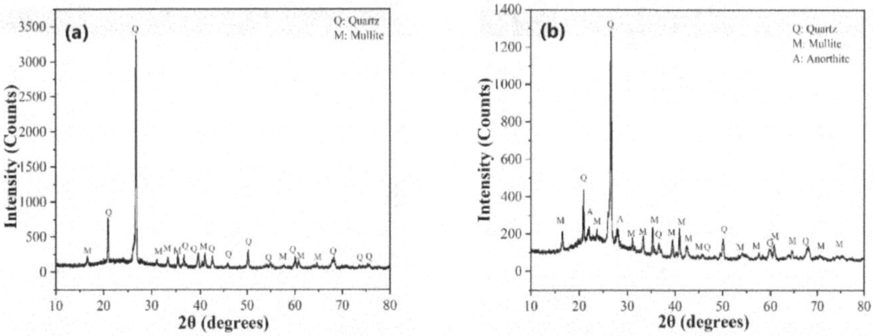

Fig. 11. XRD pattern of porcelain sample (a) without TTPOFA (P0) and (b) with 15 wt. % TTPOFA

SEM characterization was used to examine the fractured surface morphologies of samples P0 and P15. The incorporation of TTPOFA into the porcelain formulation had notable microstructural effects. Figure 12a showed that P0 had many irregularly shaped pores, which led to the formation of interconnected pores (Fig. 12b) that fused during sintering. The highly porous structure observed in P0 sample accounted for its high water absorption rate, as shown in Fig. 9. This can be the result of the decomposition of thermally unstable materials present in the raw materials. In contrast, the microstructure of P15 sample shows uniformly distributed pores that are small, as shown in Fig. 13a. The pores are more circular with a smoother and denser surface (Fig. 13b). These microstructural findings agree with the bulk density results, as P15 was more densified, which implies that they are more vitrified and have a high glassy phase [38]. The microstructural changes are also indicative of the potential mechanical improvements observed in the porcelain, which contribute to the overall enhancement of the properties and demonstrate the viability of TTPOFA as a valuable additive in porcelain production.

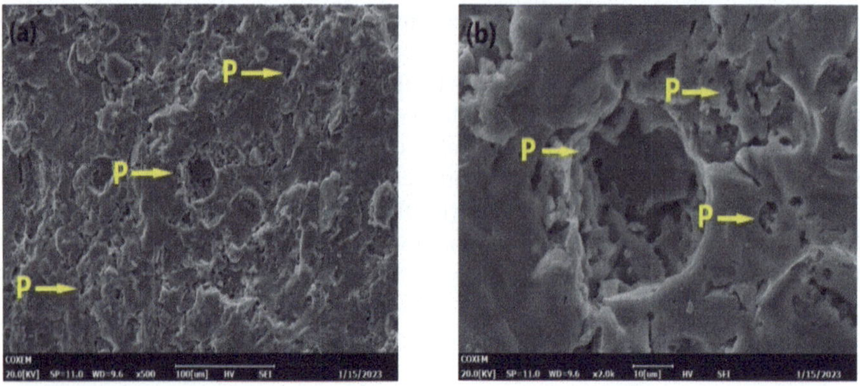

Fig. 12. SEM micrographs of fractured surface of porcelain sample without TTPOFA (P0) at (a) × 500 and (b) × 2000 (P: Pore).

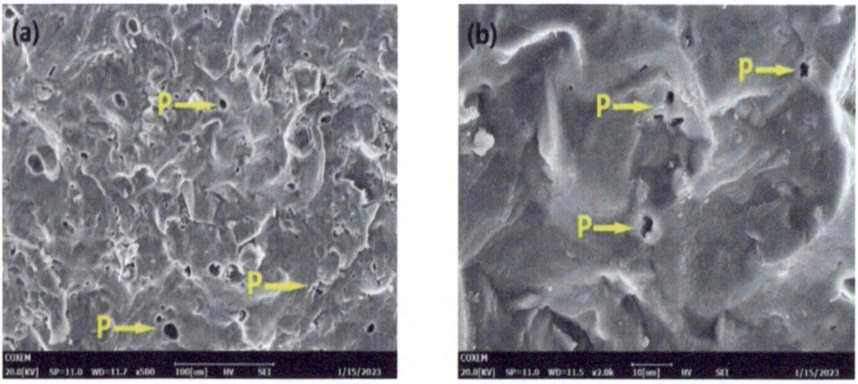

Fig. 13. SEM micrographs of the fractured surface of porcelain sample without TTPOFA (P0) at (a) × 500 and (b) × 2000 (P: Pore).

4 Conclusion

This study investigated the use of TTPOFA as a replacement for quartz in the production of triaxial porcelain and their effect on physical, mechanical, and microstructural properties with compositional variations. The addition of 15 wt. % TTPOFA showed good results in terms of both physical and mechanical properties. A bulk density of 2.44 g/cm^3, with a high compressive strength of 247 MPa and very low water absorption of 0.024%, was achieved at 15 wt. % TTPOFA. Therefore, the successful purification of POFA through thermal treatment can open opportunities for its better utilization, promoting the development of sustainable and environmentally friendly materials such as porcelain. Other methods of purification, such as chemical treatments, could be explored for further research.

References

1. Owoeye, S.S., Toludare, T.S., Isinkaye, O.E., Kingsley, U.: Influence of waste glasses on the physico-mechanical behavior of porcelain ceramics. Bol. Soc. Esp. Ceram. **58**, 77–84 (2019)
2. Ochen, W., D'ujanga, F.M., Oruru, B., Olupot, P.W.: Physical and mechanical properties of porcelain tiles made from raw materials in Uganda. Results Mater. **11**, 100195 (2021)
3. Sawadogo, Y., Zerbo, L., Sawadogo, M., Seynou, M., Gomina, M., Blanchart, P.: Characterization and use of raw materials from Burkina Faso in porcelain formulations. Results Mater. **6**, 100085 (2020)
4. Mukhopadhyay, T.K., Ghosh, S., Ghosh, J., Ghatak, S., Maiti, H.S.: Effect of fly ash on the physico-chemical and mechanical properties of porcelain composition. Ceram. Int. **36**, 1055–1062 (2010)
5. Tabit, K., Hajjou, H., Waqif, M., Saâdi, L.: Effect of CaO/SiO2 ratio on phase transformation and properties of anorthite-based ceramics from coal fly ash and steel slag. Ceram. Int. **46**, 7550–7558 (2020)
6. Khattab, R.M., Sadek, H.E.H., Taha, M.A., El-Rafei, A.M.: Recycling of silica fume waste in the manufacture of β-eucryptite ceramics. Mater Charact **171**, 110740 (2021)
7. Hassan, U.J., Maharaz, M.N.: Influence of addition of rice husk ash on porcelain composition. Sci. World J. **10**, 7–16 (2015)
8. Dana, K., Das, S., Das, S.K.: Effect of substitution of fly ash for quartz in triaxial kaolin–quartz–feldspar system. J. Eur. Ceram. Soc. **24**, 3169–3175 (2004)
9. Haque, M.J., Munna, A.H., Rahman, S., Rahman, M.: Fabrication and performance analysis of porcelain stoneware tiles incorporated with labware glass cullet. Trans. Indian Ceram. Soc. **80**, 64–70 (2021)
10. Hamada, H.M., Yahaya, F., Muthusamy, K., Humada, A.: Comparison study between POFA and POCP in terms of chemical composition and physical properties-Review paper. IOP Conf. Ser. Earth Environ. Sci. **365**, 1–8 (2019)
11. Alnahhal, M.F., Alengaram, U.J., Jumaat, M.Z., Alqedra, M.A., Mo, K.H., Sumesh, M.: Evaluation of industrial byproducts as sustainable pozzolanic materials in recycled aggregate concrete. Sustainability **9**, 767 (2017)
12. Rajesh, C., Sameer, G.N., Reddy, M.S.M., Jagarapu, D.C.K., Jogi, P.K.: Consumption of palm oil fuel ash in producing lightweight concrete. Mater. Today **33**, 1073–1078 (2020)
13. Ayub, M., Othman, M.H.D., Yusop, M.Z.M., Khan, I.U., Zakria, H.S.: Facile and economical, single-step single-chemical method for conversion of palm oil fuel ash waste into graphene nanosheets. Appl. Mater. Today **25**, 101193 (2021)
14. Kusumastuti, E., Desita, R., Prasetya, A.T., Sulistyaningsih, T., Sulistyani, M.: Conversion of palm oil fuel ash (POFA) into foamy geopolymer for lightweight building material application by aluminum powder addition. J. Phys. Conf. Ser. **1918**, 032001 (2021)
15. Durumin-Iya, S.G., Noh, M.Z., Adamu, M.S.: Influence of modified chemical compositions on palm oil fuel ash to the physico-mechanical properties of porcelain. Key Eng. Mater. **908**, 626–632 (2022)
16. Vegliò, F., Passariello, B., Abbruzzese, C.: Iron removal process for high-purity silica sands production by oxalic acid leaching. Ind. Eng. Chem. Res. **38**, 4443–4448 (1999)
17. Ochen, W., D'ujanga, F.M., Oruru, B.: Influence of residual stress on the mechanical behavior of ceramics with various quartz sizes. Sci. Afr. **11**, e00648 (2021)
18. Amran, M., Murali, G., Fediuk, R., Vatin, N., Vasilev, Y., Abdelgader, H.: Palm oil fuel ash-based eco-efficient concrete: a critical review of the short-term properties. Materials **14**, 332 (2021)
19. Belal, A., Shafigh, P., Ibrahim, Z., Jumaat, M.Z.: Heat-treated palm oil fuel ash as an effective supplementary cementitious material originating from agriculture waste. Constr. Build. Mater. **167**, 44–54 (2018)

20. Shaladi, R.J., Johari, M.A.M., Ahmad, Z.A., Mijarsh, M.J.A.: The influence of palm oil fuel ash heat treatment on the strength activity, porosity, and water absorption of cement mortar. Environ. Sci. Pollut. Res. **29**, 72493–72514 (2022)
21. Tai, Z.S., et al.: Influence of pretreatment temperature of palm oil fuel ash on the properties and performance of green ceramic hollow fiber membranes toward oil/water separation application. Sep. Purif. Technol. **222**, 264–277 (2019)
22. Kaiyum, M.A., Ahmed, A., Hasnat, M.H., Rahman, S.: Effect of MgO on physical and mechanical properties of dental porcelain. J. Korean Ceram. Soc. **58**(1), 42–49 (2020). https://doi.org/10.1007/s43207-020-00083-4
23. Zheng, W., et al.: Influence of Al2O3 content on mechanical properties of silica-based ceramic cores prepared by stereolithography. J. Adv. Ceram. **10**, 1381–1388 (2021)
24. Krzysztof, G.: Influence of mineralogical composition of applied ball clays on properties of porcelain tiles. Ceram. Int. **37**, 851–861 (2011)
25. Kitouni, S., Harabi, A.: Sintering and mechanical properties of porcelains prepared from Algerian raw materials. Cerâmica **57**, 453–460 (2011)
26. Hamada, H.M., Jokhio, G.A., Yahaya, F.M., Humada, A.M., Gul, Y.: The present state of the use of palm oil fuel ash (POFA) in concrete. Constr. Build. Mater. **175**, 26–40 (2018)
27. Tangchirapat, W., Jaturapitakkul, C., Chindaprasirt, P.: Use of palm oil fuel ash as a supplementary cementitious material for producing high-strength concrete. Constr. Build. Mater. **23**, 2641–2646 (2009)
28. Irfan Khan, M., Azizli, K., Sufian, S., Man, Z., Khan, A.S.: Simultaneous preparation of nano silica and iron oxide from palm oil fuel ash and thermokinetics of template removal. RSC Adv. **5**, 20788–20799 (2015)
29. Jarnthong, M., Malawet, C., Liao, L., Li, P., Peng, Z.P.: Preparation and characterization of ultrafine oil palm ash powder by ultrasonication and alkaline treatment for its evaluation as reinforcing filler in natural rubber. Polymers **13**, 1–22 (2021)
30. Tasnim, S., Rahman, M.E., Ahmadi, R.: Mechanical performance of modified cement paste made with micro fine POFA in ammonium nitrate environment. Constr. Build. Mater. **162**, 534–542 (2018)
31. Njindam, O.R., Njoya, D.D., Mache, J.R., Mouafon, M., Messan, A., Njopwouo, D.: Effect of glass powder on the technological properties and microstructure of clay mixture for porcelain stoneware tiles manufacture. Constr. Build. Mater. **170**, 512–519 (2018)
32. Hassan, U.J., Mohamad, Z.N., Zainal, A.A.: Effects of palm oil fuel ash composition on the properties and morphology of porcelain-palm oil fuel ash composite. J. Teknol. **70** (2014)
33. Ozturk, Z.B., Karaca, Y., Kayali, B., Ubay, E.: The use and recycling of filter-press cake wastes in eco-friendly porcelain tile formulations. Int. J. Environ. Sci. Technol. **20**, 6307–6318 (2023)
34. Singer, F., Singer, S.S.: Industrial Ceramics, 1st edn. Springer, Dordrecht (1963)
35. Selli, N.T.: Relationship between microstructure and the impact resistance of porcelain stoneware tiles. Bol. Soc. Esp. Cerám. **61**, 210–219 (2022)
36. Romero, M., Martín-Márquez, J., Rincón, J.M.: Kinetic of mullite formation from a porcelain stoneware body for tiles production. J. Eur. Ceram. Soc. **26**, 1647–1652 (2006)
37. Carty, W.M., Udayan, S.: Porcelain—raw materials, processing, phase evolution, and mechanical behavior. J. Am. Ceram. Soc. **81**, 3–20 (1998)
38. Cheng, X., Ke, S., Wang, Q., Wang, S., Liu, P.: Fabrication and characterization of anorthite-based ceramic using mineral raw materials. Ceram. Int. **38**, 3227–3235 (2012)

Calix[6]Arene Combined 4-Tert-Butylcalix[4]Arene Coated Quartz Crystal Microbalance Sensor for High Sensitivity Gas Vapour Detection

Nur Firzanah Rosnan and Mohd Kamarulzaki Mustafa(✉)

Photonics Devices and Sensor Research Centre (PDSR), Faculty of Applied Science and Tecnology (FAST), Universiti Tun Hussein Onn Malaysia (UTHM), Kampus Cawangan Pagoh, Hub Pendidikan Tinggi Pagoh, Jalan Panchor, Pagoh, 84600 Muar, Johor, Malaysia
kamarulz@uthm.edu.my

Abstract. The Quartz Crystal Microbalance (QCM) plays a crucial role as an electronic cornerstone in gas sensor development, capable of detecting minute mass alterations on its quartz crystal resonator triggered by gas molecule adsorption or desorption. In this study, we adopt the QCM dip-coat technique, involving the application of layers comprising pairs of calix[6]arene and 4-tert-Butylcalix[4]arene to enhance sensor sensitivity. Results demonstrate that employing three pairs significantly boosts gas vapor detection compared to using two or one pair. Sensor sensitivity is validated through the testing of various concentrations of Volatile Organic Compounds (VOCs) gases, revealing a direct correlation between gas dilution and impedance levels. Observed gas detection trend indicate that toluene vapor induces higher impedance than acetone and ethanol due to its greater molecular weight. Toluene's higher molecular weight fosters stronger interactions with the quartz surface, resulting in more pronounced frequency shifts and higher impedance readings. Moreover, our findings align with the Sauerbrey Equation, confirming the reliability and accuracy of the QCM-based gas sensor developed in this study. The incorporation of calix[6]arene and 4-tert-butylcalix[4]arene in the QCM dip-coat technique enhances gas molecule entrapment by providing increased surface area, selective binding, synergistic effects, and stabilization, consequently amplifying sensor sensitivity and performance.

Keywords: QCM Gas Sensor · Volatile Organic Compounds · Calix[6]Arene combined 4-Tert-Butylcalix[4]Arene

1 Introduction

The escalating population represents the primary contributor to pollutants that continuously degrade the environment, compounded by vehicle emissions, industrial discharges, and open burning [1–4]. The adverse effects of environmental pollutants pose significant threats to human health, manifesting in respiratory issues, cardiovascular ailments, and

P. Chang et al. (Eds.): SCIEMATHIC 2024, SPPHY 427, pp. 101–112, 2025.
https://doi.org/10.1007/978-981-96-7679-8_9

other health complications [5]. Volatile Organic Compounds (VOCs) fall into the category of substances harmful to both human health and the environment [6]. VOCs, organic chemical compounds with high vapor pressure and low water solubility, swiftly evaporate into the atmosphere [7, 8]. Examples include methanol [9], benzene [10], toluene [11], and dichloromethane [12]. Some VOCs, like benzene, exhibit high toxicity and carcinogenic properties, posing risks to human health and ecosystems [13]. Prolonged exposure to VOCs may heighten cancer risks [14]. Amidst the dynamic landscape of sensing technologies, gas sensors play a pivotal role in detecting and monitoring various gases, ensuring environmental safety and industrial security [15].

Both calix[6]arene and 4-tert-Butylcalix[4]arene belong to the calixarene family, characterized by macrocyclic structures formed from phenolic rings linked by methylene bridges [16]. Calix[6]arene's cup-like structure, featuring a hydrophobic interior, facilitates the recognition and characterization of other molecules within its cavity, a trait beneficial in host-guest chemistry and supramolecular applications [17]. Consequently, calix[6]arene finds application in QCM sensors due to its unique structure, allowing selective binding with target molecules, resulting in detectable mass changes with high sensitivity, making it valuable in environmental monitoring and medical diagnostics [18, 19]. On the other hand, 4-tert-Butylcalix[4]arene, known for its macrocyclic structure with a basket-like shape, influenced by cyclic organization and tert-butyl substitutions, creates a hydrophobic cavity with steric hindrance [20]. This enables preferential interaction and binding with hydrophobic guest molecules, rendering it suitable for supramolecular chemistry, molecular recognition investigations, and QCM sensors [21, 22]. The steric hindrance caused by tert-butyl groups influences the compound's reactivity and capacity to establish specific interactions with other molecules, making it a vital member of the calixarene class [23]. Consequently, both calix[6]arene and 4-tert-Butylcalix[4]arene possess unique macrocyclic structures and are utilized in various chemical applications; their specific structures and functional groups confer distinct properties, rendering them suitable for different purposes within supramolecular chemistry, molecular recognition, and sensor technologies.

This research is dedicated to developing a low cost and mobile gas sensor, leveraging the Quartz Crystal Microbalance (QCM) and the dip coat method for efficient detection of VOCs gas vapor. The QCM stands out as a highly sensitive and versatile sensor widely utilized across diverse fields, from material science to environmental monitoring, making it suitable for gas sensor applications [24, 25]. By utilizing the piezoelectric properties of precisely cut quartz crystals, QCM detects subtle changes in mass [26]. Its versatility is evident in applications such as ultra-sensitive gas sensing, real-time chemical monitoring, and biosensing for biomolecular interactions [27].

Sauerbrey equation was established by Ernst Sauerbrey in 1959, which remains the foundation of QCM technology and describes how a quartz crystal resonator's frequency influences its surface mass [29]. The Sauerbrey equation is given by,

$$f = -\frac{2fo^2}{A\sqrt{\rho_q \mu_q}}\Delta m \tag{1}$$

where Δf is the change in resonant frequency of crystal, f_o is fundamental resonance frequency, A is area of crystal electrode, ρ_q is the density of quartz and μ_q is the shear modulus of quartz. By rearranging the Sauerbrey equation, the change in mass can be express as,

$$\Delta m = -\frac{A\sqrt{\rho_q \mu_q}}{2fo^2}\Delta f \tag{2}$$

In RLC circuit, impedance, Z is related to the resonance frequency, f by equation,

$$Z = \frac{1}{2\pi fC} \tag{3}$$

where C is capacitance of the crystal and the value of capacitance is constant (2 – 5pF). The equation can be written as,

$$Z \propto \frac{1}{f} \tag{4}$$

Therefore, as the frequency, f increases, the impedance, Z decreases. The Sauerbrey Eq. (2) states that increasing mass, Δm on a quartz surface reduces their frequency shift, Δf. In RLC circuit, a reduction in frequency, f causes an increase in impedance, Z as impedance is inversely proportional to frequency. Thus, the change in mass on the quartz surface has an inverse connection with the impedance value. This theory emphasises the sensitivity of QCM-based gas sensors, since small changes in mass cause noticeable alterations in impedance, allowing for precise gas detection and quantification due to the binding event between chemicals coated on quartz surface towards VOC's molecules.

2 Materials

Calix[6]arene ($C_{42}H_{36}O_6$, 97%) and 4-tert-Butylcalix[4]arene ($C_{44}H_{56}O_4$, 95%) were purchase from Sigma-Aldrich (M) Sdn. Bhd. All chemicals e.g., chloroform, toluene, acetone from QREC (ASIA) Sdn. Bhd. and ethanol from Progressive Scientific Sdn. Bhd. is absolute for research analysis. The Agilent 4294A Precision Impedance Analyzer is the main instrument employed in this work for measurement of impedance changes. An AT cut Quartz crystal microbalance (QCM) with 10 MHz fundamental frequency bought form Euroquartz were exploited to detect the changes of frequencies due to adding mass on quartz surface.

3 Sample Preparation and Measurement

Preparation of chemical solutions for this research involved meticulous handling of two substances, calix[6]arene (C42H36O6) and 4-tert-Butylcalix[4]arene (C44H56O4). Calix[6]arene were weighed and diluted with chloroform in beaker to form 10 mg ml^{-1} concentration of solution labelled as stock solution. The dilution step for calix[6]arene was repeated for 4-tert-Butylcalix[4]arene compound as the desired concentration for both chemical need to be achieved for 1 mg ml^{-1}. VOCs solution were prepared for detection on QCM sensor according to different ratio as stated in Table 1.

Table 1. Ratios for VOC's dilution.

Ratio	Calculation
1:100	1 ml VOCs chemical + 99 ml DI water
1:200	1 ml VOCs chemical + 199 ml DI water
1:300	1 ml VOCs chemical + 299 ml DI water

To ensure precise and reliable research results, the Quartz Crystal Microbalance (QCM) surface must be thoroughly cleaned with diluted HCl and rinsed with deionized (DI) water before use.

3.1 Control Impedance Curve

The impedance measurement conducted on the QCM in ambient air require meticulous setup, involving the connection of the Quartz Crystal Microbalance (QCM) to the Agilent 4294A Precision Impedance Analyzer. The objective of the experiment is to scrutinize the impedance properties of the QCM standard, with a specific focus on monitoring and contrasting the alterations in resonance frequency and impedance on the crystal surface before the binding event. Throughout the experiment, real-time results and an impedance curve versus frequency in air are anticipated from the impedance analyzer's display. This impedance curve elucidates how the QCM's impedance interacts with frequency, furnishing invaluable insights into its resonant behaviour. This experiment marks the initial stride towards comprehending the impedance dynamics of the QCM frequency, facilitating comparison with other impedance measurements in various solutions.

3.2 Deposition Layer by Layer (LbL) of Chemical onto Quartz Surface

The deposition process of chemical layers onto the Quartz surface is carefully conducted to fabricate QCM samples with varying layers, comprising up to three pairs of chemicals. Initially, the QCM is prepared by thorough cleaning to ensure the surface is free from contaminants. The first layer is deposited using a 1 mg/ml solution of calix[6]arene and undergoes an incubation period of 90 s followed by a drying process. This deposition process is carried out in 2 ml of calix[6]arene solution.

Subsequently, the process is repeated using a 1 mg/ml solution of 4-tert-Butylcalix[4]arene. The initial deposition sample is defined as one pair of layers. The experiment progresses to fabricate two pairs of layers by repeating the deposition of both calix[6]arene and 4-tert-Butylcalix[4]arene solutions on the QCM surface. This process is conducted twice to achieve two pairs of layers and further extends to achieve three pairs of layers by repeating the deposition process for both solutions three times. The deposition process for both chemicals utilizes the dipping method.

3.3 Impedance Measurement of Deposited QCM

Impedance measurements are conducted on Quartz Crystal Microbalance (QCM) samples with varying layers to emulate real-world conditions. These fabricated QCM samples, featuring one, two, and three pairs of layers, undergo systematic impedance analysis. For each deposited QCM sample, the impedance curve versus frequency is meticulously recorded. Subsequently, the impedance curves of the deposited QCM samples are compared with the baseline QCM standard curve to assess the influence of layer stacking. This comparative analysis aids in pinpointing deviations, resonance frequency shifts, and alterations in impedance behavior resulting from layering.

The fabricated QCM is assessed within a closed and transparent container by connecting it to the terminal and impedance analyzer. This method furnishes direct insights into how the number of layers impacts QCM sensitivity. The findings, encompassing impedance curves and other observations, are thoroughly documented for data analysis.

3.4 Impedance Measurement of Exposure QCM Sensor to VOCs Vapours

Acetone, ethanol, and toluene were selected as typical VOCs for this experiment, as they are commonly found chemicals in laboratories and can impact the health of users. The goal is to understand how vapor exposure affects the sensor's impedance behavior. QCM samples, previously fabricated with different layers, were placed inside a closed container for exposure to VOCs vapor. This closed container ensures a controlled environment, facilitating precise measurement of the impact of VOCs on QCM impedance. The volumes of acetone, ethanol, and toluene were consistently set at 50 ml to maintain uniform conditions across the experiment.

QCM samples were simultaneously exposed to ethanol, acetone, and toluene vapors in the closed container to maintain consistent and controlled exposure conditions for VOCs. This setup guarantees stable vapor concentrations throughout the experiment, minimizing the influence of external factors on the results. The closed system enables accurate measurements and reliable comparisons between different VOCs. This exposure was conducted concurrently with the impedance measurement setup, allowing for real-time monitoring of changes in the sensor's impedance properties during VOC exposure over a 10-min duration. The closed container technique confines and manages the VOCs environment, resulting in precise and repeatable findings.

Using a Precision Impedance Analyzer, the impedance of QCM samples exposed to VOCs vapors was measured. This step aimed to capture any variations in the impedance curves resulting from exposure to different VOCs. Real-time monitoring of impedance during exposure provides crucial data for understanding how the sensor responds to specific gas vapors. The collected impedance data undergoes rigorous analysis, focusing on identifying shifts in frequencies and changes in impedance magnitude. Interpreting these variations provides insights into the specific effects of acetone, ethanol, and toluene vapors on the QCM sensor.

4 Result and Discussion

4.1 QCM Impedance Analysis

Figure 1 illustrates a distinctive trend in the data in air, wherein the peak of the QCM frequency reach the peak at 10 MHz. This behavior is attributable to the QCM's resonance frequency, which is determined by its intrinsic physical properties, including the crystal's mass and flexibility. At resonance, the crystal vibrates most effectively, typically measured at a given frequency, around 10 MHz for this specific QCM device.

As the frequency approaches the resonance point, the crystal's ability to transmit energy efficiently improves, leading to an impedance peak. This phenomenon occurs because the crystal reacts most strongly to changes in mass or viscoelastic characteristics at its resonance frequency. Thus, the observed rise in the peak of the typical QCM at 10 MHz reflects the crystal's optimal resonance state, indicating heightened sensitivity to alterations in mass or material qualities. Understanding and leveraging these impedance characteristics are crucial for interpreting QCM data and optimizing sensor performance across different medium (Table 2).

Table 2. Ratios for VOC's dilution.

Ratio	Calculation
1:100	1 ml VOCs chemical + 99 ml DI water
1:200	1 ml VOCs chemical + 199 ml DI water
1:300	1 ml VOCs chemical + 299 ml DI water

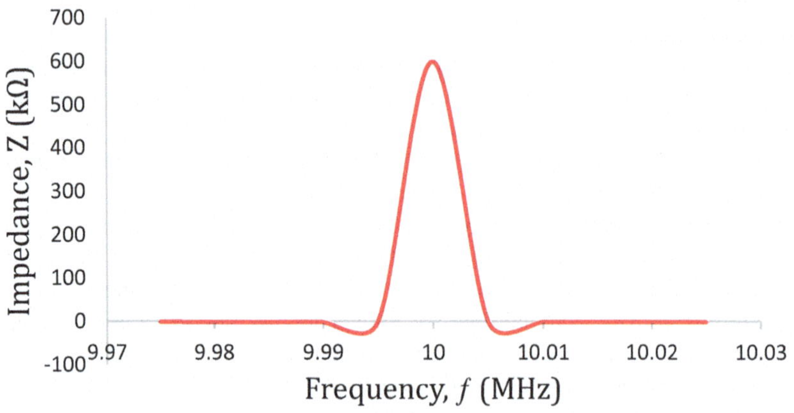

Fig. 1. The profile of impedance value in air

4.2 Impedance Analysis of Fabricated QCM

QCM samples were constructed through repeated depositions of calix[6]arene and 4-tert-Butylcalix[4]arene, forming up to three pairs of layers. Calix[6]arene, a macrocyclic molecule comprising six phenol units linked by methylene bridges, boasts a spacious, bowl-shaped structure capable of accommodating guest molecules. This structural feature enables calix[6]arene to selectively bind certain gas molecules through non-covalent interactions such as hydrogen bonding, van der Waals forces, and π-π stacking. On the other hand, 4-tert-butylcalix[4]arene, another macrocyclic molecule, features a smaller cavity formed by four phenol units. Despite its smaller size compared to calix[6]arene, 4-tert-butylcalix[4]arene also exhibits strong molecular recognition capabilities. The tert-butyl groups attached to the phenol units can sterically hinder the entry of certain guest molecules into its cavity, allowing for the selective binding of specific gas molecules that can fit into the available space. The combination of these two molecules offers complementary advantages for gas molecule detection: calix[6]arene provides a larger cavity suitable for accommodating bulkier gas molecules, while 4-tert-butylcalix[4]arene offers selectivity towards smaller gas molecules due to its steric hindrance effects. This synergistic combination enhances the overall sensitivity and selectivity of the detection system.

Figure 2 illustrates the acquired results, revealing consistent patterns in the impedance response. The observed increase in impedance, Z, with a decreasing number of layer pairs can be attributed to the cumulative impact of mass loading. Each new biochemical layer contributes to the cumulative total mass on the QCM surface, enhancing mass sensitivity and leading to a higher impedance response as the crystal becomes more sensitive to changes in mass.

Furthermore, concentration effects can be inferred from the observed pattern in which one pair of layers exhibited greater impedance values than two and three pairs. The mass loading is more concentrated with one pair of layers, resulting in a more pronounced influence on the resonance frequency and, consequently, a greater impedance peak. Conversely, the cumulative mass loading is dispersed across a greater area when more layers are added, resulting in comparatively lower impedance values.

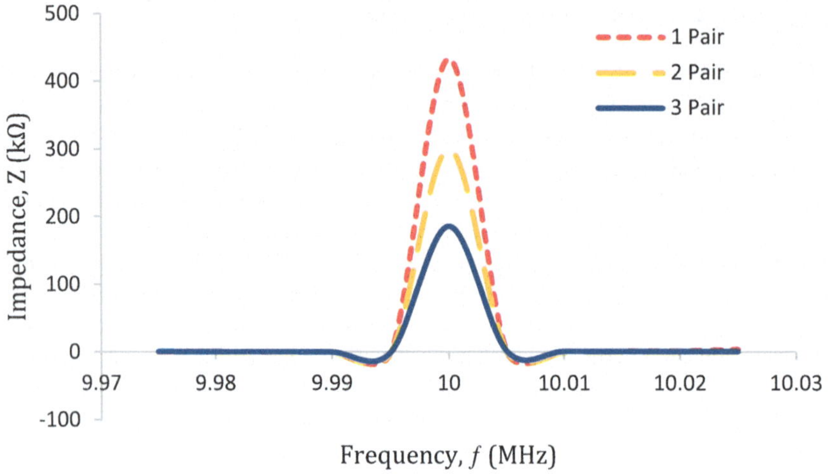

Fig. 2. The profile of impedance towards varies pair of calix[6]arene and 4-tert-Butylcalix[4]arene.

4.3 Impedance Measurement Towards Different Concentration of VOCs

Impedance measurement proves to be a valuable technique for detecting VOCs in diverse environments. The results in Table 3 unveil a consistent trend: at 100% concentration, each type of VOC exhibits the lowest impedance values compared to diluted ratios such as 1:100, 1:200, and 1:300. This trend underscores the reliability and predictability of impedance measurement in assessing VOC concentrations, with higher concentrations consistently yielding lower impedance values. Furthermore, the investigation explores the influence of different combinations of calix[6]arene and 4-tert-Butylcalix[4]arene on impedance values.

Notably, the results pertaining to three pairs of these compounds demonstrate higher impedance values compared to configurations with one or two pairs. This observation highlights the significance of the molecular architecture of calix[6]arene in enhancing the sensitivity of impedance measurement for VOC detection, with specific configurations offering superior performance.

4.4 Sensitivity of QCM Sensor Towards Different Types of VOCs

The observed pattern, Ethanol exhibited the lowest impedance values compared to Acetone and Toluene, can be attributed to the distinct molecular characteristics of each VOC. Ethanol, being a smaller and lighter molecule, interacts differently with the QCM surface, resulting in a less pronounced influence on the impedance. Conversely, Toluene induces a more significant shift in mass loading due to its larger molecular size and unique chemical properties, resulting in higher impedance values.

Table 3. Impedance value based on different VOC concentration

Type of VOCs	Impedance value in different concentration (kΩ)			
(a) One pair of deposition chemical onto QCM				
	100%	1:100	1:200	1:300
Ethanol	10.39	11.27	19.90	39.30
Acetone	12.19	91.84	205.33	361.90
Toluene	32.20	159.40	379.62	520.857
(b) Two pairs of deposition chemical onto QCM				
	100%	1:100	1:200	1:300
Ethanol	22.43	31.17	38.94	51.43
Acetone	32.85	109.42	567.87	629.22
Toluene	58.88	320.66	640.80	854.57
(c) Three pairs of deposition chemical onto QCM				
	100%	1:100	1:200	1:300
Ethanol	30.56	54.25	63.48	71.84
Acetone	54.74	383.22	850.19	943.74
Toluene	71.05	642.60	887.52	1084.46

Figure 3, depicting an increase in impedance following exposure of the QCM sample to increasingly diluted vapors, underscores the sensor's sensitivity to concentration fluctuations. As the concentration diminishes, the QCM becomes more responsive to minor changes in mass or viscoelastic properties, yielding larger impedance peaks. The results revealed that pure VOCs of Ethanol, Acetone, and Toluene exhibited the lowest impedance values compared to diluted VOCs, affirming that higher VOC concentrations exert a greater influence on the QCM surface, thus resulting in lower impedance values.

Regarding hazard assessment, the graph and findings indicate that Toluene elicits higher impedance levels than Ethanol and Acetone, suggesting a more pronounced impact on the QCM surface. This aligns with Toluene's recognized hazardous nature, underscoring the sensor's capability to discern between various VOCs based on their molecular attributes and concentrations. Ultimately, the QCM sensor's sensitivity to diverse VOCs is intricately linked to molecular characteristics, concentration levels, and dilution effects. Grasping these nuances enhances the sensor's efficacy in distinguishing between different VOCs, rendering it a potent tool for selective and precise gas detection across various applications.

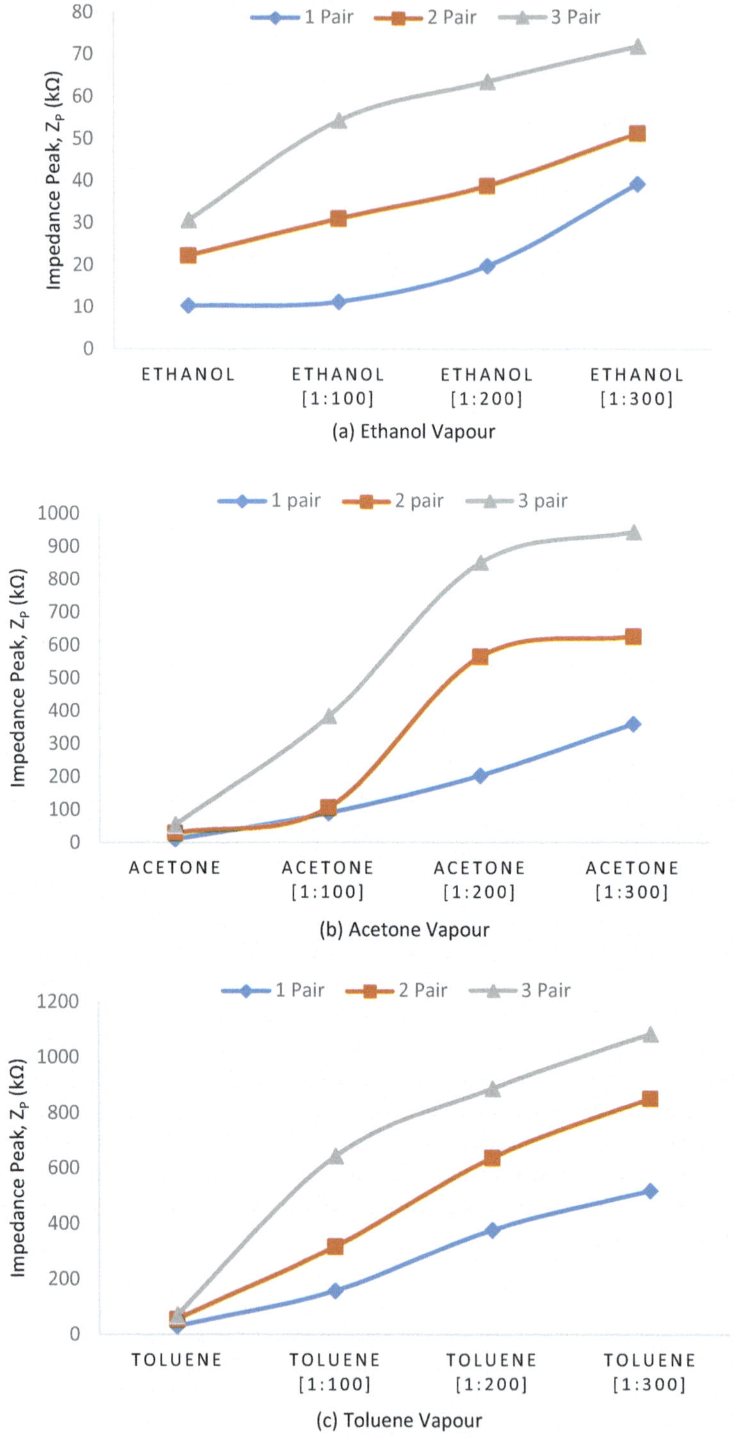

Fig. 3. Impedance peak profile of VOCs vapour (a) Ethanol; (b) Acetone; (c) Toluene

5 Conclusion

In conclusion, this study underscores the potential of combining calix[6]arene and 4-tert-Butylcalix[4]arene in detecting VOCs vapor using the dipping method on QCM. The efficacy of depositing calix[6]arene and 4-tert-Butylcalix[4]arene onto QCM enhances sensor sensitivity. These molecules possess intricate structures with cavities and functional groups that effectively absorb gas molecules. By combining them, the surface area available for gas absorption increases, facilitating more interactions with the sensor. Moreover, the combination exhibits selective binding properties towards certain gas molecules due to their specific structural features and functional groups, enhancing the sensor's sensitivity.

The synergistic effects resulting from combining calix[6]arene and 4-tert-Butylcalix[4]arene could lead to increased absorption capacity and improved detection capabilities for a broader range of gas molecules. Furthermore, this combination may contribute to stabilizing the absorbed gas molecules on the sensor's surface, preventing their desorption and ensuring prolonged interaction and detection. This stabilization effect enhances the sensor's sensitivity and reliability over time.

The positive correlation between chemical characteristics and QCM sensitivity provides a basis for customization. Utilizing three pairs of calix[6]arene and 4-tert-Butylcalix[4]arene yields higher sensitivity compared to one and two pairs. The consistency of sensitivity measurements is validated by testing different concentrations of defected gas at ratios of 100, 200, and 300. Results indicate that more diluted gas yields higher impedance values, with toluene vapor registering higher readings compared to acetone and ethanol due to its higher molecular weight. These findings underscore the sensitivity of a simple QCM-based gas sensor for common VOCs exposure in environments such as laboratories or chemical stores.

References

1. Briffa, J., Sinagra, E., Blundell, R.: Heavy metal pollution in the environment and their toxicological effects on humans. Heliyon **6**(9), e04691 (2020)
2. Dhall, S., Mehta, B.R., Tyagi, A.K., Sood, K.: A review on environmental gas sensors: materials and technologies. Sens. Int. **2**, 100116 (2021)
3. Mohd Shafie, S.H., Mahmud, M., Mohamad, S., Rameli, N.L.F., Abdullah, R., Mohamed, A.F.: Influence of urban air pollution on the population in the Klang Valley, Malaysia: a spatial approach. Ecol. Process. **11**(1) (2022)
4. Munsif, R., Muhammad Zubair, A.A.: Industrial air emission pollution: potential sources and sustainable mitigation. Intech **11**(tourism), 13 (2016)
5. Manisalidis, I., Stavropoulou, E., Stavropoulos, A., Bezirtzoglou, E.: Environmental and health impacts of air pollution: a review. Front. Public Health **8**(Feb), 1–13 (2020)
6. David, E., Niculescu, V.C.: Volatile organic compounds (Vocs) as environmental pollutants: occurrence and mitigation using nanomaterials. Int. J. Environ. Res. Public Health **18**(24) (2021)
7. Epa, U.: What are volatile organic compounds (VOCs)? | US EPA. United States Environmental Protection Agency, 1 (2022)
8. Khan, A.A., Bahadar, A., Hussain, M., Ullah, F., Ullah, A., Rasheed, S.: Analytical evaluation of polymeric CNTs/CuO nanocomposite electrode for the room temperature detection of volatile organic compounds (VOCs). Results Chem. **5**(Feb), 100928 (2023)

9. Yeoman, A.M., Shaw, M., Carslaw, N., Murrells, T., Passant, N., Lewis, A.C.: Simplified speciation and atmospheric volatile organic compound emission rates from non-aerosol personal care products. Indoor Air **30**(3), 459–472 (2020)
10. Sonne, C., Xia, C., Dadvand, P., Targino, A.C., Lam, S.S.: Indoor volatile and semi-volatile organic toxic compounds: need for global action. J. Build. Eng. **62**, 105344 (2022)
11. Pirola, C., Mattia, M.: Purification of air from volatile organic compounds by countercurrent liquid gas mass transfer absorption process. Int. J. Thermofluids **9** (2021)
12. Public Health Division. VOCs and well water: What you need to know
13. Mirzaei, A., Leonardi, S.G., Neri, G.: Detection of hazardous volatile organic compounds (VOCs) by metal oxide nanostructures-based gas sensors: a review. Ceram. Int. **42**(14), 15119–15141 (2016)
14. American cancer society. Benzene and cancer risk. Am. Cancer Soc. 1–8 (2016)
15. Nikolic, M.V., Milovanovic, V., Vasiljevic, Z.Z., Stamenkovic, Z.: Semiconductor gas sensors: materials, technology, design, and application. Sensors **20**(22), 1–31 (2020)
16. Guérineau, V., et al.: The synthesis and characterization of giant Calixarenes. Nat. Commun. **10**(1), 1–14 (2019)
17. Wang, C., Xu, L., Jia, Z., Loh, T.-P.: Recent applications of macrocycles in supramolecular catalysis. Chin. Chem. Lett. 109075 (2023)
18. Edwards, N.Y., Schnable, D.M., Gearba-dolocan, I.R., Strubhar, J.L.: Sensors (2021)
19. Quaglio, D., et al.: Calixarene: a versatile scaffold for the development of highly sensitive biosensors. Supramol. Chem. **33**(7), 345–369 (2021)
20. Widanelage, L., Priyadarshana, V.: Developing a Basket-Shaped Host Molecule Based on P-Tert- Butylcalix[4]arene Featuring Urea Groups for Encapsulating Volatile Guest Molecules, Master Thesis, Eastern Illinois University, US (2019)
21. Malinska, M.: Temperature- And solvent-induced crystal-form transformations of the pyridine@p-tert-butylcalix[6]arene host-guest system. Cryst. Growth Des. **21**(2), 1103–1112 (2021)
22. Moris, S., Galdámez, A., Jara, P., Saitz-Barria, C.: Synthesis of novel p-tert-butylcalix[4]arene derivative: Structural characterization of a methanol inclusion compound. Crystals **6**(9) (2016)
23. Lo, P.K., Wong, M.S.: Extended calix[4]arene-based receptors for molecular recognition and sensing. Sensors **8**(9), 5313–5335 (2008)
24. NanoScience Instruments. What is a Quartz Crystal Microbalance - QCM What is a Quartz Crystal Microbalance – QCM (2012)
25. Zhou, X., Zhou, X., Wang, C., Zhou, H.: Environmental and human health impacts of volatile organic compounds: a perspective review. Chemosphere **313**(Feb) (2023)
26. Biologic. QCM : History and principles (2023). https://www.biologic.net/topics/quartz-crystal-microbalance-measurement-principles/
27. Akgönüllü, S., Özgür, E., Denizli, A.: Recent advances in quartz crystal microbalance biosensors based on the molecular imprinting technique for disease-related biomarkers. Chemosensors **10**(3) (2022)
28. Vashist, S.K., Vashist, P.: Recent advances in quartz crystal microbalance-based sensors. J. Sens. (2011). https://doi.org/10.1155/2011/571405

Label Free Optical Biosensor of Cancer Cells Using Two Single-Slot GaAs Microring Resonators

Faezeh Pakfetrat[1], Mohd Kamarulzaki Mustafa[2(✉)], and Mahdi Bahadoran[1]

[1] Department of Physics, Shiraz University of Technology, 31371555 Shiraz, Fars, Iran
bahadoran@sutech.ac.ir

[2] Photonics Device and Sensor Research (PDSR), Faculty of Applied Sciences and Technology, Universiti Tun Hussein Onn Malaysia, Pagoh Educational Hub, 86400 Muar, Johor, Malaysia
kamarulz@uthm.edu.my

Abstract. In this article, a refractive index-based biosensor using a microring resonator structure is presented. The sensor configuration consisted of a circular ring resonator placed inside a racetrack ring resonator. Both resonators fabricated from a single-slot core from GaAs on a substrate layer of Al0.5Ga0.5As. The simulations were performed using the variational finite difference time domain (var-FDTD) method. An extended free spectral range of 129 nm was measured for the layout output signals. This sensor stands out with its ability to detect even the smallest changes in refractive index, achieving a sensitivity of 267.7 nm/RIU. This high level of precision makes it highly effective in identifying various cancer cells, including HeLa cells. Such remarkable sensitivity highlights its strong potential for advancing biomedical diagnostics and improving early disease detection.

Keywords: refractive index · microring resonators · variation finite difference time domain

1 Introduction

Cancer is a major public health problem worldwide. This disease is characterized by uncontrolled growth and expansion of abnormal cells [1]. These malignant cells can invade nearby tissues and lead to severe health complications and death if not treated effectively [2]. There are more than 100 different types of cancer, including the common types of breast cancer, lung cancer, prostate cancer, colon cancer, etc. [3]. Despite advances in medical science, cancer remains one of the leading causes of death worldwide. Cancer detection sensors are at the forefront of medical diagnostics, offering innovative ways to identify cancer at its earliest and most treatable stages. These sensors utilize various technologies to detect biomarkers associated with cancer, such as specific proteins, DNA mutations, or cellular changes [4]. Additionally, advancements in optical and electrochemical sensors have enabled the development of portable and cost-effective devices, which can be used for point-of-care testing and early cancer screening, thereby significantly improving patient outcomes [5].

P. Chang et al. (Eds.): SCIEMATHIC 2024, SPPHY 427, pp. 113–120, 2025.
https://doi.org/10.1007/978-981-96-7679-8_10

Today several optical-based approaches offer valuable insights into biomolecular interactions and are essential for scientific research and clinical diagnostics. These techniques include plasmonic biosensing, electro-optical techniques, electrochemical surface Plasmon resonance (EC-SPR) [6], electrochemical optical waveguide light mode spectroscopy (EC-OWLS) [7] and silicon-based electro-photonic (SBEP) approach. Plasmonic biosensors utilize surface plasmons that are collective oscillations of free electrons in metal (typically gold, silver and aluminum) at the interface with a liquid medium. These sensors are successful for label-free detection and characterization of biomolecular interactions [8].

EC-SPR blends electrochemical methods with Surface Plasmon Resonance (SPR) to study and analyze electrochemical processes happening at metal surfaces. In this approach, the movement of free electrons at the metal-dielectric boundary, known as surface plasmons, is affected by shifts in the refractive index of the surrounding medium. This allows for the detection and measurement of molecular interactions without the need for labeling. EC-SPR is especially valuable in creating sensitive biosensors for biomolecular interactions, providing real-time, label-free analysis with exceptional sensitivity [9]. EC-OWLS is an advanced technique that combines evanescent-field optical sensing with electrochemical control to study molecular adsorption processes on surfaces. In EC-OWLS, a thin layer of indium tin oxide (ITO) serves both as a high fractive-index waveguide and a conductive electrode. This setup allows for real-time monitoring of molecular interactions without the need for labeling. A pilot study demonstrated that applying potentials between 0 and 1.5 V to the ITO waveguide facilitated the adsorption of poly(L-lysine)-grafted-poly(ethylene glycol) (PLL-g-PEG), with adsorption increasing from 125 to 475 ng/cm^2. This technique holds promise for developing sensitive biosensors capable of label-free, real-time analysis with high sensitivity [7]. EC-SPR combines optical and electrochemical information to detect low concentrations of biomolecules and study molecular interactions in real time [10]. EC-OWLS provides label-free detection and detailed measurements of biomolecular processes using optical waveguides. SBEP approach is technique that integrates silicon technology for biosensing applications [10].

Ring resonators are considered as waveguide-based sensors that emerge as a powerful tool in the early detection and diagnosis of cancer [11]. Due to the manufacturing method in accordance with the existing technologies in electronics, the cheap manufacturing price, the linear response and narrow resonance peaks, the ring resonator-based systems have found many applications in medicine [12], optical switches [13], optical sensors [14], and optical telecommunications [15]. The operation of optical waveguides is based on total internal reflection, which directs the incoming light modes to the destination with minimal loss. In addition to telecommunications, the systems consisted of optical waveguides have used in sensors, lasers [16], tools for light confinement [17]. These optical sensors operate based on the principle of resonant frequency shifts when specific biomolecules bind to the sensor surface. The high sensitivity and specificity of ring resonator sensors make them particularly effective in detecting cancer biomarkers at very low concentrations [18]. They can identify molecular interactions with exceptional precision, which is crucial for early cancer diagnosis. The compact size and integration capabilities of these sensors allow for their incorporation into lab-on-a-chip devices, facilitating point-of-care testing and real-time monitoring of cancer biomarkers in clinical settings [19]. Advancements in ring resonator technology have significantly enhanced their performance and application in cancer diagnostics. This means that multiple cancer biomarkers can be detected simultaneously, providing a comprehensive diagnostic profile [20].

The gallium arsenide (GaAs) semiconductor material can be placed on an aluminum gallium arsenide (AlGaAs) optical buffer layer, which performs optimally in the wavelength range of 0.9–25 μm (MIR) [21]. This structure can have various applications, including in chemical and biological sensors [21].

In this work, we designed and simulated a ring resonator-based biosensor using a layout of two circular and racetrack ring resonator fabricated from a single-slot waveguide from GaAs on an Al0.5Ga0.5As substrate. The simulations results showed a linear response of sensor to small changes in the refractive indices that contributed to a sensitivity of 267.7 nm/RIU for detection of several cancer cells including Jurkat and HeLa cells.

2 Theory and Result

In this paper, we study a cancer microring resonator sensor consists of two single-slot GaAs ring resonators on an Al0.5Ga0.5As substrate with a refractive index of 3.374 and 3.184 at wavelength of 1550 nm, respectively. The sensor structure consists of an inner and outer rims of single-slot with radii of R1 and R2 which is shown in Fig. 1(a). The mode of the electric field of the entire sensor structure is shown in Fig. 1-b, which indicates the propagation of light through all the slots and rings that satisfy resonance conditions.

$$n_{eff} L = m\lambda \quad (1)$$

Here n_{eff} is effective index of waveguide, L shows the total circumference of rings, m represents the resonant mode number and λ is wavelength of input light.

The dimensions that are very important in the structure are shown schematically in the y-z direction in Fig. 2(a) and in the x-z direction in Fig. 2(b).

The structure of ring resonators has a great impact on its sensing, so the dimensions should be optimized in such a way that it has a good sensitivity and quality factor. The optimal dimensions are shown in Table 1. All parameters in Table 1 is optimized one by one in such a way that for each structural parameter we simulated a range of dimensions and considered the maximum transmission. In the same way, we optimized each parameter individually. After optimizing the structure, the refractive index of different cells was applied in var-FDTD.

Gallium Arsenide (GaAs) was chosen for the ring resonator sensor due to its excellent physical and electronic properties. A high refractive index is possessed by GaAs, making it highly sensitive to changes in the surrounding environment, a key factor for precise optical sensors. It is also known for its superior performance at high frequencies and its ability to transmit signals with minimal loss, which makes it ideal for sensitive applications like cancer detection. In addition, compared to materials like silicon,

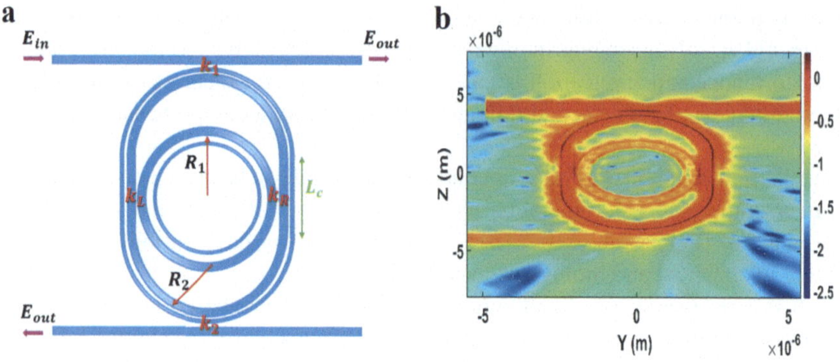

Fig. 1. Schematic diagram of a (a) proposed microring resonator sensor, (b) the electric field mode due to the propagating light via structure.

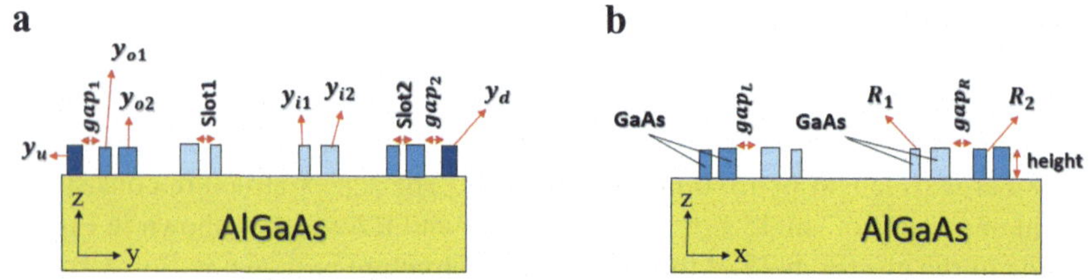

Fig. 2. Cross section of waveguides used in microring resonator sensor (a) y-z direction, (b) x-z direction.

GaAs is preferred for use in challenging conditions, such as high temperatures, making it particularly suitable for medical and diagnostic applications.

The simulation was done using the variational finite difference time domain (var-FDTD) method. In this structure, an extended free spectral range as wide as 129 nm and FWHM equal to 4.81 nm have been obtained. To check the sensing characteristics, the refractive index of Table 2 was used. Here, the phosphate buffer saline (PBS) with refractive index of 1.330 [22] was used as a base-line of sensing. Jurkat cells are an immortalized line of human T lymphocyte cells that can be used to explore cancer drug susceptibility, radiation effects, and interleukin-2 production. The refractive index of Jurkat cells is 1.390 at 1550 nm. HeLa cells are an immortalized cell line derived from cervical cancer cells and are widely used in scientific research due to their unique properties: they can divide and proliferate indefinitely in vitro. HeLa cells have played a crucial role in diverse studies, including investigations related to viruses, cancer, and human genetics [23]. The refractive index of HeLa cells was reported to be 1.392 at 1550 nm [22]. Michigan Cancer Foundation-7 (MCF-7) is another breast cancer cell line those express expresses estrogen receptor alpha (ERα) and is widely utilized in breast cancer research. Scientists investigate hormone effects, drug resistance, and breast

Table 1. Dimensions and parameters used in microring resonator-based sensor.

Geometric parameter	Size of sensor (μm)
R1	1.83
R2	4.59
yu	0.4
yd	0.4
yo1	0.23
yo2	0.37
yi1	0.2
yi2	0.33
Slot1	0.15
Slot2	0.05
height	0.1
gap1	0.1
gap2	0.1
gapL	0.1

cancer-related molecular mechanisms using MCF-7 cells [24]. The index of MCF-7 is 1.401 at 1550 nm [22].

The transmission diagram in terms of wavelength is shown in Fig. 3-a, the spectral shift of resonance peak based on the change of refractive index is demonstrated in Fig. 3(b), and the sensing line is plotted in Fig. 3-c. After the parameter optimization was completed, the refractive index in var-FDTD was adjusted according to the cancer cell properties to obtain the wavelength shift and the sensitivity was calculated. However, the refractive index of each cancer cell was set individually to generate the calibration line (Fig. 3c). The wavelength shift is equal to 18.74 nm for the resonance peak at a wavelength of about 1600 nm that it is equivalent to 267.7 nm/RIU sensitivity that corresponds to a detection limit of 7.5×10^{-5}.

Table 2. The refractive index types of cancer.

Types of cancer	Refractive index
PBS	1.330
Culture Medium	1.350
Jurkat	1.390
HeLa	1.392
Breast cancer (MCF7)	1.401

-NoValue-Jurkat, HeLa and breast cancer cell lines are commonly used in cancer research because of their medical significance, unique optical properties, and ease of growth in the lab. HeLa cells, taken from cervical cancer, are the first immortal human cell line and have become a standard model in cancer studies. Jurkat cells, which come from

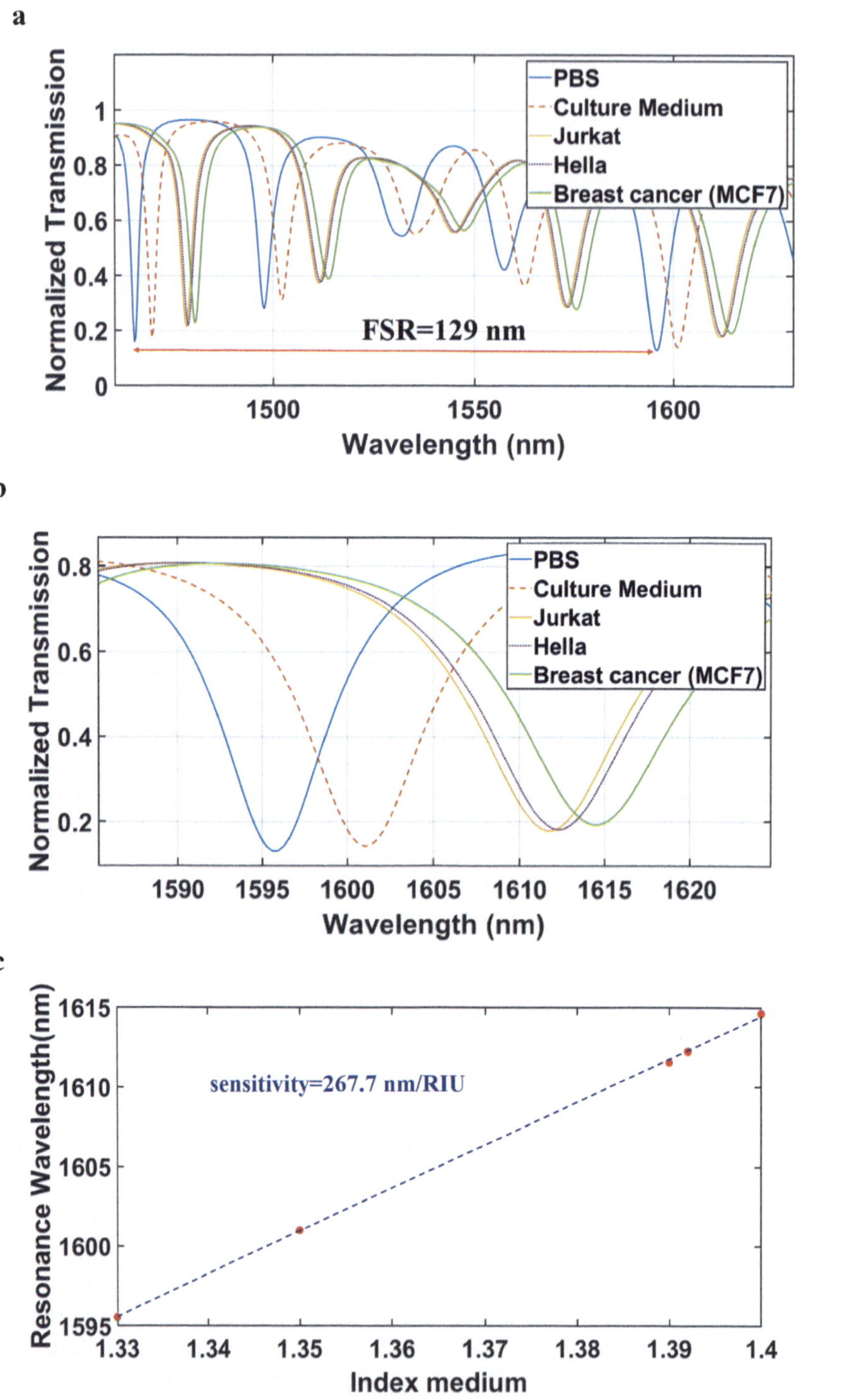

Fig. 3. (a) light transmission versus wavelength for various cancerous samples, (b) the spectral shift of resonance peaks, (c) sensing line of sensor.

T-cell leukemia, are crucial for researching blood cancers and immune-based treatments. Breast cancer cell lines like MCF-7 help scientists understand tumor progression and test new cancer drugs. Their well-known refractive index values also make them great candidates for optical biosensing, especially in developing sensors for cancer detection. Our designed biosensor achieves a sensitivity of 267.7 nm/RIU and a detection limit of 7.5×10^{-5} RIU, offering better performance than the biosensor reported in [25], which reached only ~100 nm/RIU and a detection limit of 10^{-6} RIU These results show that the proposed sensor can detect cancer biomarkers with greater accuracy. The combination of sensitivity and detection limit makes this sensor a promising candidate for biomedical diagnostics.

3 Conclusion

In this article, two single-slot resonator rings were presented as biosensors. Results were simulated using parameter settings for different geometrical data. Proposed sensor has a linear response to very small changes in refractive index. With this configuration, we achieved a free spectral range of 129 nm and a sensitivity of 267.7 nm/RIU with a detection limit of 7.5×10^{-5}. Future work will focus on further optimizing the sensor's design to enhance sensitivity and detection limits, as well as exploring its integration with microfluidic systems for real-time, point-of-care cancer diagnostics.

Acknowledgments. The authors would like to acknowledge the Shiraz University of Technology, Iran and Faculty of Applied Sciences and Technology, University Tun Hussein Onn Malaysia for providing the research facilities.

References

1. Siegel, R.L., Miller, K.D., Fuchs, H.E., Jemal, A.: Cancer statistics, 2022. CA Cancer J. Clin. **72** (2022)
2. Bray, F., Ferlay, J., Soerjomataram, I., Siegel, R.L., Torre, L.A., Jemal, A.: Global cancer statistics 2018: GLOBOCAN estimates of incidence and mortality worldwide for 36 cancers in 185 countries. CA Cancer J. Clin. **68**, 394–424 (2018)
3. Hanahan, D., Weinberg, R.A.: Hallmarks of cancer: the next generation. Cell **144**, 646–674 (2011)
4. Chikkaveeraiah, B.V., Bhirde, A.A., Morgan, N.Y., Eden, H.S., Chen, X.: Electrochemical immunosensors for detection of cancer protein biomarkers. ACS Nano **6**, 6546–6561 (2012)
5. Song, F.-X., et al.: Recent progress in nanomaterial-based biosensors and theranostic nanomedicine for bladder cancer. Biosensors **13**, 106 (2023)
6. Blidar, A., Feier, B., Tertis, M., Galatus, R., Cristea, C.: Electrochemical surface plasmon resonance (EC-SPR) aptasensor for ampicillin detection. Anal. Bioanal. Chem. **411**, 1053–1065 (2019)
7. Bearinger, J., Vörös, J., Hubbell, J., Textor, M.: Electrochemical optical waveguide lightmode spectroscopy (EC-OWLS): a pilot study using evanescent-field optical sensing under voltage control to monitor polycationic polymer adsorption onto indium tin oxide (ITO)-coated waveguide chips. Biotechnol. Bioeng. **82**, 465–473 (2003)

8. Herrera-Domínguez, M., Morales-Luna, G., Mahlknecht, J., Cheng, Q., Aguilar-Hernández, I., Ornelas-Soto, N.: Optical biosensors and their applications for the detection of water pollutants. Biosensors **13**, 370 (2023)
9. Jungnickel, R., Balasubramanian, K.: Electrochemical surface plasmon resonance sensing using a van der waals heterostructure. Adv. Sens. Res. **3**, 2400028 (2024)
10. Chen, C., Wang, J.: Optical biosensors: an exhaustive and comprehensive review. Analyst **145**, 1605–1628 (2020)
11. Fan, X., White, I.M., Shopova, S.I., Zhu, H., Suter, J.D., Sun, Y.: Sensitive optical biosensors for unlabeled targets: a review. Anal. Chim. Acta **620**, 8–26 (2008)
12. Bahadoran, M.: Analysis of ingaasp-inp double microring resonator using signal flow graph method. J. Optoelect. Nanostruct. **3**, 41–52 (2018)
13. Bahadoran, M., et al.: Modeling and analysis of a microresonating biosensor for detection of Salmonella bacteria in human blood. Sensors **14**, 12885–12899 (2014)
14. Babakhani-Fard, M.-M., Daraei, A., Hassani, S.: Bio-medical sensing by ultra-high quality-factor/modal-volume (Q/V) multi-slotted SOI Bragg grating cavity. Eur. Phys. J. Plus. **137**, 648 (2022)
15. Bahadoran, M., Afroozeh, A., Ali, J., Yupapin, P.P.: Slow light generation using microring resonators for optical buffer application. Opt. Eng. **51**, 044601 (2012)
16. Bogaerts, W., et al.: Silicon microring resonators. Laser Photon. Rev. **6**, 47–73 (2012)
17. Sanati, P., Shafiee, A., Bahadoran, M., Akbari, E.: Label-free biosensor array comprised of Vernier microring resonator and 3×3 optical coupler. Eur. Phys. J. Plus. **135**, 1–14 (2020)
18. Chao, C.-Y., Guo, L.J.: Biochemical sensors based on polymer microrings with sharp asymmetrical resonance. Appl. Phys. Lett. **83**, 1527–1529 (2003)
19. Armani, A.M., Kulkarni, R.P., Fraser, S.E., Flagan, R.C., Vahala, K.J.: Label-free, single-molecule detection with optical microcavities. Science **317**, 783–787 (2007)
20. Washburn, A.L., Luchansky, M.S., Bowman, A.L., Bailey, R.C.: Quantitative, label-free detection of five protein biomarkers using multiplexed arrays of silicon photonic microring resonators. Anal. Chem. **82**, 69–72 (2010)
21. Haas, J., Artmann, P., Mizaikoff, B.: Mid-infrared GaAs/AlGaAs micro-ring resonators characterized via thermal tuning. RSC Adv. **9**, 8594–8599 (2019)
22. Liang, X., Liu, A., Lim, C., Ayi, T., Yap, P.: Determining refractive index of single living cell using an integrated microchip. Sens. Actuators A **133**, 349–354 (2007)
23. Rahbari, R., Sheahan, T., Modes, V., Collier, P., Macfarlane, C., Badge, R.M.: A novel L1 retrotransposon marker for HeLa cell line identification. Biotechniques **46**, 277–284 (2009)
24. Lee, A.V., Oesterreich, S., Davidson, N.E.: MCF-7 cells—changing the course of breast cancer research and care for 45 years. J. Natl. Cancer Inst. **107**, djv073 (2015)
25. De Goede, M., et al.: Al2O3 microring resonators for the detection of a cancer biomarker in undiluted urine. Opt. Express **27**, 18508–18521 (2019)

The Study of Physical and Chemical Properties Turmeric (Curcuma Longa) Added with Copper Particles - A Preliminary Study

Engku Noradila Zulaika Engku Razali and Siti Amira Othman(✉)

Faculty of Applied Sciences and Technology, Universiti Tun Hussein Onn Malaysia, 84600 Pagoh, Johor, Malaysia
sitiamira@uthm.edu.my

Abstract. The study of turmeric's physical and chemical properties (Curcuma Longa) added with copper particles has drawn significant interest for its environmentally benign and long-lasting properties as a reducing and stabilising agent. Utilizing Ultraviolet-visible Spectroscopy (UV-Vis), Fourier Transform Infrared Spectroscopy (FTIR) and Scanning Electron Microscopy (SEM), the synthesised copper was characterised. The active ingredients found in Curcuma Longa reduced copper ions in an aqueous solution to create copper particles. In UltraViolet Visible Spectroscopy (UV-Vis), increasing concentration led to absorbance fluctuation, attributed to varying solubility of the particle mixture within the wavelength range between 420 nm to 550 nm. The Fourier Transform Infrared Spectroscopy (FTIR) study revealed the molecular composition of copper particles' transmittance, functional group, and chemical bonds. A broad peak around 3200–3600 cm^{-1} indicated stretching vibrations of hydroxyl (O-H) groups covering alcohols and phenols. Peaks around 1700–1800 cm^{-1} indicated stretching vibrations of the carbonyl ($C = O$) group covering ketones, aldehydes, and carboxylic acids; meanwhile, peaks around 3300–3500 cm^{-1} indicated stretching vibrations of the amino (N-H) group cover amines and amides bonds. The surface morphology of the synthesised copper particles was examined using Scanning Electron Microscopy (SEM), which provided visual proof of production, revealing that the particle size of Curcuma Longa rhizome extract increases with concentration due to particle mixture insolubility illustrates a smoother copper particle morphology at 3M compared to a rougher texture at 5M. Copper impacts biological systems when the characterisation shows copper particles' good physical and chemical properties. In conclusion, it enhanced several applications such as catalytic, drug delivery systems and antibacterial effects. Their environmentally benign synthesis process enables a wide range of applications, significantly advancing the safe application of copper in several industries.

Keywords: Turmeric · Curcuma Longa · Copper · Morphology

P. Chang et al. (Eds.): SCIEMATHIC 2024, SPPHY 427, pp. 121–132, 2025.
https://doi.org/10.1007/978-981-96-7679-8_11

1 Introduction

With uses in many sectors, copper is a valuable metal due to its unique blend of conductivity, malleability and corrosion resistance. Historically, Curcuma Longa has been used for culinary and medical purposes, mainly because of its active ingredient, curcumin. According to recent research, curcumin may impact metals' chemical and physical characteristics. Curcuma Longa can alter and improve the properties of metal surfaces, which has drawn attention to its use with copper. Curcumin, the active ingredient, possesses anti-inflammatory and antioxidant qualities [1]. Researchers have looked at its interaction with metal surfaces and its biological impacts. According to research by [2], curcumin may strengthen copper resistance to corrosion by inhibiting corrosion on the metal in acidic environments. Researchers [3] investigated how curcumin can suppress copper corrosion in chloride solutions.

This substance, well known for its antibacterial, anti-inflammatory, and antioxidant effects, offers a chance to give the copper particles beneficial biological characteristics and provide opportunities for use in various industries. Turmeric's biocompatibility has become vital in biomedical applications where it has been used medicinally, and adding it to particle formulations improves their overall biocompatibility, which opens up new possibilities for the use of the final particles in pharmaceutical and medical applications. Turmeric contributes to the stability of the copper particles by shielding them from deterioration through its various components. This characteristic increases the shelf life of particles and improves their functionality in particular situations, making them more dependable for long-term uses. Turmeric's vivid yellow hue has a valuable function in experimental settings. It makes it simple to follow and characterise the particles visually, examining them at different phases of their creation and use. The usefulness of the particles is increased by the possible health advantages of curcumin, which is present in turmeric and is especially useful in medication delivery and antibacterial applications. The study investigates how adding curcumin to copper particles improves their therapeutic benefits and increases the range of applications for these particles. Overall, the study explores a variety of applications, including biomedical, stability, visibility and health-related ones, that might benefit from the varied and possibly significant synergy between turmeric and copper particles.

There is a great deal of interest in further exploring this cutting-edge and ecologically advantageous technology due to the quick advancement in manufacturing metal particles utilizing plant extracts. This proposal discusses the study of the physical and chemical properties of turmeric (Curcuma Longa) added with copper particles. Zetasizer, Ultraviolet-Visible Spectroscopy (UV-Vis), Fourier Transform Infrared Spectroscopy (FTIR) and Scanning Electron Spectroscopy (SEM) are employed. The absorbance of copper particles in these rhizomes is determined using Ultraviolet-Visible Spectroscopy (UV-Vis) imaging. Fourier Transform Infrared (FTIR) spectra show the molecular composition of copper particles' transmittance, functional group and chemical bonds. The Scanning Electron Spectroscopy (SEM) picture clearly shows the surface morphology process of forming particles. The yield and stability of copper are remarkably high [4]. Any material must be used entirely for the benefit of humans, medicines and the healthcare industry. The generated particles must be characterised before usage. Examining the distinguishing characteristics of materials, such as size, shape, size distribution, surface area, form, solubility, and aggregation, is important before assessing toxicity or bio-compatibility. Creating value-added Curcuma Longa products for the biomedical and technology sectors will be the most crucial result of the projects.

Due to its many applications, copper one of the noble metal particles, is of great interest. Due to their distinctive features, copper has attracted much attention in various sectors. However, further research is needed to determine if they are harmful and how they could affect the environment [5]. This work uses a natural plant extract to study the physical and chemical properties of turmeric (Curcuma Longa) added with copper particles. The findings of this study will aid in developing safer and more environmentally friendly methods for producing and using copper particles while also shedding light on any possible hazards to human health and the environment.

Malaysia has a more extensive Curcuma Longa plantation. The greatest outcome would be using natural resources as much as possible instead of Curcuma Longa plantations. Numerous applications are expanding in scope as they play increasingly significant roles in daily life. Copper's remarkable physical and chemical properties, such as its high conductivity, malleability and corrosion resistance, are responsible for its widespread use in various industries. Nevertheless, the traditional approaches to improving these qualities frequently call for using materials or procedures that are harmful to the environment. Other methods of enhancing the properties of copper are required, especially with the growing focus on sustainable practices.

Curcuma Longa has gained attention as a possible tool for altering metals' chemical and physical characteristics, especially copper. While initial research has demonstrated encouraging outcomes concerning the ability of curcumin to inhibit copper corrosion [2, 3], a thorough examination of Curcuma Longa, which has broader effects on the different aspects of copper properties, is lacking. The following problem statement is formulated in response to this research gap: What specific effects does Curcuma Longa have on the chemical and physical characteristics of copper particles, and how can this naturally occurring substance be used to improve the performance and sustainability of copper? Solving this issue could lead to the creation of environmentally friendly methods for enhancing the qualities of copper and improve our knowledge of the possible uses of Curcuma Longa in material science. Therefore, a thorough investigation into the various aspects of copper treated with Curcuma Longa is imperative to close this knowledge gap and promote long-term progress in the field.

Due to the wide range of uses it may be used, copper, one of the noble metal particles, is of great interest. Urban planning and policy choices to foster sustainable development and enhance public health outcomes for toxicity studies can be informed by research on herbal plantations like Curcuma Longa [6]. Extracts from the rhizomes of Curcuma Longa can create very stable copper particles in solution.

Copper, one of the noble metal particles, is of tremendous interest due to their diverse uses. Copper has gained significant attention in various fields due to its unique properties, but its potential toxicity and environmental impact must be thoroughly investigated [5]. Curcuma Longa is a natural plant extract known for its antioxidant and anti-inflammatory properties [7]. Utilising Curcuma Longa to explore its potential as a green and eco-friendly approach for synthesising and characterising copper also evaluates the toxicity of these particles. This study aims to contribute to understanding copper characterization using a natural plant extract and compare the physical and chemical properties of copper using Curcuma Longa. The results obtained from this research will help develop safer and more sustainable approaches for the synthesis and application of copper, while also providing insights into their potential risks to human health and the environment.

Examining the impact of Curcuma Longa on the physical and chemical characteristics is highly significant in several fields. The research advances sustainable material science by helping to alter the paradigm in favor of sustainable material science. In line with the worldwide movement for ecologically friendly technology, if the interaction between Curcuma Longa and copper is successful, it may open the door for environmentally benign methods of improving metal characteristics [1]. It also reduces environmental impact by conventional ways of increasing metal characteristics sometimes requiring the use of materials that are harmful to the environment. If Curcuma Longa is shown to be successful, it provides a natural and biodegradable substitute, lessening the harm that traditional therapies cause to the environment [2].

It also opens doors for innovative applications: If this study's findings are successful, copper may find new uses in various sectors. Applications for improved quality might arise in electronics, building, and transportation, producing more robust and effective goods. Curcumin, the main ingredient in Curcuma Longa, is well-known for its advantages to health and safety [1]. Curcuma Longa may help produce copper materials with additional health advantages if its application to copper surfaces is safe and effective. This is especially true when copper antibacterial capabilities are already being used. The discovery adds to our understanding of how natural substances and metals interact, which advances scientific knowledge. The study contributes significantly to material science and corrosion studies by thoroughly characterising the effect of Curcuma Longa added with copper particles [8].

Copper particles with important biological properties are enhanced by turmeric, which is well known for its antibacterial and antioxidant properties. Copper particles have several uses in different sectors. The biocompatibility of turmeric plays a critical role in biomedical applications, enhancing its general suitability for usage in pharmaceutical and medical settings. Its stability contribution increases the functionality and shelf life of the particles, guaranteeing dependability for extended uses. In test environments, the bright yellow colour facilitates visual tracking. The health advantages of curcumin are increased copper particle usage, especially for antibacterial and medication delivery applications.

Encouragement of additional studies and utilization of positive results might lead to more investigation into the relationship between Curcuma Longa and other metals. Furthermore, the results might stimulate the creation of new materials with improved qualities by using natural components [3]. In conclusion, this study is critical because it has the potential to improve environmentally friendly material science practices, lessen the impact on the environment, find new uses for copper, improve health and safety, add to scientific knowledge and inspire more innovation and research in the field. Subsequent paragraphs, however, are indented.

2 Methodology

The materials used in this experiment included Curcuma Longa, which was purchased from a local market in Malaysia. An extract of the Curcuma Longa plant was created, and Curcuma Longa was washed, cut, weighed using a weighting balance, and punched using mortar and pestle of a 50g sample. Curcuma Longa was then mixed with 150ml of distilled water. It is heated and stirred magnetically using a hot plate and magnetic stirrer until the solution is completely dissolved. The colour of the solution changed from watery to pale yellow colour. The extract is separated by filtration using Whatman No. 1 filter paper and filter funnel. The forms exist for the solution and the precipitate, which were then split into liquid and powder. The precipitate form was heated in an oven at 73 °C for 24 h. The synthesis of copper involved Copper Sulphate solution (CuSO4), which was diluted manually into five different molarities of 1M, 2M, 3M, 4M and 5M using 249.69 M Copper Sulphate Solution (CuSO4) powder concerning 30 ml of distilled water. In this case, it involved M1V1 = M2V2 formulation. 10 ml of the Curcuma Longa extract was added into 20 ml of Copper Sulphate solution (CuSO4) each to five different molarities. Copper sulphate solution (CuSO4) acts as a copper precursor of each molarity, respectively, heated and stirred magnetically using a hot plate and magnetic stirrer until the solution dissolves completely.

The colour of the extracted change from straw yellow to sea green indicates the reduction of copper. This is monitored by measuring the Ultraviolet-Visible spectrum for absorbance detection where distilled water is used as a blank solution. Characterization of copper of the surface morphology of the prepared samples is evaluated using a Scanning Electron Microscope (SEM) with a dynamic light scattering of sputtering coater technique where the sample was dried overnight in an oven at a temperature of 110 °C. Gold (Au) coated copper grid was inserted into the stub with the coated side upwards for 15 min. Air-dried on a tape sample on the stub to ensure the sample completely sticks and is immovable. The Fourier Transform Infrared studies were carried out using Fourier Transform Infrared Spectroscopy (FTIR). Each sample was placed in an FTIR spectrometer to analyse the molecular composition of copper particles' transmittance, functional group and chemical bonds. Each characterization involved the usage of acetone (C_3H_6O) and Kim Wipes as a cleaning agent for an accurate result.

3 Result and Discussion

The size and form of the synthesised copper particles were analysed using Zetasizer. Figure 1 displays a Zetasizer image of biosynthesised copper particles. The microscale particle sizes of the synthesised copper particles were observed around the peak, respectively. The content of plant extract significantly influences the size of the synthesised copper particles. It has been demonstrated that some particle sizes of Curcuma Longa rhizome extract fluctuated as the concentration did due to the different solubility of the particle mixture of the solution. The peak observed for each result is 31.9 for 1M, 25.3 for 2M, 36.6 for 3M, 36.8 for 4M and 21.7 for 5M, respectively. This was observed as microparticles since the size of the particles ranged between 100 nm and 1000 nm (Table 1).

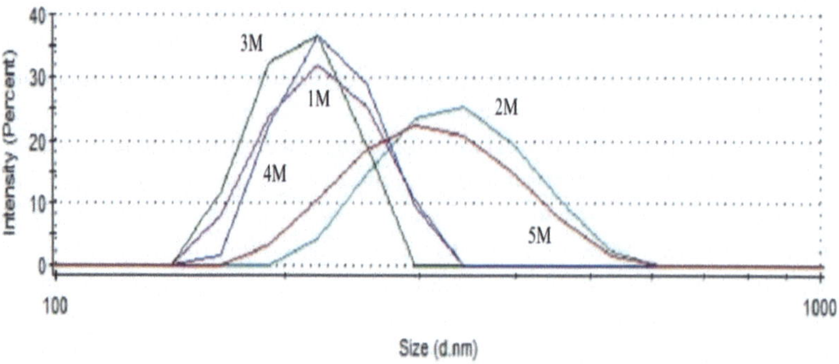

Fig. 1. Particle Size of Copper for 1M, 2M, 3M, 4M and 5M of Curcuma Longa Extract

Table 1. Peak Size of Copper using Zetasizer

Molarity (M)	**1**	**2**	**3**	**4**	**5**
Peak Size	0.0	0.0	0.0	0.0	0.0
	8.0	0.0	11.7	1.7	0.0
	23.7	0.0	32.4	22.6	3.4
	31.9	4.4	36.6	36.8	10.8
	25.6	14.6	19.3	29.2	18.4
	10.8	23.5	0.1	9.7	22.4
	0.0	25.3	0.0	0.0	21.0
	0.0	19.6	0.0	0.0	15.0
	0.0	10.1	0.0	0.0	7.4
	0.0	2.5	0.0	0.0	1.4
	0.0	0.0	0.0	0.0	0.0

A combination of Curcuma Longa and Copper Sulphate (CuSO4) may clump at high concentrations for several reasons, mainly related to their chemical interactions. In chemical reactions, the polyphenol curcumin, which has been shown to have antioxidant and chelating properties, is present in Curcuma Longa. Conversely, Copper Sulphate (CuSO4) is a copper salt. Besides, the effects of pH on different solubility of substances can be affected by the pH of the mixture—natural substances found in Curcuma Longa. Often, it can change pH. Precipitation may occur if the pH is such that less soluble chemicals are more likely to form.

Furthermore, at saturation point, the solution may become saturated with the constituents at high concentrations, decreasing their solubility [9]. Excess components may begin to precipitate or form solid particles if the solubility limit is reached. In addition, ionic strength involves the stability of colloidal systems, which can be affected by the amount of ions in the solution. Particle clumping or aggregation may result from variations in ionic strength brought on by adding Copper Sulphate (CuSO4). Effects of temperature also could be one of the factors where temperature variations can affect the compound's solubility. Solid particles may develop if solubility is reduced as the concentration rises.

3.1 Scanning Electron Microscopy (SEM)

The size and form of the synthetic copper particles were analysed using Scanning Electron Microscopy (SEM). The figure displays a Scanning Electron Microscope (SEM) morphology image of biosynthesised copper particles according to different concentrations of 3M, 4M and 5M. The content of plant extract significantly influences the size of the synthesised copper particles. It has been demonstrated that the particle size of Curcuma Longa rhizome extract increases as the concentration does due to the insolubility of the particle mixture in solution. Figure 2 shows a smooth texture of the morphology of copper particles in 3M compared to the rough texture of the morphology in 5M. The morphology texture was rougher, ranging from 3M, 4M and 5M.

(a) (b) (c)

Fig. 2. Scanning Electron Microscopy (SEM) Images of (a) 3M (b) 4M (c) 5M Copper Particles derived from Curcuma Longa Extracts

In addition, when the concentration of Copper Sulphate (CuSO4) rises, the particle size distribution widens. The particle sizes displayed a heterogeneous distribution at 3M copper sulphate concentration. This implies that the kinetics of copper particle crystallisation may be significantly influenced by the concentration of copper sulphate (CuSO4). Turmeric quickly produced primary nuclei at low Copper Sulphate (CuSO4) concentrations, considerably decreasing the concentration of copper ions and inhibiting the creation of a secondary nucleus [10]. In this instance, the particle development period was constrained, and producing tiny, well-uniform particles was easy. In contrast, a longer nucleation duration was seen at high concentrations of Copper Sulphate (CuSO4) [11]. Thus, repeated heterogeneous phase nucleation eventually led to a heterogeneous particle distribution. Scanning Electron Microscopy (SEM) observations of clumping of a combination of Curcuma Longa and Copper Sulphate (CuSO4) in high concentration can be attributed to several reasons relating to the materials' morphology.

Particle agglomeration occurs when individual Curcuma Longa and Copper Sulphate (CuSO4) particles come into close contact at high concentrations, resulting in clumping or agglomeration. Van der Waals forces and electrostatic interactions are two examples of the attractive forces between particles that might cause this [9].

The observed agglomerate may be attributed to surface morphological abnormalities, roughness or unique characteristics on the surfaces of Curcuma Longa and Copper Sulphate (CuSO4) particles that favour their interlocking or joining. In addition, if Copper Sulphate (CuSO4), Curcuma Longa or both are hygroscopic and tend to absorb moisture from the air, this may cause moisture-induced bridges to develop between the particles, encouraging agglomeration. Furthermore, the development of new substances or modifications to the surfaces of the particles may result from interactions and chemical reactions between the components in the mixture, such as the reactions between the chemicals in Curcuma Longa and Copper Sulphate (CuSO4), which might affect the particle's propensity to clump.

Ultraviolet-Visible Spectroscopy (UV-Vis). Ultraviolet-visible spectroscopy (UV-Vis) analysis is being used to determine the absorbance of copper by looking for copper ion reduction. The creation of copper from Curcuma Longa extract was verified by the shifted absorption peak ranging between 420 nm to 550 nm wavelength at concentrations of 1M, 2M, 3M, 4M and 5M, respectively shown in Figs. 3. Over time, the effects of copper absorbance fade due to the oxidation of the newly formed copper particles. The wavelength range between 420nm to 550nm falls within the visible light spectrum commonly used in Ultraviolet-Visible Spectroscopy because most Curcuma Longa exhibit absorbance in this region. As concentration increases, some absorbance fluctuates due to the different solubility of the particle mixture in solution at the wavelength range between 420 nm to 550 nm. The peak observed for 475 nm of wavelength is around 2.699 for 1M, 2.603 for 2M, 2.511 for 3M, 3.094 for 4M and 2.947 for 5M, respectively.

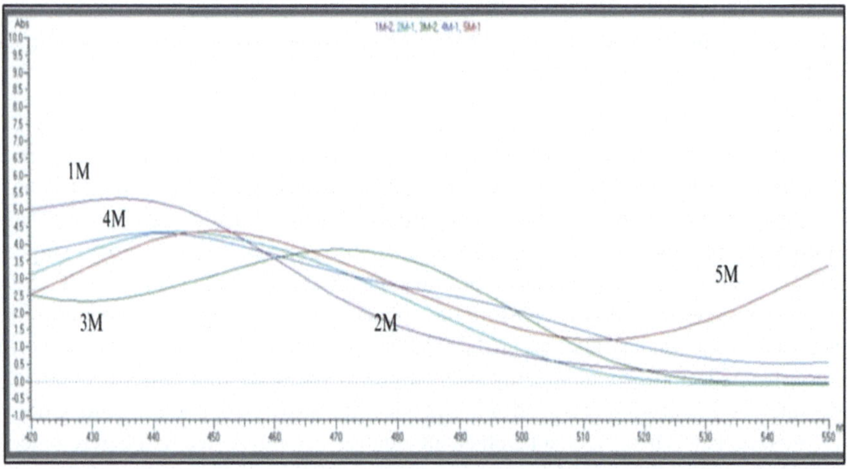

Fig. 3. UV Spectrum of Copper of 1M, 2M, 3M, 4M and 5M of Curcuma Longa Extracts

The high concentration of Curcuma Longa and Copper Sulphate (CuSO4) that clumps or precipitates may be connected to the absorbance characteristics of the UV-

Vis (ultraviolet-visible) spectrum constituents. The high peak was observed at 4M of 3.094, while the low peak was at 3M of 2.511. Absorbance in UV-Vis spectroscopy is a measurement of the amount of light that a sample containing a mixture of turmeric and Copper Sulphate (CuSO4) absorbs at the wavelengths range between 420 nm to 550 nm within the visible light spectrum. Due to interactions between the molecules, the concentration of the components in the combination may result in higher and lower absorbance fluctuation.

Based on UV-Vis. Absorbance, aggregation or precipitation particles may precipitate or form aggregates when Curcuma Longa and Copper Sulphate (CuSO4) concentrations are high [9]. Higher absorbance values might arise from light scattering or blocking as a result. Besides, due to colour interactions, Curcuma Longa and Copper Sulphate (CuSO4) have different hues, and when they combine in large amounts, they may produce intricate colour variations. These modifications could impact the absorbance spectrum and lead to precipitation or clumping. Regarding chemical reactions, new compounds or complexes may be created when Curcuma Longa and Copper Sulphate (CuSO4) are combined in high quantities [12]. These compositional alterations may have an impact on the UV-Vis—absorbance spectrum.

Fourier Transform Infrared Spectroscopy (FTIR). Fourier Transform Infrared Spectroscopy (FTIR) analysis was used to analyse the molecular composition of transmittance, functional group and chemical bond of copper particles. Figure 4 shows Fourier Transform Infrared Spectroscopy (FTIR) observation of Curcuma Longa of the copper particle formation. Fourier Transform Infrared Spectroscopy (FTIR) may detect peaks at 552 nm wavelength around 164.513 cm^{-1} for 1M, 56.012 cm^{-1} for 2M, 1.000 cm^{-1} for 3M, 2.308 cm^{-1} for 4M and 1.000 cm^{-1} for 5M. The absorption of each peak may be ascribed to the Ammonium Hydroxide (NH3OH) and Isopropyl Alcohol (C3H8O) bond, whereas the peak can be related to OH stretching in alcohols and phenolic chemicals. Fourier Transform Infrared Spectroscopy (FTIR) was used to identify various functional groups such as alcohol (OH stretch, H-bonded), alkenes ($C = C$ stretching) and primary amines (NH bending) [11]. In copper particles, the positions and dimensions of the absorption bands have changed a little. The investigation of Fourier Transform Infrared Spectroscopy (FTIR) spectra leads us to conclude that each copper was bonded with Ammonium Hydroxide (NH3OH). This demonstrates that a layer that acts as a capping agent and inhibits particle aggregation develops on top of copper particles. Each molarity of the sample in FTIR spectra shows that the positions and dimensions of the absorption bands have changed a little. Absorption peaks that provide information about the molecular composition and structure of the sample show a broad peak around 3200–3600 cm^{-1} and 1600–1800 cm^{-1}. A broad peak around 3200–3600 cm^{-1} indicated stretching vibrations of hydroxyl (O-H) groups covering alcohols and phenols. Peaks around 1700–1800 cm^{-1} indicated stretching vibrations of the carbonyl ($C = O$) group covering ketones, aldehydes, and carboxylic acids; meanwhile, peaks around 3300–3500 cm^{-1} indicated stretching vibrations of the amino (N-H) group cover amines and amides.

Previous studies have demonstrated that some organic polar functional ligand groups allow them to adsorb on particle surfaces [13–15]. Curcuma Longa has a chemical component called curcumin, which is recognised for being polyphenolic. Because of

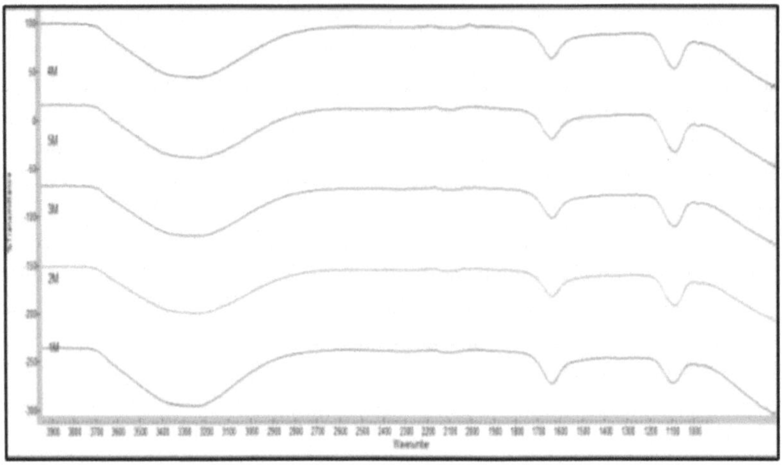

Fig. 4. FTIR Measurement of Copper Particles Formation of Curcuma Longa Extract

its functional groups, curcumin may create a variety of interactions, including hydrogen bonds. Conversely, hydrated crystals of Copper Sulphate (CuSO4) are usually seen, with water molecules establishing coordination connections with copper ions. There may be more interactions between these substances at high doses. One possible explanation for the clumping is the creation of new chemical connections or improved interactions between already-existing bonds.

Curcumin from Curcuma Longa contains functional groups such as keto-enol tautomeric forms and phenolic OH groups. These groups can engage in additional interactions, such as hydrogen bonding. Sulphate ions are present in Copper Sulphate (CuSO4), and interactions between the water molecules in the crystal lattice may also be necessary. The clumping may be caused by the distinct functional groups in Copper Sulphate (CuSO4) and curcumin that may create complexes or attractive forces.

Functional groups in curcumin of Curcuma Longa would show distinctive peaks at particular wavenumbers in terms of transmittance in FTIR. These peaks may move or alter in strength in response to concentration or interaction changes. Concurrently, Copper Sulphate (CuSO4) exhibits comparable characteristics with distinct peaks. Because clumping alters the molecular environment and vibrational modes of the functional group, it may impact the transmittance in FTIR. In conclusion, changing chemical interactions between curcumin in Curcuma Longa and Copper Sulphate (CuSO4), including hydrogen bonds or other attractive factors, may cause the mixture to clump. The molecular structures of the substances in the mix may vary due to concentration variations, which might also impact the FTIR spectrum.

4 Conclusion

This research has been successfully observed by preparation of Curcuma Longa via plant extraction, analyzing data of the characterization of Curcuma Longa added with copper particles by using Zetasizer, UltraViolet-Visible Spectroscopy (UV-Vis), Fourier

Transform Infrared Spectroscopy (FTIR) and Scanning Electron Microscopy (SEM) and compare the physical and chemical properties of Curcuma Longa added with copper particles. Zetasizer, UltraViolet Visible Spectroscopy (UV-Vis), Fourier Transform Infrared Spectroscopy (FTIR) and Scanning Electron Microscopy (SEM) were among the instruments utilized in this experiment to characterize copper particles. A Zetasizer was used to determine the size and peak of the particles. Due to the varying solubility of the particle mixture of the solution, it has been shown that some particle sizes of Curcuma Longa rhizome extract drop and rise as the concentration does. By routinely taking samples from the reaction mixture and utilizing UltraViolet Visible Spectroscopy (UV-Vis), where distilled water is used as a control blank solution, it was possible to detect the absorbance of the copper particles. Because of the variable solubility of the particle mixture in solution at the range of wavelengths between 420 nm to 550 nm, some absorbance fluctuates as the concentration rises. The surface morphology of the generated samples was assessed using a Scanning Electron Microscope (SEM) and a dynamic light scattering of sputtering coater technique that uses gold (Au). It has been shown that the insolubility of the particle combination in solution causes some of the particle size of Curcuma Longa rhizome extract to rise as concentration does. The surface morphology of copper particles in 3M has a smoother texture than those in 5M, which have a rougher texture. One could observe that the morphological texture became rougher as one moved from 3M, 4M and 5M. An FTIR spectrometer was used in the Fourier Transform Infrared studies to find the molecular composition of transmittance, functional group and chemical bonds in each molarity. We conclude from the Fourier Transform Infrared Spectroscopy (FTIR) spectra analysis that each peak indicated the stretching of different functional groups. A broad peak around 3200–3600 cm^{-1} indicated stretching vibrations of hydroxyl (O-H) groups covering alcohols and phenols. Peaks around 1700- 1800 cm^{-1} indicated stretching vibrations of the carbonyl ($C = O$) group covering ketones, aldehydes, and carboxylic acids; meanwhile, peaks around 3300–3500 cm^{-1} indicated stretching vibrations of the amino (N-H) group cover amines and amides. This illustrates how a layer forms on top of copper particles that function as a capping agent and prevent particle aggregation.

There are several potential uses for copper particles. These particles may have special catalytic capabilities in the real of catalysis, which is one possible use. Because copper is known to have antibacterial qualities, it may also be used as an antimicrobial agent or in drug delivery systems. In the catalytic process, catalytic activity is well-known for copper. The larger surface area can improve catalytic efficiency when the material is in the form of microparticles. These microparticles may be applied to chemical processes in which copper acts as a catalyst to encourage changes in the chemical composition. Curcumin in turmeric may have a supportive function or alter the surface characteristics. In drug delivery systems, medicinal substances may be released under regulated conditions. The possible interplay between the components of copper and turmeric may impact drug encapsulation, release kinetics or enhance the stability of drug carriers. In antimicrobial substances, copper's antibacterial qualities are well-established as it may strengthen their antibacterial properties. This combination may be investigated as a component in formulations intended to inhibit the development of microorganisms or as a coating for surfaces and medical equipment. These uses demonstrate how adaptable the copper microparticle combination is in various academic and professional settings.

References

1. Li, X., et al.: Curcumin as a promising candidate for treating chronic wounds: a review of its potential mechanisms of action. Biomed. Pharmacother. **103**, 58–65 (2018)
2. Ramanathan, S., et al.: Evaluation of corrosion inhibition effect of curcumin on copper in acidic media. J. Mol. Liq. **242**, 395–403 (2017)
3. Zhang, L., Webster, T.J.: Nanotechnology and nanomaterials: promises for improved tissue regeneration. Nano Today **25**, 66–85 (2019)
4. Singh, P., Kim, Y.J., Zhang, D., Yang, D.C.: Biological synthesis of nanoparticles from plants and microorganisms. Trends Biotechnol. **34**(7), 588–599 (2016)
5. Liu, J., Hurt, R.H.: Ion release kinetics and particle persistence in aqueous nanosilver colloids. Environ. Sci. Technol. **44**(6), 2169–2175 (2010)
6. Dwivedi, A.D., Dubey, S.P., Kumar, V., Singh, S.C.: Characterization of silver nanoparticles synthesized using Urtica dioica Linn. leaves and their synergistic effects with antibiotics. J. Radiat. Res. Appl. Sci. **3**(2), 197–202 (2010)
7. Hewlings, S.J., Kalman, D.S.: Curcumin: a review of its effects on human health. Foods **6**(10), 92 (2017)
8. Bhuyan, B., et al.: Effect of curcumin on the corrosion behavior of copper in sulfuric acid. J. Market. Res. **8**(5), 4716–4727 (2019)
9. Zahirullah, S., Prince, J., Rajendran, S.: Influence of curcumin Extract on the characteristics of copper sulphide thin film prepared by silar method. Int. J. Adv. Eng. Technol. **VII**, 1086–1091 (2016)
10. Lu, Z., Wang, S., Ma, H., Ma, S., Xu, D., Guo, Y.: Size-controlled synthesis of copper nanoparticles in supercritical water. Chem. Eng. Res. Des. **98**, 36–43 (2015)
11. Sathiyabama, M., Indhumathi, M., Amutha, T.: Preparation and characterization of curcumin functionalized copper nanoparticles and their application enhances disease resistance in chickpea against wilt pathogen. Biocatal. Agric. Biotechnol. **29**, 101823 (2020)
12. Jayarambabu, N., Akshaykranth, A., Venkatappa Rao, T., Venkateswara Rao, K., Kumar, R.R.: Green synthesis of Cu nanoparticles using Curcuma longa extract and their application in antimicrobial activity. Mater. Lett. **259**, 126813 (2020)
13. Rangappa, D., Ohara, S., Umetsu, M., Naka, T., Adschiri, T.: Synthesis, characterization and organic modification of copper manganese oxide nanocrystals under supercritical water. J. Supercrit. Fluids **44**(3), 441–445 (2008)
14. Lu, J.F., Minami, K., Takami, S., Shibata, M., Kaneko, Y., Adschiri, T.: Supercritical hydrothermal synthesis and in situ organic modification of indium tin oxide nanoparticles using continuous flow reaction system. Appl. Mater. Interfaces **4**(1), 351–354 (2012)
15. Mousavand, T., et al.: Hydrothermal synthesis and in situ surface modification of boehmite nanoparticles in supercritical water. J. Supercrit. Fluids **40**(3), 397–401 (2007)

Influence of Magnetic Ions Doped on Physical Properties of Perovskite Manganite $Pr_{0.75}Na_{0.2}Ag_{0.05}MnO_3$

T. W. Lin, S. Shamsuddin(✉), and A. Azizi

Ceramic and Amorphous Group, Faculty and Applied Sciences and Technology, Universiti Tun Hussien Onn Malaysia, 86400 Pagoh, Johor, Malaysia
suhadir@uthm.edu.my

Abstract. These studies report the influence of Cr and Co-doped at Mn-site in perovskite manganite $Pr_{0.75}Na_{0.2}Ag_{0.05}MnO_3$, which were prepared using the solid-state method. The parent compound, Cr-doped sample, and Co-doped sample are called Pr-1, Pr2, and Pr-3 respectively. The focus of this research work was to elucidate how the Cr and Co affected the Jahn-Teller effect by examining the crystalline phase, surface morphology, and electrical properties. These behaviors of perovskite manganite were analyzed by the powder X-ray diffraction (XRD), Scanning electron microscope (SEM) equipped with Energy-dispersive X-ray spectroscopy (EDX), and four-point probe. All samples showed an orthorhombic structure with the Pnma space group according to the X-ray diffraction pattern. Scanning Electron Microscope (SEM) images show that the samples were not regular particles. The grain size decreases from 7.239 μm, 5.957 μm, and 5.576 μm with the replacement of a smaller ionic radius of Cr and Co-doped respectively accompanied by the values of resistivity. The existence of the element substitution of samples illustrated the spectrum of EDX. Apart from that, the value of resistivity shows a continuous decrease from 113.1 Ω, 108.3 Ω, and 106.7 Ω respectively, with the magnetic ion concentration suggestively due to the iso-electronic of Cr-doped ions and Co-doped ions, which improved the conductivity of the samples. Hence, Cr and Co-doped play a crucial role in influencing the physical properties that alter the lattice arrangement and reduce the unit cell volume of all samples for the enhancement of magnetic sensor devices.

Keywords: Perovskite Manganite · Magnetic Ions · Jahn-Teller Effect

1 Introduction

Recent high sustainability in green technology has received remarkably fair attention due to the escalated energy demands [1]. Among the green technology, magnetic refrigeration technology (MRT) has been given more attention due to utilization in the refrigeration field via the magnetocaloric effect (MCE) of magnetic materials such as La (Fe, Si)$_{13}$ [2], Gd-Si-Ge [3], MnAs alloys [4], and perovskite manganite materials [5]. The perovskite manganite materials categorized in the perovskite structure, ABO_3, where A is a lanthanide rare earth element and B is a trivalent manganite ion, are promising candidates

P. Chang et al. (Eds.): SCIEMATHIC 2024, SPPHY 427, pp. 133–139, 2025.
https://doi.org/10.1007/978-981-96-7679-8_12

for advanced technologies due to their interesting properties that can be finely tuned by the doping effect significantly alter the electronic, magnetic, and electrical behaviors as well as the colossal magnetoresistance (CMR) effect. These signature studies on perovskite manganite relate the lattice distortion may affect the Jahn-Teller effect (JT), charge ordering (CO), and double exchange mechanism (DE) [6–10]. Further research revealed that a doped perovskite manganite shows a mixed valence Mn^{3+}/Mn^{4+} oxidation state that can variously change the physical properties [7].

Although perovskite manganite has an antiferromagnetic (AFM) ordering, specific features can be studied using a doping procedure, either doped at A-site or B-site and by manipulating the sintering temperatures. It has been observed that many researchers focus on doping at the A-site to study the behaviors of magnetic and electrical transport properties [8]. For example, the Ag-doped of $Pr_{0.75}Na_{0.2}Ag_{0.05}MnO_3$ exhibited a ferromagnetic-metallic (FMM) phase as the DE mechanism, causing the CO state to become weakened [9]. For instance, the $La_{0.7}Y_{0.05}Ba_{0.25}Mn_{1-x}Cr_xO_3$ (x = 0 and 0.15) revealed a magnetization transition from a ferromagnetic state to a paramagnetic state near the Curie temperature (T_C) with the significant reduction suggested due to the formation of antiferromagnetic coupling between Cr^{3+} ions result in the formation of antiferromagnetic coupling between Cr^{3+}-Cr^{3+}, Cr^{3+}-Mn^{3+}, and Mn^{4+}-Mn^{4+} [10]. However, limited studies have been conducted on comparing the doping effect at the B-site for crystalline structure, surface morphology, and elemental composition analysis of the samples of monovalent-doped manganite.

Several reports showed that the direct substitution at the B-site with a small amount of transition-metal element has influenced the physical properties due to a change in structure brought about by a shift in bond angle, bond length, and an increase in induced distortion (tolerance factor). It demonstrated remarkable features, such as the ability to produce transition metal-insulator (MI) in the CO compound, resulting in the significant CMR effect, especially Cr^{3+} ions and Co^{3+} ions. For example, $La_{0.7}Ca_{0.3}Mn_{1-x}Cr_xO_3$ and $Nd_{0.7}Sr_{0.3}Mn_{1-x}Cr_xO_3$ showed the electrical transport properties that the MI transition modified to a higher temperature due to the amplification of the DE mechanism between Mn^{3+}-O-Mn^{3+} [11]. Besides that, the $YMn_{1-x}Co_xO_3$ (x = 0.20 and 0.40) exhibited lower grain and grain boundary energies when Co-doped into the sample. The dielectric constants of Co-doped manganite showed enhancement at 100 °C due to the formation of charge carriers of Mn^{3+}- Mn^{4+} and Co^{2+}-Co^{3+} pairs [12]. Thus, these critical behaviors can be optimized by the doping mechanism with different magnetic ions to enhance the physical properties of the material. However, a comparative study of elemental substitution on monovalent doped manganite at the Mn-site of $Pr_{0.75}Na_{0.2}Ag_{0.05}MnO_3$ has still not been extensively studied. In addition, to the best of our knowledge, the doping effect on the $Pr_{0.75}Na_{0.2}Ag_{0.05}MnO_3$ on reporting the physical properties such as structural, surface morphology, and electrical resistivity properties have not been previously reported.

The effect of Cr and Co-doped at the Mn-site on the physical properties of the $Pr_{0.75}Na_{0.2}Ag_{0.05}MnO_3$ (Pr-1), $Pr_{0.75}Na_{0.2}Ag_{0.05}Mn_{0.98}Cr_{0.02}O_3$ (Pr-2), and $Pr_{0.75}Na_{0.2}Ag_{0.05}Mn_{0.98}Co_{0.02}O_3$ (Pr-3) have been reported in this work. Cr and Co dopants are chosen in this work due to their remarkable features, as the Cr^{3+} and Co^{3+} ions have the same oxidation state as the Mn^{3+} ion but a smaller ionic radius than the Mn^{3+} ion. In addition, the porosity with density values were also compared and discussed.

2 Experiment Investigation

The formation of perovskite manganite of Pr-1, Pr-2, and Pr-3 samples was synthesis from the grinding of the stoichiometric mixture of Pr_6O_{11}, Na_2CO_3, Ag_2O, MnO_2, Co_3O_4, and Cr_2O_3 fine powders (>99.99%) which prepared by the solid-state synthesis method. These powders were pulverized and calcined using the Protherm Furnace PLF130/15 at 1000 °C with a heating rate and cooling rate of 15 °C/min and 1 °C/min, respectively. The samples were then pressed into pellet form with the 5 tons for 15 min by Carver Hydraulic Presser Equipment. The pelleted samples were continued to be heated at 1200 °C for the sintering process. Powder X-ray diffraction (XRD) was carried out to investigate the structural properties of all samples using Bruker D2 phaser with CuKα radiation measured in the range of $20° \leq 2\theta \leq 80°$ at room temperature. Scanning Electron Microscope (SEM) COXEM EM-30AX PLUS model equipped with Energy Dispersive X-ray Spectroscopy (EDX) was used to determine the surface morphology and identification of elemental substitution of Gold-coated samples with the magnification of 2 kX. The four-point probe measurement (SR4-4PP) was also used for the resistivity data analysis by applying 0.1 A current at room temperature. Bulk density was determined by employing Archimedes' principle using propanol as the buoyant medium and measured by a digital electronic balance EJ-OP-13 density determination kit. Figure 1 shows the overall flow process of the perovskite samples mechanism and investigation.

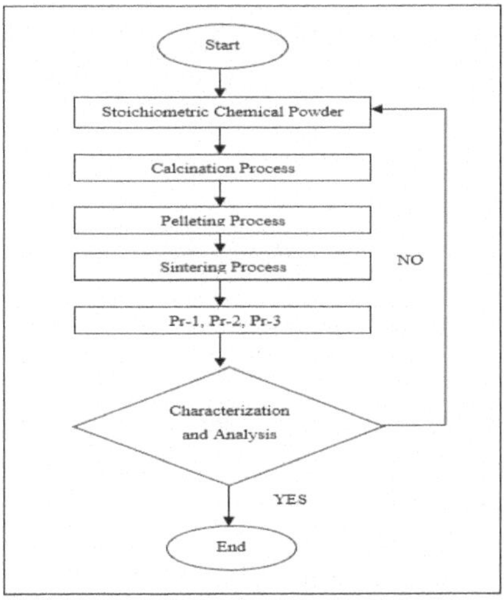

Fig. 1. Overall process flow of perovskite samples

3 Results and Discussion

Figure 2 presents the XRD pattern for the monovalent doped Pr-1, Pr-2, and Pr-3 samples. All samples have been detected as an orthorhombic crystalline structure with the space group of Pnma, which is consistent with the previous study reported by S. Shamsuddin et al., 2022 [13] and A. Hossain et al., 2023 [14]. Table 1 shows the lattice parameters, unit cell volume (V), density (D), porosity, and resistivity for the Pr-1, Pr-2, and Pr-3 manganite. The Pr-1, Pr-2, and Pr-3 samples were shown with the V values of 232.3 $Å^3$, 231.9 $Å^3$, and 2311.6 $Å^3$ with the porosity 13.73%, 16.34%, and 15.45% respectively. The lattice parameters have been shown to be decreased when the replacement of smaller ionic radius of Cr^{3+} (0.615Å) and Co^{3+} (0.610 Å) ions doped into the larger Mn^{3+} (0.645Å) ions. In addition, the V also showed a similar pattern where the V decreased from 232.3Å (Pr-1) to 231.9 Å (Pr-2) and 231.6Å (Pr-3), suggested to be due to the replacement of smaller ions that existing the lattice distortion. Based on the findings, it is shown that the replacement of larger ionic radius Mn^{3+} ions ($e_{2g}{}^1$ $t_g{}^3$) led to the bond length being compressed due to the different energy levels of d-orbitals as the result of increasing density of samples aligned with the previous study [15, 16].

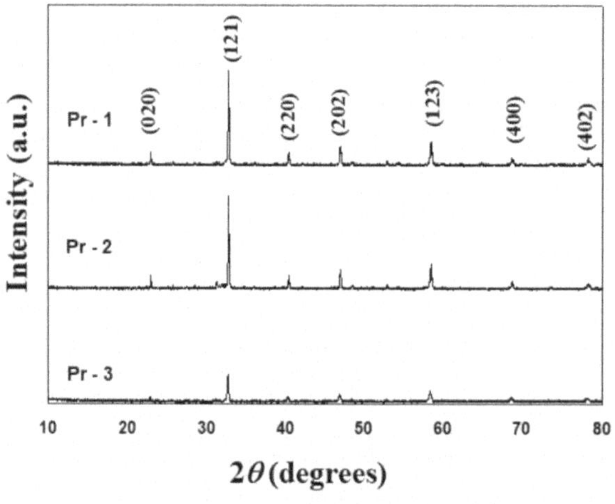

Fig. 2. X-ray diffraction pattern (XRD) of Pr-1, Pr-2, and Pr-3 manganites

Figure 3 shows the SEM images of all samples with the 2kX magnification. The surface morphology of all samples was the same pattern. However, the Pr-2 and Pr-3 images show that the samples become compact and have well-defined grain boundaries. The grain size was decreased from 7.239 μm, 5.957 μm, and 5.576 μm for Pr-1, Pr-2, and Pr-3 samples respectively. The decreasing grain size for all samples suggested that the minor concentration of smaller ions (Cr and Co) doped rearranged the lattice as the larger Mn^{3+} ions replaced but did not change the crystal structure aligned with the literature study [17, 18]. Figure 4 illustrates the EDX analysis, which identifies the element in all the samples. Interestingly, the Cr and Co elements for Pr-2 and Pr-3 samples were

Table 1. The lattice parameter, unit cell volume (V), porosity, and resistivity for the sample Pr-1, Pr-2, and Pr-3

Doped Sample	Lattice Parameter (±0.001)			V (Å)3	D (g/cm^3)	Porosity (%)	Resistivity (Ω)
	a (Å)	b (Å)	c (Å)	(±0.1)	(±0.001)	(±0.01)	(±0.1)
Pr-1	5.474	7.753	5.474	232.3	5.889	13.73	113.1
Pr-2	5.743	7.749	5.467	231.9	5.332	16.34	108.3
Pr-3	5.470	7.747	5.466	231.6	5.600	15.45	106.7

detected respectively as the presence of Cr^{3+} and Co^{3+} ions in the compound. In contrast, Pr-1 detected the elements of Pr, Na, Ag, Mn, Co, Cr, and O without any impurity as proof of proper sample preparation.

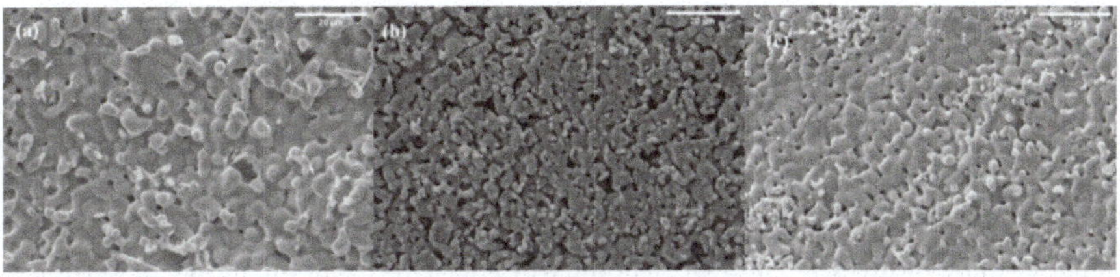

Fig. 3. Scanning Electron Microscope (SEM) images for (a) Pr-1, (b) Pr-2, and (c) Pr-3 manganites.

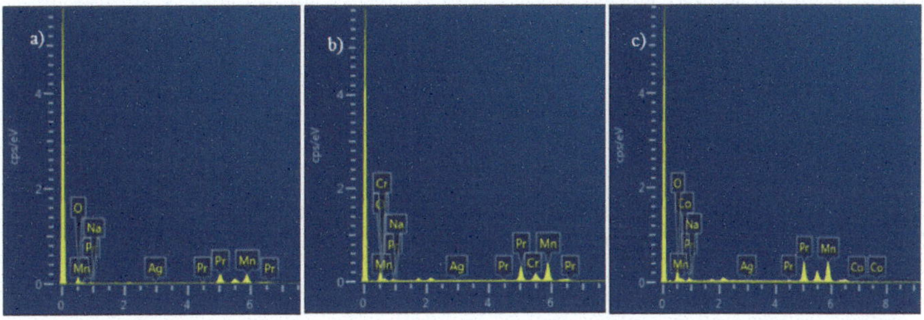

Fig. 4. Energy Dispersive X-ray Spectroscopy (EDX) Spectrum for (a) Pr-1, (b) Pr-2, and (c) Pr-3 manganite

Besides that, the resistivity of the samples was reduced from 113.1 Ω (Pr-1) to 108.3 Ω (Pr-2) and 106.7 Ω (Pr-3) as the iso-electronic of Cr^{3+} ions and Co^{3+} ions doped into the Mn-site. This is because the substitution on the Mn-site will reduce the ratio of Mn^{3+} ions to reduce the Mn^{3+}–O–Mn^{4+} bond lengths. This compression will indicate the interaction of electrons between Mn ions becomes easier and increases electron hopping [19, 20]. Based on the XRD and SEM results, the lattice parameters, V, porosity, resistivity, and average grain size were changed when a dopant doped into the Mn-site. Due to the smaller ionic radius of Cr^{3+} and Co^{+3} doped at the Mn-site, the replacement of the Mn^{3+} ion altered the lattice arrangement and presence of lattice distortion. Based on the XRD and SEM results, it was shown that the Mn^{3+} ions were replaced, and the ratio of Mn^{3+}/Mn^{4+} ions was affected because of lattice distortion. Based on these findings, it was suggested that the Mn-O bond length was decreased due to the minor ion replacement because of the weakening Jahn-Teller effect signal [21].

4 Conclusion

In conclusion, the influence of Cr and Co-doped at the Mn-site on the physical properties of $Pr_{0.75}Na_{0.2}Ag_{0.05}MnO_3$ perovskite manganite has been investigated. The XRD results showed that the samples were crystal in the orthorhombic pattern with the Pnma space group. The minor ion substitution of Cr^{3+} and Co^{3+} ions reduced the unit cell volume, V, as the presence of lattice distortion. The substitution of the higher energy level of Mn^{3+} ions will lead to a decrease in the Mn–O bond length and an increase in porosity. The SEM images, such as the reduction of grain size with the Cr and Co-doped, can also supported since the ratio of Mn^{4+} ions increased and compressed the bond length. The four-point probe analysis demonstrated the reduced signal with the Cr and Co-doped due to the decreased Mn–O bond length, causing the electron interaction between Mn to be faster and indicating the CO state change in the samples. Thus, substitution with the smaller ionic radius Cr^{3+} and Co^{3+} at the Mn-site will reveal the lattice distortion as well as the weakening of the JT effect signal to increase the conductivity and density of future devices.

Acknowledgments. This research is supported by E15203 Fund from Faculty of Applied Sciences and Technology (FAST) Universiti Tun Hussien Onn Malaysia (UTHM).

References

1. Chahar, T., Somvanshi, A., Khan, A.: Structural and surface morphological investigation of Ni-doped NdMnO3 perovskite nanomaterial. Mater. Today Proc. (2023)
2. Jiang, X., et al.: Pr/Co co-doping endows perovskite manganite with enhanced magnetic refrigeration capacity. J. Alloy. Compd. **968**, 172124 (2023)
3. Xie, Z., et al.: Changes in microstructures, magnetocaloric effect and critical behaviors of La0.8Sr0.2MnO3 manganites by K/Co doping. Chin. J. Phys. (2024)
4. Mabhouti, K., Norouzzadeh, P., Taleb-Abbasi, M.: Effects of Fe, Co, or Ni substitution for Mn on La0.7Sr0.3MnO3 perovskite: structural, morphological, and optical analyses. J. Non-crystalline Solids **610**, 122283 (2023)
5. Khairulzaman, N.N.: Influence of Ag-doped on magnetic and electrical transport properties in Pr0.75Na0.25MnO3 manganites. Doctoral dissertation, Universiti Tun Hussien Onn Malaysia (2019)
6. Rajmi, R., Ibrahim, N., Shamsuddin, S., Salleh, S.M., Othman, M.H.: Effect of Pr3+ Substitution at the A-site on the structural and electrical properties of Hole-Doped La-based manganites. Sci. Res. J. **21**(1), 67–82 (2024)

7. Shamsuddin, S., Ibrahim, N., Mayzan, M.Z.H., Esa, F., Razali, S.A.: Analysis of electrical transport properties in Nd0.75Na0.25-xKxMnO3 manganites. In: Proceedings of the 7th International Conference on the Application of Science and Mathematics 2021. Sciemathic 2021, pp. 139–145. Springer, Singapore (2022)
8. Zafar, A., Petkov, V., Abeykoon, A.M.: Local lattice distortions and electronic phases in perovskite manganite Pr0.5Sr0.5MnO3. J. Phys. Mater. **7**(3), 035007 (2024)
9. Shamsuddin, S., Ibrahim, A.B.M., Yahya, A.K.: Effects of Cr substitution and oxygen reduction on elastic anomaly and ultrasonic velocity in charge-ordered Nd0.5Ca0.5Mn1-xCr1-xO3-δ ceramics. Ceram. Int. **39**, S185–S188 (2013)
10. Mnefgui, S., et al.: Effect of Cr substituition on structural, magnetic, magnetocaloric, and critical behaviors of La0.7Y0.5Ba0.25MnO3 perovskite manganite. J. Solid-State Chem. **330**, 124481 (2024)
11. Shamsuddin, S., Mayzan, M.Z.H., Esa, F., Hao, L.K.: Crystalline structure, surface morphology of perovskite manganites Nd0.75Na0.25Mn1-yFeyO3. In: Proceedings of the 7th International Conference on the Application of Science and Mathematics 2021: Sciemathic 2021, pp 147–154. Springer, Singapore (2022)
12. Polat, O., et al.: A comprehensive investigation of the structural, chemical, and dielectric properties of Co-dopes YMnO3 multiferroic component. Appl. Phys. A **130**(3), 166 (2024)
13. Mannan, A., Nabilah, N., Razali, S.A., Shamsuddin, S., Noh, M.Z., Mohommed, Z.: Effect of Co3+ substitution on electro-magnetic properties of Pr0.75Na0.25MnO3 and Nd0.75Na0.25MnO3 manganites. Int. J. Eng. Technol. **7**(4.3), 389–392 (2018)
14. Hossain, A., et al.: Tuning of the optical and magnetic properties of Nd2Ni1-xCoxMnO6-δ (0.2≤ x ≤ 0.5) perovskite by cobalt doping. Ceram. Int. **49**(17), 29229–29236 (2023)
15. Altan, B., Ak, Y.S., Coskun, A.: Exploring the influence of B-site dopants on structural and magnetic properties in La0.8Ag0.2Mn1-xCoxO3 (x: 0 and 0.1) compounds. Usak Univ. J. Eng. Sci. **7**(1), 54–62 (2024)
16. Suresh, S., Vindhya, P.S., Kavitha, V.T.: A comprehensive study of dielectric, magnetic, and anticancerous properties of lanthanum manganite perovskite nanoparticles. J. Alloy. Compd. **976**, 173222 (2024)
17. Zaidi, N., Khadhraoui, S., Cherif, A.: The effect of Cr-doped on structural, magnetic, and magnetocaloric properties of La0.8Sr0.15Na0.05Mn1-xCrxO3 (x=0.00, 0.15, and 0.20) compounds. J. NanoScience Adv. Mater. **3**(1), 32–41 (2024)
18. Zhang, W., Xie, Z., Zou, Z., Jiang, X., Xu, C., Feng, M.: Eu/Ni doping on the structure, magnetocaloric effect, and critical behavior of La0.65Sr0.35MnO3 ceramics. Ceram. Int. **50**(3), 4921–4935 (2024)
19. Nadig, P.R., Murari, S., Daivajna, M.D.: Influence of heat sintering on the physical properties of bulk La0.67Ca0.33MnO3 perovskite manganite: role of oxygen in tuning the magnetocaloric response. Phys. Chem. Chem. Phys. **26**(6), 5237–5252 (2024)
20. Rueda, D.A., et al.: Multifunctional features of perovskite-type ferro-manganite La0.5Ca0.5Mn0.5Fe0.5O3. J. Mol. Struct. **1314**, 138821 (2024)
21. Estemirova, S.K., Mitrofanov, V.Y., Uporoy, S.A., Bykov, V.A., Kozhine, G.K.: Influence of A-site ionic radii on the structural, magnetic, and magnetocaloric properties in Nd0.85Me0.15MnO3 (Me=Ca, Sr, Ba) manganites. J. Magn. Magn. Mater. **605**, 172355 (2024)

Chemistry

Effect of Lime Addition and Storage Period on the Stability of Aloe Vera-Lime Juice

Zalilah Murni Yunus[1(✉)] and Wan Norizzati Wan Seman[2]

[1] Department of Physics and Chemistry, Faculty of Applied Sciences and Technology, Universiti Tun Hussein Onn Malaysia, Pagoh Educational Hub, 86400 Pagoh, Malaysia
zalilah@uthm.edu.my

[2] Department of Technology and Natural Resources, Faculty of Applied Sciences and Technology, Universiti Tun Hussein Onn Malaysia, Pagoh Educational Hub, 86400 Pagoh, Malaysia

Abstract. Aloe vera and lime (Citrus aurantifolia) nutritional contents are known as highly beneficial plants that contain various functional compounds. This study is aimed to formulate aloe vera – lime (AV-L) juice and characterizes the physicochemical of the fruit juice mixture. Five samples were prepared which were reference sample (control) from commercialized aloe vera drink, Formulation 1 (80% aloe vera + 20% lime), Formulation 2 (75% aloe vera + 25% lime), Formulation 3 (70% aloe vera + 30% lime) and Formulation 4 (65% aloe vera + 35% lime). Physicochemical properties, microbiological analysis and sensory of the formulations were determined. Formulation 3 resulted in the highest pH value at the end storage with pH 2.13. Titratable acidity analysis shows that Formulation 4 has the highest acidity with 1.82%, total soluble solids (TSS) of 13.4°Brix. Color analysis shows that Formulation 3 has high value of L* (37.54), a* (−0.73) and b* (2.93) at day 28 which gives the juice darker, greener and yellower color than the initial. Formulation 3 shows the highest antioxidant, total phenolic content and total flavonoid content of 75.32%, 1.153% mgGAE/g, and 1.044% mgQE/g respectively. Formulation 3 has the lowest microbial growth with 5.0 x 105 cfu/ml after 28 days of storage. The sensory evaluation resulted in the acceptance of Formulation 3 is the highest with an overall acceptability score of 8.6. Thus, this study has concluded that the addition of lime in aloe vera can be improvised to produce a longer shelf life fruit juice drink.

Keywords: aloe vera · antioxidant · lime · physicochemical · stability

1 Introduction

The advancement of the juice industry has led to the development of diverse innovations, including a range of beverages and juices. Consumers perceive juice as one of the most accessible methods to integrate fruits and vegetables into their daily diet while enjoying a pleasant flavor. Nonetheless, fruit juice has become increasingly inconvenient over time as consumers demand beverages with extended storage capabilities without undesirable alterations. During the storage of fruit juice, alterations in flavor fractions can lead to

P. Chang et al. (Eds.): SCIEMATHIC 2024, SPPHY 427, pp. 143–157, 2025.
https://doi.org/10.1007/978-981-96-7679-8_13

the loss of freshness and original taste, and the gradual development of unpleasant odors compromises the original taste of the fruit juice [1]. Therefore, researchers have shown interest in developing functional fruit juice as a means to satisfy consumer demand for convenience and to offer improved concepts of ready-to-drink options with extended shelf life.

Functional fruit juice is defined as a product being formulated with ingredients such as raw fruits, herbs, vitamins, minerals, amino acids, and other bioactive compounds that provide specific health benefits. With a growing interest in fruit juice, manufacturers have endeavored to produce mixed fruit juice. According to a previous study, mixing two or more types of fruit can result in a product that offers more vitamins and minerals and exhibits different sensory, color, and flavor characteristics compared to the raw materials [2–4]. This suggests that two or more types of fruits may be formulated to yield more nutritional fruit juice. Aloe vera and Aloe barbadensis Miller are recognized for their high nutraceutical content and benefits in healing burns, oral health, skincare, and digestive health. Citrus aurantifolia, or lime fruit, is a citrus fruit that is a valuable source of soluble and insoluble fiber with numerous benefits, including the mitigation of toxic effects in the body [5]. Additionally, lime fruits are rich in flavonoids, terpenes, phytonutrients, and a variety of phenolic compounds, vitamin C, and carotenoids [6]. Consequently, a combination of aloe vera and lime is proposed to produce functional juices with superior nutritional and nutraceutical value.

Considering the aforementioned, this study focuses on formulating aloe vera-lime (AV-L) juice, characterizing the physicochemical properties of AV-L juice, and thus determining and examining the effect of lime addition and storage period on the physicochemical properties and microbiological stability. The acceptance method was employed to characterize the sensory evaluation of the juice.

2 Materials and Methods

2.1 Materials

Aloe vera (Aloe barbadensis Miller), lime fruits, sugar, citric acid, 1,1-diphenyl-2-picryl-hydrazyl (DPPH) (Sigma Aldrich, Malaysia), distilled water, aluminium chloride ($AlCl_3$) (Bendosen, Malaysia), methanol (CH_3OH) (R & M Chemical, Malaysia), phenolphthalein indicator, quercetin (Sigma Alridch, Malaysia), sodium carbonate ($CaCO_3$) (Bendosen, Malaysia), sodium hydroxide (NaOH) (QRec, Malaysia) and sodium nitrate ($NaNO_2$) (QRec, Malaysia), Folin-Ciocalteu reagent (Sigma-Aldrich, Malaysia) and gallic acid solution (Sigma-Aldrich, Malaysia). Peptone water (HiMedia, India) and plate count agar (PCA) (HiMedia, India). All of the solutions and solvents were of analytical grade.

The instruments used in this study were UV-VIS spectrophotometer (BIOMATE 3S, Malaysia), pH - meter (EUTECH Instruments pH 700, USA), refractometer (ATAGO hand refractometer, ATAGO CO., LTD, Japan), blender (MX-GM1011, Japan), colorimeter (MiniScan EZ 4000L Spectrophotometer, Hunter Associates Laboratory, Inc., United States), biosafety level cabinet (ESCO, Malaysia), autoclave (HIRAYAMA, Japan), colony counter count (Rocker Galaxy 230 Taiwan, Chiller > 10 °C (ELBA, Malaysia), incubator (Memmert Incubator Oven INB200, Germany).

2.2 Sample Preparation

The aloe vera (AV) leaves were meticulously harvested from the plants. The AV gel was extracted, homogenised using a blender, and filtered through a muslin cloth. The lime juice underwent filtration through a muslin cloth to eliminate extraneous materials. The filtered AV-L juice was combined in accordance with the specified mixture ratios and recipes to produce a ready-to-serve beverage by integrating calculated quantities of blended AV-L, sugar syrup, and water [7]. Prior to pasteurisation, sugar syrup was incorporated into the AV-L juice, which was then packaged in bottles and stored at temperatures ranging from 4–5 °C [8]. The juice was subjected to muslin cloth filtration prior to utilisation in analyses conducted at 7-day intervals over a 28-day period. The formulations of AV-L juice are presented in Table 1.

Table 1. Formulation ratio of aloe vera and lime in juice

Samples	Ratio	
	Aloe vera gel	Lime juice
Reference sample	*commercialized fruit juice drink	
Control 1	100%	0
Control 2	0	100%
Formulation 1	80.0%	20.0%
Formulation 2	75.0%	25.0%
Formulation 3	70.0%	30.0%
Formulation 4	65.0%	35.0%

2.3 pH, and Total Soluble Solids (TSS)

The pH of the AV-L juice was determined by introducing 10 ml of the juice into a beaker in accordance with the AOAC method [9]. Prior to this measurement, the pH meter was calibrated using phosphate buffer solutions of pH 4, 7, and 10. The total soluble solids (TSS) were assessed by applying a few drops of AV-L juice onto the scale of a hand refractometer at a temperature range of 0–32 °C. The mean of three readings was utilized to represent the TSS content of the AV-L juice, which was expressed in degrees Brix as a percentage [10].

2.4 Titratable Acidity (TA)

The titratable acidity (TA) of the AV-L juice was assessed following a method previously described [11], with modifications. Specifically, approximately 5 ml of the sample was introduced into a conical flask, diluted with 25 ml of distilled water, and subsequently titrated with 0.1N NaOH until reaching a pH of 8.1. The 0.1 N NaOH solution was prepared by dissolving 0.4 g in 100 ml of distilled water. Titratable acidity was quantified as grams of citric acid per milliliter. The percentage of citric acid in the sample was calculated using the following equation (Eq. 1) [12]:

$$\% \, citric \, acid = \frac{V(ml) \times N(N) \times citric \, acid \, MW(\frac{g}{mol})}{S(ml) \times 10} \tag{1}$$

where V = volume of NaOH used, N = NaOH normality, MW = molecular weight of citric acid (192.12 g/ml) and S = sample volume.

2.5 Color Measurement Analysis

AV-L juice was transferred into a Petri dish until the dish was entirely covered. The coloration was assessed utilizing a Hunterlab colorimeter, with the results expressed in terms of L*, a*, and b* units. Here, L* denotes luminosity or brightness, a* represents greenness () or redness (+), and b* signifies blueness () or yellowness (+) [13].

2.6 Antioxidant Properties

Determination of Antioxidant Properties Using 1, 1-dyphenyl-2-picrylhydrazyl (DPPH) Radical Scavenging Activity. The antioxidant capacity of the juice to scavenge DPPH free radicals was assessed utilizing the methodology outlined in [14]. A DPPH solution was prepared in 100 ml of methanol, and absorbance was measured at 517 nm. The fruit juice sample (0.5 ml) was combined with 3 ml of the prepared DPPH solution and incubated in the dark for 1 h. Absorbance was then measured at 517 nm. The experiment was conducted in triplicate. The percentage inhibition of DPPH by the samples was computed using the specified formula (Eq. 2):

$$\% \, Inhibition = \frac{A - B}{A} \times 100 \tag{2}$$

where, A = the absorbance of pure DPPH and B = the absorbance of the sample.

Total Phenolic Content. The total phenolic content was quantified employing the Folin-Ciocalteu reagent methodology, as delineated in a prior established protocol [15]. A standard solution of gallic acid was formulated by dissolving 1 g of gallic acid in 100 ml of CH3OH. In place of the standard solution, 0.1 ml of a diluted AV-L juice sample (1:5 v/v) was combined with 0.1 ml of Folin-Ciocalteu reagent and allowed to react for a duration of 6 min. Subsequently, 1 ml of 7% NaCO3 solution and 0.5 ml of distilled water were incorporated into the mixture. The absorbance of the resultant solution was measured after 90 min at a wavelength of 760 nm using a UV-visible spectrophotometer. The same methodology was uniformly applied to all AV-L juice formulations in a triplicate manner. Gallic acid served as the calibration standard, with results articulated in terms of gallic acid equivalents (mg GAE/g) as per the calibration curve developed with the identical standard.

Total Flavonoid Content. The total flavonoid content (TFC) was determined according to the method outlined in [16], with slight modifications, and was expressed as milligrams of quercetin per gram of plant tissue. Quercetin served as the standard and was prepared by dissolving 1 g of quercetin in 100 ml of methanol. A standard curve was constructed utilizing the standard solution. A 0.1 ml sample of AV-L juice was combined with 0.5 ml of distilled water, followed by the addition of a 0.1 ml 5% NaNO2 solution, and was allowed to react for 5 min. Subsequently, 0.2 ml of 10% AlCl3 was added, and the reaction proceeded for an additional 5 min before introducing 0.2 ml of 1M NaOH to the mixture. The absorbance of the mixture was immediately ascertained at 510 nm using a UV-vis spectrophotometer. The total flavonoid content was computed as quercetin equivalents (mg QE/g). All procedures were executed in triplicate.

2.7 Microbiological Analysis

Microbial analysis was conducted to ascertain the total plate count (TPC) of the AV-L juice in accordance with the method delineated in [17]. The analysis was executed on days 0, 7, 14, 21, and 28 of storage. Serial dilutions of the sample were prepared in buffered peptone water and diluted to 10–6. TPC was determined utilizing the spreading plating technique as per the standard method on plate count agar (PCA) and incubated at 30 °C for 48 h. All plates were scrutinized for growth on the culture medium.

2.8 Sensory Evaluation

A sensory evaluation of AV-L juice was conducted to evaluate the acceptability of its formulation. Each panelist was provided with three samples. The sensory assessment employed the 9-point hedonic scale, where a score of 1 indicates the least liked attribute and a score of 9 indicates the most liked attribute [18]. The evaluated attributes in the sensory analysis included the appearance, flavor, aroma, texture, and overall acceptability of the AV-L juice.

2.9 Statistical Analysis

All physicochemical experiments were performed in triplicate. All results are expressed as mean ± standard deviation. The mean, standard deviation, and significant difference ($p < 0.05$) were determined using the two–way analysis (ANOVA) by IBM SPSS Statistics version 20.

3 Results and Discussion

3.1 PH Value and Total Soluble Solids (TSS)

The pH of the AV-L juice samples exhibited a declining trend from days 0, 7th, 14th, 21st, and 28th as depicted in Table 2. This pattern is attributed to the augmentation of organic acids within the juice, thereby resulting in a general decrease in pH with the extension of the storage duration across all formulations. As documented in the study, the pH of the juice sample incorporating 40% AV gel commenced at 3.14 on day 0 and progressively diminished to a pH of 3.05 by day 60. Corresponding with the findings of Masood [19], all formulations demonstrated a reduction in pH as the number of storage days increased. This is due to the process of acid hydrolysis, leading to a lower pH value of the formulation samples as examined by [20].

Table 2. pH value for the formulation of the aloe vera-lime juice drink

Samples	pH				
	Day 0	Day 7	Day 14	Day 21	Day 28
Reference sample	3.69 ± 0.01	3.56 ± 0.04	3.40 ± 0.01	3.22 ± 0.04	3.07 ± 0.01
Control 1	5.20 ± 0.03	4.70 ± 0.08	4.44 ± 0.06	4.30 ± 0.02	4.08 ± 0.03
Control 2	2.30 ± 0.01	2.24 ± 0.03	2.18 ± 0.01	1.98 ± 0.02	1.86 ± 0.02
Formulation 1	2.89 ± 0.02	2.71 ± 0.07	2.43 ± 0.04	2.26 ± 0.04	2.08 ± 0.01
Formulation 2	3.00 ± 0.02	2.83 ± 0.01	2.76 ± 0.04	2.44 ± 0.01	2.13 ± 0.02
Formulation 3	2.83 ± 0.01	2.67 ± 0.02	2.61 ± 0.02	2.33 ± 0.01	1.99 ± 0.01
Formulation 4	2.85 ± 0.01	2.68 ± 0.01	2.51 ± 0.01	2.24 ± 0.03	1.93 ± 0.03

*Values are expressed as mean ± standard deviation ($p < 0.05$)

The preliminary observations indicated that the pH of all samples ranged around 3.69. Additionally, it was determined that the proportion of AV-L within the formulations significantly influenced the pH value. The minimum pH was recorded in Formulation 3 (comprising 65% AV gel and 35% lime juice), whereas Formulation 2 exhibited the maximum pH value among the formulations, with a pH of 3.00 on day 0. This variation is attributed to the lime juice content in each formulation, as it affects the pH levels. A reduced quantity of lime juice corresponded to a higher pH value, whereas an increased proportion of lime juice in the formulation resulted in a lower pH value. Table 3 outlines a significant increase ($p < 0.05$) in Total Soluble Solids (TSS) content throughout the storage period. Notably, the formulation containing 70% AV and 30% lime juice achieved the highest total soluble solids, ranging from 12.2 to 13.3 Brix. Over the storage period, Formulation 3 exhibited a continuous increase until day 28, while Formulation 1 demonstrated the lowest TSS, ranging from 11.5 to 12.3 Brix from day 0 to day 28. This finding substantiates that formulations with a higher proportion of lime juice have elevated TSS levels, resulting from the ongoing breakdown of long-chain carbohydrate compounds into soluble sugar compounds [21, 22].

Table 3. Total Soluble Solids Value for Aloe Vera-lime Juice Drink Formulation

Samples	Total Soluble Solids (TSS) ((Brix)				
	Day 0	Day 7	Day 14	Day 21	Day 28
Reference sample	11.0 ± 0.0	11.0 ± 0.0	11.1 ± 0.0	11.2 ± 0.173	11.3 ± 0.26
Control 1	0.70 ± 0.0	0.70 ± 0.01	0.73 ± 0.05	0.80 ± 0.01	0.80 ± 0.05
Control 2	7.20 ± 0.05	7.30 ± 0.01	7.50 ± 0.04	7.70 ± 0.01	7.80 ± 0.03
Formulation 1	11.5 ± 0.1	11.9 ± 0.10	12.0 ± 0.0	12.4 ± 0.208	12.3 ± 0.28
Formulation 2	11.9 ± 0.1	12.4 ± 0.30	12.4 ± 0.23	12.6 ± 0.11	12.9 ± 0.05
Formulation 3	12.2 ± 0.15	12.6 ± 0.26	12.6 ± 0.05	12.9 ± 0.0	13.3 ± 0.05
Formulation 4	11.9 ± 0.1	12.3 ± 0.15	12.3 ± 0.057	12.7 ± 0.15	13.4 ± 0.3

* Values are expressed as mean ± standard deviation ($p < 0.05$)

3.2 Titratable Acidity (TA)

The titratable acidity (TA) of the AV-L juice samples was observed to exhibit an inverse relationship with the pH value of the juice. As illustrated in Table 4, Formulation 4 (comprising 65% AV and 35% lime juice) demonstrated the highest TA at 1.63% on day 0, with a continual increase by day 28. In contrast, Formulation 2 (consisting of 75% AV and 25% lime juice) exhibited the lowest TA among all formulations, starting at 0.80% on day 0 and experiencing an increase by day 28. In a related study on acidity [13], the TA of aloe vera-papaya beverages varied from 0.27% to 0.62% during the storage period from day 0 to day 150 with statistical significance ($p < 0.05$). The fluctuations in acidity may also be explained by the chemical interactions between the organic components of the juice, mediated by enzymatic activity [23].

Table 4. Titratable Acidity Value for Aloe Vera-lime Juice Drink Formulation

Samples	Titratable acidity (% citric acid)				
	Day 0	Day 7	Day 14	Day 21	Day 28
Reference sample	0.26 ± 0.01	0.27 ± 0.01	0.27 ± 0.01	0.28 ± 0.01	0.28 ± 0.01
Control 1	0.15 ± 0.02	0.17 ± 0.02	0.23 ± 0.03	0.25 ± 0.01	0.28 ± 0.02
Control 2	5.65 ± 0.04	5.73 ± 0.03	5.71 ± 0.05	5.76 ± 0.03	5.80 ± 0.03
Formulation 1	0.80 ± 0.063	0.88 ± 0.04	0.90 ± 0.05	0.94 ± 0.02	0.92 ± 0.08
Formulation 2	0.80 ± 0.040	0.82 ± 0.04	0.88 ± 0.00	0.91 ± 0.05	1.00 ± 0.00
Formulation 3	1.32 ± .023	1.34 ± 0.00	1.39 ± 0.05	1.43 ± 0.02	1.48 ± 0.02
Formulation 4	1.63 ± 0.023	1.65 ± 0.00	1.66 ± 0.02	1.74 ± 0.02	1.82 ± 0.05

* Values are expressed as mean ± standard deviation ($p < 0.05$)

The increase in titratable acidity (TA) may be attributed to the hydrolysis of polysaccharides and the conversion of non-reducing sugars into reducing sugars, both contributing to these alterations. Furthermore, titratable acidity can vary among formulations owing to differing rates of acid degradation. Acids, such as ascorbic acid and citric acid, are commonly utilized to assess the acidity of juice; the formulation with the highest lime content has been demonstrated to exhibit the greatest acidity during storage..

3.3 Color Measurement

The elevated L* value indicates the lightness of the sample color; it is near absolute white, which corresponds to a 0 value [24]. Formulations 1 and 2 exhibit lighter colors in comparison to Formulations 3 and 4, potentially due to the introduction of lime juice, which imparts a greenish hue. A negative a* value signifies a more greenish color, whereas a positive a* value denotes a red color. It is noted that the L* color value consistently declines over the storage period from day 0 to day 28, signifying that AV-L juice samples become less light as storage progresses. Formulations 3 and 4, containing greater quantities of lime juice, result in lower L* values, thus appearing darker than other formulations. On day 0, the L* values for Formulations 1, 2, 3, and 4 were 47.68 ± 0.24, 45.51 ± 0.60, 41.84 ± 1.62, and 43.75 ± 0.44, respectively.

In contrast, the commercialized AV drink sample exhibited a lighter color than all the formulation samples..

On the initial day, Formulation 1 recorded the most negative value of a*, demonstrating the highest degree of greenness among all formulations. This greenness is attributed to the formulation's minimal lime content juxtaposed with a maximal concentration of AV. Conversely, a noticeable deterioration in the red hue was observed across all samples during the storage period. Upon conclusion of the storage duration, Formulation 1 sustained a superior a* value relative to the other formulations, continuing to exhibit a greener appearance.

The negative b* value signifies a bluish hue, whereas a positive b* value signifies a yellowish hue. The reference sample, a commercialized aloe vera-apple beverage, exhibited the highest b* value, signifying that it possesses the most pronounced yellowish coloration among the juices examined. Formulation 1 recorded the highest b* value amongst all formulations and exhibited a continuous increase, culminating in a yellowish tint in the juice by the conclusion of the storage period. Consequently, the a* values for all formulations demonstrated an increase from day 0 to day 28, attributable to the lime content present in the juice. The acidic nature facilitates nonenzymatic browning, which accounts for the observed chromatic alterations in the juice throughout storage [25]. The elevated proportion of lime content in AV-L juice accelerates the degradation of sugars, which subsequently participate in browning reactions, either through polymerization or interaction with amino acids, resulting in browning compounds. Furthermore, the degradation of color is exacerbated by the heightened acidity.

3.4 Antioxidant Properties

Free Radical Scavenging (DPPH) Activities. The antioxidant capacity of all formulations demonstrated a declining trend subsequent to storage, as depicted in Table 5. Formulation 2 initially exhibited the highest antioxidant capacity; nevertheless, degradation was observed following 28 days of storage. Conversely, Formulation 3 attained the highest antioxidant capacity of 75.32% after a 28-day storage period. The antioxidant capacity of the reference sample decreased by 25% from its initial level on day 0. Formulation 1 exhibited a reduction of 14% in antioxidant capacity, whereas Formulation 2 underwent a decline of 15.8%. Significantly, Formulation 3 manifested the least antioxidant loss at 11.9%, whereas a 15.11% reduction was observed in Formulation 4 after 28 days of storage. According to Klimczak et al. (2007) [26] and Plaza et al. [27], the reduction in antioxidants in fruit juice during storage might be ascribed to the degradation of ascorbic acid. Furthermore, Castro-Lopez et al. [28] have postulated that alterations in antioxidants and other active compounds and constituents in food may be associated with the oxidation process. Oxidation is characterized by an imbalance between the generation of free radicals and the neutralizing capacity of antioxidants. Collectively, the reduction in antioxidant levels was statistically significant ($p < 0.05$) throughout the storage duration of the juice. Lourenco [29] indicated that the modifications in antioxidants and other active compounds and components in food could be linked to the oxidation process; consequently, Formulation 3 was determined to be the most stable in terms of antioxidant properties in comparison to other formulations.

Table 5. Antioxidant value for aloe vera-lime juice drink formulation

Samples	Antioxidant (% inhibition)				
	Day 0	Day 7	Day 14	Day 21	Day 28
Reference sample	65.94 ± 3.33	63.67 ± 3.41	60.48 ± 6.14	61.51 ± 7.36	48.94 ± 7.73
Control 1	75.17 ± 6.73	73.96 ± 2.93	72.43 ± 1.73	71.89 ± 4.02	71.11 ± 3.42
Control 2	75.78 ± 3.71	72.33 ± 2.05	71.74 ± 1.11	70.57 ± 3.21	68.88 ± 6.11
Formulation 1	85.78 ± 1.63	85.26 ± 1.11	84.02 ± 1.09	77.69 ± 5.72	73.51 ± 7.25
Formulation 2	88.05 ± 0.58	87.63 ± 0.46	86.34 ± 0.23	82.38 ± 2.84	74.08 ± 4.99
Formulation 3	85.57 ± 0.62	85.31 ± 0.70	84.34 ± 0.38	77.22 ± 3.15	75.32 ± 6.92
Formulation 4	82.12 ± 2.29	80.16 ± 3.86	80.57 ± 5.26	78.10 ± 9.40	69.71 ± 5.10

* Values are expressed as mean ± standard deviation ($p < 0.05$)

Total Phenolic Content. A prior investigation [24] documented that the total phenolic content of fruit juice ranged from 0.92% to 0.45% over a storage duration of 60 days. Table 6 details the analyzed total phenolic content across various formulations. On day 0, Formulation 2 exhibited the highest total phenolic content, whereas the reference sample and Formulation 3 demonstrated lower levels. By the conclusion of the storage period, Formulation 3 presented the highest total phenolic content at 1.153%. It was observed that the total phenolic content of the AV-L juice sample diminished as the storage period increased. Formulation 4 displayed the greatest reduction in total phenolic content, with a 12.96% decrease, declining from 1.259% to 1.103% between day 0 and day 28. Conversely, Formulation 3 maintained the most stability in total phenolic content, with a loss of 7.98%. The degradation observed in the total phenolic content in a juice beverage is primarily attributed to the transformation of monomeric anthocyanin into polymeric compounds, which is identified as the principal factor in phenolic content diminution within the drink [30].

Table 6. Total Phenolic Content value for aloe vera-lime juice drink formulation

Samples	Total Phenolic Content (mg GAE/g)				
	Day 0	Day 7	Day 14	Day 21	Day 28
Reference sample	1.226 ± 0.010	1.209 ± 0.013	1.150 ± 0.007	1.110 ± 0.016	1.088 ± 0.004
Control 1	0.700 ± 1.14	0.667 ± 0.050	0.636 ± 0.031	0.627 ± 0.012	0.611 ± 0.010
Control 2	0.960 ± 7.50	0.861 ± 0.021	0.752 ± 0.022	0.739 ± 0.010	0.711 ± 0.010
Formulation 1	1.259 ± 0.001	1.220 ± 0.003	1.170 ± 0.021	1.149 ± 0.011	1.103 ± 0.002
Formulation 2	1.286 ± 0.024	1.265 ± 0.016	1.205 ± 0.009	1.174 ± 0.01	1.131 ± 0.019
Formulation 3	1.253 ± 0.015	1.260 ± 0.012	1.228 ± 0.012	1.198 ± 0.008	1.153 ± 0.001
Formulation 4	1.258 ± 0.006	1.217 ± 0.016	1.150 ± 0.010	1.131 ± 0.005	1.098 ± 0.003

* Values are expressed as mean ± standard deviation ($p > 0.05$)

Total Flavonoid Content. According to Table 7, Formulation 3 demonstrated the highest total flavonoid content (TFC), with a value of 1.0579% on day 0 and 1.0440% at the conclusion of the storage interval. Conversely, Formulation 4 exhibited the lowest TFC at the outset, with a measurement of 1.0530% on day 0. Nevertheless, at the termination of the storage duration, the minimal total phenolic content was observed in Formulation 1 and the reference sample on day 28. Throughout the storage period, Formulation 1 experienced a 1.30% reduction in TFC, starting with an initial percentage of 1.0536%. This indicates that Formulation 4 maintained the greatest stability in terms of TFC decrement, despite not holding the highest TFC among all the formulations analyzed.

Table 7. Valuation of total flavonoid content for formulation of aloe vera-lime juice drink

Samples	Total Flavonoid Content (mg QE/g)				
	Day 0	Day 7	Day 14	Day 21	Day 28
Reference sample	1.050 ± 0.001	1.0502 ± 0.001	1.0434 ± 0.001	1.0418 ± 0.0004	1.0337 ± 0.002
Control 1	0.3500 ± 0.970	0.3371 ± 0.033	0.3115 ± 0.0531	0.2710 ± 0.0003	0.1887 ± 0.010
Control 2	0.6400 ± 2.800	0.5781 ± 0.043	0.5132 ± 0.0021	0.4771 ± 0.003	0.4532 ± 0.210
Formulation 1	1.0536 ± 0.002	1.0522 ± 0.001	1.0459 ± 0.001	1.0432 ± 0.002	1.0399 ± 0.002
Formulation 2	1.0565 ± 0.002	1.0532 ± 0.002	1.0474 ± 0.002	1.0424 ± 0.003	1.0400 ± 0.003
Formulation 3	1.0579 ± 0.007	1.0517 ± 0.010	1.0571 ± 0.004	1.0498 ± 0.009	1.0440 ± 0.007
Formulation 4	1.0530 ± 0.001	1.0497 ± 0.001	1.0443 ± 0.001	1.0432 ± 0.001	1.0401 ± 0.002

* Values are expressed as mean ± standard deviation ($p > 0.05$)

3.5 Microbiological Analysis

Table 8 presents the total plate count (TPC), representing the total colonies of bacteria, yeast, molds, and fungi, which commence proliferation on the 14th day of storage for Formulations 1, 2, 3, and 4. These findings corroborate a preceding study asserting the absence of growth of bacteria, coliforms, and yeast and molds on the initial day of preparation [31].

Table 8. Total plate count of aloe vera lime juice

Samples	Total plate count (cfu / ml)				
	Day 0	Day 7	Day 14	Day 21	Day 28
Reference sample	0	0	0	0	0
Control 1	0	0	$3.1 \times 10^5 \pm 1.1 \times 10^5$	$3.7 \times 10^5 \pm 2.4 \times 10^5$	$7.5 \times 10^5 \pm 1.7 \times 10^5$
Control 2	0	0	$4.0 \times 10^5 \pm 2.5 \times 10^5$	$5.1 \times 105 \pm 1.1 \times 10^5$	$6.3 \times 10^5 \pm 2.0 \times 10^5$
Formulation 1	0	0	$4.5 \times 10^5 \pm 2.1 \times 10^5$	$4.0 \times 10^5 \pm 1.4 \times 105$	$8.0 \times 10^5 \pm 9.8 \times 105$
Formulation 2	0	0	$1.5 \times 10^5 \pm 7.1 \times 10^4$	$3.5 \times 10^5 \pm 7.1 \times 10^4$	$5.0 \times 10^5 \pm 2.8 \times 10^5$
Formulation 3	0	0	$1.0 \times 10^5 \pm 1.4 \times 10^5$	$2.0 \times 10^5 \pm 1.4 \times 10^5$	$5.0 \times 10^5 \pm 2.8 \times 10^5$
Formulation 4	0	0	$2.5 \times 10^5 \pm 2.1 \times 10^5$	$4.5 \times 10^5 \pm 3.5 \times 10^5$	$9.0 \times 10^5 \pm 2.8 \times 10^5$

* Values are expressed as mean ± standard deviation ($p < 0.05$)

The proliferation of these microorganisms was examined in each formulation over a 28-day storage period. Formulation 3 uniquely exhibited the presence of the smallest colony on the plate by day 14. However, it was observed that microbial growth increased in Total Plate Count (TPC) as the storage period progressed. Among all formulation samples, Formulations 2 and 3 demonstrated superior microbiological shelf-life stability compared to other samples. This is attributable to the fact that Formulations 2 and 3 contained lime juice in a higher ratio than Formulation 1. Consequently, the incorporation of lime into the formulations can significantly decrease its pH, inactivate certain indigenous microorganisms, and suppress the growth of those that persist. Notably, Formulation 4, despite containing the highest lime juice content, exhibited greater microbiological growth. This indicates that the efficacy of lime juice as a bio preservative with antimicrobial properties within the juice mixture is limited at certain concentrations. Formulation 3 displayed the most effective microbial stability, attributable to the antimicrobial capabilities of lime juice.

3.6 Sensory Evaluation

The reference sample was excluded from sensory evaluation. Onyekwelu [32] assessed mixed fruit juice using color, flavor, taste, mouthfeel, and general acceptability, linked to total soluble solids (TSS) and acidity in Table 9.

Table 9. Sensory evaluation for three formulations of Aloe vera-lime juice drink

Samples	Sensory evaluation				
	Appearance	Flavor	Smell	Texture	Overall
Formulation 1	5.6 ± 0.54	7.4 ± 0.54	6.0 ± 0.70	5.0 ± 0.70	5.8 ± 0.83
Formulation 2	6.6 ± 0.54	6.6 ± 1.14	6.8 ± 0.44	5.4 ± 1.14	6.8 ± 0.44
Formulation 3	8.0 ± 0.70	8.8 ± 0.44	8.2 ± 0.83	7.0 ± 0.70	8.6 ± 0.54

* Values are expressed as mean ± standard deviation ($p > 0.05$)

As depicted in Fig. 1 and associated with Table 9, Formulation 3 emerged as the most favored among the study participants for overall appeal. This indicates that when evaluating various attributes such as color, flavor, taste, mouthfeel, and general acceptability, Formulation 3 received the highest preference in terms of general acceptability compared to other formulations. This preference is likely influenced by the total soluble solids (TSS) and acidity levels assessed in the study. Therefore Formulation 3 was most preferred in general acceptability.

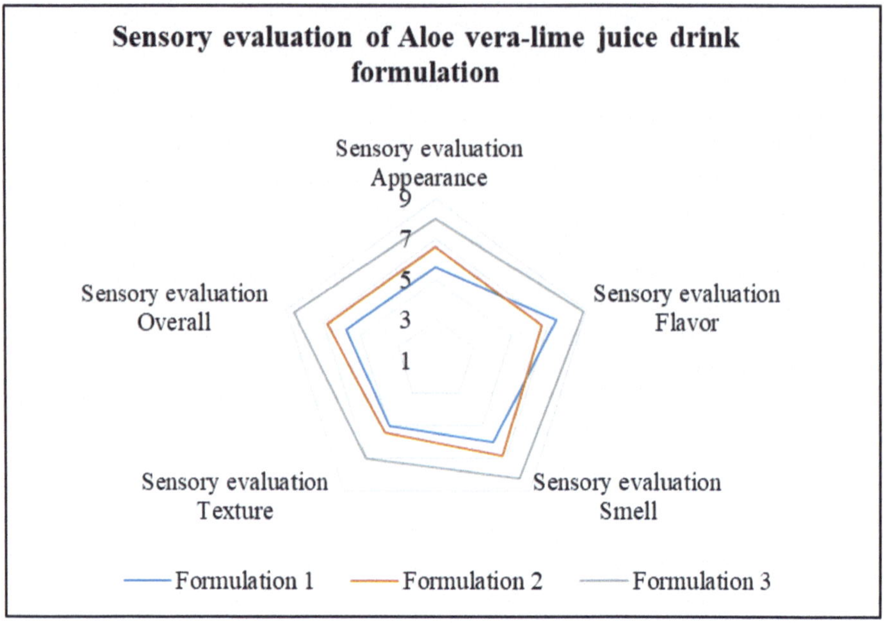

Fig. 1. Sensory evaluation of preference for aloe vera-lime juice

4 Conclusion

The research has demonstrated that Formulation 4 initially had a notably low pH of 2.84 on day 0 and experienced a 32% reduction in pH, leading to a pH of 1.93, the lowest among all formulations by the end of storage. Formulation 4 also recorded the highest titratable acidity (day 0, 1.63%; day 28, 1.82%). To sum up, the pH and titratable acidity in Formulation 4 impacted the total soluble solids of the juice sample, which showed the highest total soluble solids (13.4% on day 28).

The L*, a*, and b* values exhibited variability across all formulations. The resultant color of the formulations was observed to be darker, greener, and yellower than that of the initial juice. Formulation 3 demonstrated the highest levels of DPPH radical scavenging activities, as well as the total flavonoid content. In contrast, the total phenolic content was most abundant in Formulation 4. The elevated levels of antioxidants, flavonoids, and phenolics were associated with a higher concentration of lime juice. The microbiological analysis indicated an increase in microbial growth with prolonged storage time, with Formulation 3 displaying superior stability compared to other formulations. Sensory evaluation revealed a preference for Formulation 3 of the AV-L juice samples. This study suggests that the AV-L juice formulation exhibited enhanced shelf-life stability, functioning as a viable functional beverage, comparable to the control sample of raw aloe vera and lime juice. Nonetheless, future studies are recommended to undertake a more comprehensive investigation into the processing conditions and packaging of the juice mixture.

Acknowledgement. The authors would like to give gratitude to Universiti Tun Hussein Onn Malaysia (UTHM) and, Faculty of Applied Sciences and Technology (FAST) for their support.

References

1. Erkmen, O.: Practice 35 - analysis of fruit juices and concentrates. In: Microbiological Analysis of Foods and Food Processing Environments, pp. 385–391 (2022). https://doi.org/10.1016/B978-0-323-91651-6.00045-8
2. Cristina Pedro, A.L., Maciel, G.M., Lima, N.F., Ribeiro, I.S., Pinheiro, D.F.: Valorization of bioactive compounds from juice industry waste: Applications, challenges, and future prospects. Trends Food Sci. Technol. **152** (2024). https://doi.org/10.1016/j.tifs.2024.104693
3. Zeece, M.: Food additives. In: Introduction to the Chemistry of Food, pp. 251–311 (2020). https://doi.org/10.1016/B978-0-12-809434-1.00007-4
4. Tahosin, A., et al.: Production and evaluation of quality characteristics of ready-to-drink Aloe vera juice incorporation with ginger and lemon. Food Hum. **3** (2024). https://doi.org/10.1016/j.foohum.2024.100324
5. Mandha, J., Shumoy, H., Matemu, A., Raes, K.: Characterization of fruit juices and effect of pasteurization and storage conditions on their microbial, physicochemical, and nutritional quality. Food Biosci. **51**, 102335 (2023). https://doi.org/10.1016/j.fba. ISSN 2212-4292
6. Hong, M., Zhou, L., Zhang, H., Huang, L., He, M.: Unraveling the chemical composition, antibacterial, antioxidant and cytotoxic potential of Citrus reticulata Blanco dropped fruit extracts. Postharvest Biol. Technol. **222** (2025). https://doi.org/10.1016/j.postharvbio.2025.113400
7. Singh, O., Choro, H.K., Kanwar, J.S., Dwivedi, S.K., Singh, R.: Formulation, nutraceutical profile and storage stability of aloe gel and ginger juice functional beverage blend. Pharma Innov. J. **6**(12), 373–379 (2017)
8. Kumar, M., Jha, A., Bharti, K., Mirsha, B.: Extraction, structural properties, and applications of aloe mucilage. Natural Gums **13**(3), 319–337 (2023)
9. A.O.A.C.: Official method of analysis of Association of Official Analytical Chemists International. Maryland, USA (2000)
10. AOAC: Official methods of analysis of the association of official analytical chemists, 20th edn., pp. 1058–1059 (2016)
11. Noor, A., Moyle, P.M., Malik, A., Ziora. Z.M., Pant, K.K.: Transformative upcycling of fruit-vegetable waste for nutraceutical and pharmaceutical breakthroughs and circular economy evolution. Process Saf. Environ. Protect. **187**, 1022–1036 (2024). https://doi.org/10.1016/j.psep.2024.05.002
12. Munankarmi, N.N., et al.: Characterization of the genetic diversity of Citrus species of Nepal using simple sequence repeat (SSR) markers. S. Afr. J. Bot. **156**, 192–201 (2023). https://doi.org/10.1016/j.sajb.2023.03.014
13. Martínez-Burgos, W.J., et al.: Aloe vera: from ancient knowledge to the patent and innovation landscape – a review. S. Afr. J. Bot. **147**, 993–1006 (2022). https://doi.org/10.1016/j.sajb.2022.02.034
14. Alim, M.A., et al.: Study on stability of antioxidant activity of fresh, pasteurized, and commercial fruit juice during refrigerated storage. Food Hum. **1**, 1117–1124 (2023). https://doi.org/10.1016/j.foohum.2023.09.008
15. Chiara Roberta Girelli, C.R., et al.: Metabolomic NMR analysis and organoleptic perceptions of pomegranate wines: influence of cultivar and yeast on the product characteristics. Heliyon **9**(6) e16774 (2023). https://doi.org/10.1016/j.heliyon.2023.e16774

16. Wang, H., et al.: Lactic acid bacteria fermentation improves physicochemical properties, bioactivity, and metabolic profiles of Opuntia ficus-indica fruit juice. Food Chem. **453**, 139646 (2024). https://doi.org/10.1016/j.foodchem.2024.139646
17. Alvarado-Morales, G., Minjares-Fuentes, R., Contreras Esquivel, J.C., Julio Montañez, J., Meza-Velázquez, A., Femenia, A.: Application of thermosonication for Aloe vera (Aloe barbadensis Miller) juice processing: impact on the functional properties and the main bioactive polysaccharides. Ultrason. Sonochem. **56**, 125–133 (2019). https://doi.org/10.1016/j.ultsonch.2019.03.030
18. Arya, S.S., et al.: Effect of hydrodynamic cavitation processing on orange juice physicochemical and nutritional properties. J. Agri. Food Res. **14**, 100781 (2023). https://doi.org/10.1016/j.jafr.2023.100781
19. Shaik, M.I., Hamdi I.H., Sarbon N.M.: A comprehensive review on traditional herbal drinks: Physicochemical, phytochemicals and pharmacology properties. Int. Food Chem. Adv. **3**, 100460 (2023). https://doi.org/10.1016/j.focha.2023.100460
20. Lan, T., et al.: Effects and impacts of technical processing units on the nutrients and functional components of fruit and vegetable juice. Food Res. Intern. **168**, 112784 (2023). https://doi.org/10.1016/j.foodres.2023.112784
21. Adeoye, L., Adegbite, O., Fashogbon, A., Layade, A.: Consumer purchasing behavior for tomatoes. Intern. J. Veg. Sci. **22**, 259–265 (2016)
22. Astuti, S.D., Salengke, S., Laga, A., Bilang, M., Mochtar, M.H., Waris, A.: Influences of AC frequency and electric field strength on changes in bioactive compounds in Ohmic heating of pomelo juice. Inn. Food Sci. Emerging Technol. **72**, 102754 (2021). https://doi.org/10.1016/j.ifset.2021.102754
23. Cassiet, M.D., Kozono, M,L., Andreone, A., Schenk, M., Guerrero, S.: Effect of texturization on microbial inactivation in UV-C treated orange juice: predictive modeling and analysis of Alicyclobacillus acidoterrestris physiological and morphological changes. Intern. J. Food Microb. **428**, 110979 (2025)
24. Afonso, T., et al.: UV-Vis and CIELAB based chemometric characterization of Manihot esculenta carotenoid contents. J. Integr. Bioinform. **14**(4), 56–64 (2017)
25. Gupta, A.K., et al.: Current and emerging applications in detection and removal of bitter compounds in citrus fruit juice: a critical review. Food Bioci. **55**, 102995 (2023). https://doi.org/10.1016/j.fbio.2023.102995
26. Habibi, F., et al.: Postharvest storage temperature strategies affect anthocyanin levels, total phenolic content, antioxidant activity, chemical attributes of juice, and physical qualities of blood orange fruit. Food Chem. Adv. **4**, 100722 (2024). https://doi.org/10.1016/j.focha.2024.100722
27. Plaza, L., Sánchez-Moreno, C., Elez-Martinez, P., De Ancos, B.: Comparing the antioxidant properties and volatile compounds of carrot-orange juice blend processed through varied chemical, pasteurization and ultrasound conditions. Ultras. Sonochem. **98**, 106534 (2023). https://doi.org/10.1016/j.ultsonch.2023.106534
28. Van de Velde, F., et al.: Intestinal and colonic bioaccessibility of phenolic compounds from fruit smoothies as affected by the thermal processing and the storage conditions. Food Res. Intern. **155**, 111086 (2022). https://doi.org/10.1016/j.foodres.2022.111086
29. Fernandes, F.A., et al.: Valorization of natural resources - development of a functional plant-based beverage. Food Chem. **472**, 142813 (2025). https://doi.org/10.1016/j.foodchem.2025.142813
30. Cakmak, H., Ilyasoglu-Buyukkestelli, H., Sogut, E., Ozyurt, H.V., Gumus-Bonacina, C.E., Simsek, S.: A review on recent advances of plant mucilages and their applications in food industry: extraction, functional properties and health benefits. Food Hydrocol. Health **3**, 100131 (2023). https://doi.org/10.1016/j.fhfh.2023.100131

31. Oladunjoye, A.O., Awani-Aguma, E.U.: Effect of thermosonication on physicochemical, bioactive, sensory and microbiological qualities of African mango fruit (Irvingia gabonensis) juice. Meas.: Food. **11**, 100103 (2023). https://doi.org/10.1016/j.meafoo.2023.100103
32. Suo, H., Xiao, S., Wang, B., Cai, Y., Wang, J.: Insights into physicochemical and sensory properties and metabolomic changes of antioxidant-rich jackfruit juice fermented by lactic acid bacteria. Food Bioscie. **58**, 103815 (2024). https://doi.org/10.1016/j.fbio.2024.103815

Physicochemical Properties of Rhamnolipid Synthesis by *Burkholderia thailandensis* E264 Grown on Miscible and Immiscible Carbon Source

Nadirah Arifin[1], Zainatul 'Asyiqin Samsu[1(✉)], and Sharifah Soplah Syed Abdullah[2]

[1] Section of Bioengineering Technology, Universiti Kuala Lumpur Malaysian Institute of Chemical and Bioengineering Technology, 78000 Alor Gajah, Melaka, Malaysia
zainatul@unikl.edu.my

[2] Section of Food Engineering Technology, Universiti Kuala Lumpur Malaysian Institute of Chemical and Bioengineering Technology, 78000 Alor Gajah, Melaka, Malaysia

Abstract. The widespread use of *Pseudomonas aeruginosa* for RL production is hampered by its pathogenicity. *Burkholderia thailandensis* E264 presents a safer and potentially more cost-effective alternative, yet research on its RL production, particularly regarding the influence of carbon source type and chain length, remains limited. This study addresses this gap by investigating the effects of different carbon sources on the physicochemical properties of RLs produced by *B. thailandensis* E264. *B. thailandensis* E264 was cultured in Mineral Salt Medium (MSM) supplemented with glycerol (miscible) and oleic acid (immiscible) at concentrations of 2%, 4%, and 6% (w/v). These concentrations were chosen based on considerations of solubility, growth inhibition, and cost-effectiveness. Bacterial growth was monitored using optical density (OD600) and cell dry weight measurements. RL production was assessed using Fourier-Transform Infrared Spectroscopy (FTIR) analysis, oil displacement tests, and emulsification index (EI_{24}) measurements. Maximum cell dry weight was achieved at 4% (w/v) for both glycerol (0.2290 g/L) and oleic acid (0.1590 g/L). FTIR analysis confirmed RL production, evidenced by characteristic C-O stretching bands that show the existence of bonds between carbon atoms and hydroxyl groups in the chemical structure of the rhamnose ring (e.g. 1045.10 cm^{-1} for glycerol, 1044.9 cm^{-1} for oleic acid). Oil displacement was highest at 4% (w/v) glycerol (1.70 ± 0.99 cm) and 6% (w/v) oleic acid (3.60 ± 1.22 cm). The emulsification index (EI_{24}) peaked at 50% with 6% (w/v) glycerol and 30% with 2% (w/v) oleic acid. Statistical analysis (ANOVA) confirmed significant differences ($p < 0.05$) between carbon sources except there is no significant difference between various concentrations of glycerol (p-value = 0.55) and oleic acid for the oil displacement test (p-value = 0.27). However, these results suggest that the type and concentration of the carbon source significantly influence RL production and its physicochemical properties. Thus, this study only provides preliminary data demonstrating the potential of *B. thailandensis* E264 for RL production. Future research should focus on a wider range of carbon sources and employ HPLC-MS for detailed congener analysis to optimize RL production by *B. thailandensis* E264 for industrial applications.

P. Chang et al. (Eds.): SCIEMATHIC 2024, SPPHY 427, pp. 158–177, 2025.
https://doi.org/10.1007/978-981-96-7679-8_14

Keywords: Rhamnolipid · *Burkholderia thailandensis* E264 · physicochemical properties · carbon source

1 Introduction

Rhamnolipids have gained prominence in the last decade as a very promising category of biosurfactants because of their notable characteristics exhibited in various industrial sectors such as food, detergent, agriculture, cosmetics, and bioremediation. *P. aeruginosa* is the most effective producer strain of RL but, its pathogenicity poses significant safety and health risks for large-scale applications [1, 2]. Consequently, this led to the discovery of *B. thailandensis* E264 as a safer strain for producing RLs [3]. *B. thailandensis* E264 has emerged as a safer and more cost-effective alternative [4–6]. Its non-pathogenic nature (biosafety level 1) makes it suitable for large-scale production, and its ability to utilize substrates like glycerol and canola oil [4] offers potential cost savings using agricultural byproducts.

However, research on RL production using *B. thailandensis* E264 is limited. Efficient biosurfactant production requires optimizing growth factors and using cost-effective, sustainable resources (agricultural waste) is crucial for reducing production costs. The type and concentration of the carbon source significantly impact RL yield and quality. Common carbon sources include hydrocarbons, carbohydrates, and vegetable oils, with some microorganisms utilizing both hydrophilic and hydrophobic sources [7]. While *P. aeruginosa* can utilize a wide range of carbon sources [8], the utilization of various carbon sources by *B. thailandensis* E264 is less well understood.

The purpose of this study was to investigate the effect of different carbon sources (glycerol and oleic acid, representing miscible and immiscible sources) on the physicochemical properties of RLs produced by *B. thailandensis* E264. The goal is to understand the relationship between the type of carbon sources, and the resulting physicochemical properties to optimize production for various applications. Optimizing RL production requires understanding how different nutrients, especially carbon sources, affect both yield and physicochemical properties. The choice of carbon source impacts the final product's characteristics, directly affecting its performance in different applications [9–13]. Cost-effective and sustainable carbon sources, like agricultural waste, are essential for large-scale production.

The growth profiles of RLs produced by strain *B. thailandensis* E264 were characterized when supplied with varied concentrations (w/v) of glycerol or oleic acid. The physicochemical characteristics of RLs produced from two distinct substrates are being evaluated and compared. A study was conducted to examine the relationship between the structure and activity of RLs. The ratio of homologues has a crucial role in shaping the properties of the RL. Research has shown that even slight alterations in the composition of the combination can have a substantial effect on its physicochemical characteristics [14]. The variety of choices in physiochemical characteristics of RL is vital towards various applications such as cosmetics, pharmaceuticals, biotechnology, agriculture, food, and oil industries [15]. Understanding carbon source impact on RL production allows companies to develop unique RL products with tailored properties, giving them a competitive edge in various industries. Furthermore, by utilizing sustainable carbon sources and optimizing production for specific applications, industries can reduce their environmental impact and promote green chemistry practices.

2 Methodology

2.1 Activation and Inoculum Preparation

A modified mineral salts medium (MSM) with glycerol or oleic acid as the carbon source was used as a fermentation medium. The MSM compositions were 2 g/L of sodium nitrate ($NaNO_3$), 0.9g/L of disodium phosphate (Na_2HPO_4), 0.7g/L of monopotassium phosphate (KH_2PO_4), 0.4 g/L of magnesium sulfate ($MgSO_4$), 0.1 g/L of calcium chloride ($CaCl_2$), with 0.1% of trace element solution (v/v) and 2% glycerol or oleic acid as the carbon source. Trace element solution consists of 0.7 g/L of zinc sulfate ($ZnSO_4$), 0.5 g/L of copper sulfate ($CuSO_4$), 0.5g/L of manganese sulfate ($MnSO_4$), 0.26 g/L of boric acid (H_3BO_3), 0.06 g/L of sodium molybdate ($MoNa_2O_4$) and 0.001 g/L of ferrous sulfate ($FeSO_4$) [2]. To prevent precipitation during sterilization, MSM was initially prepared with a carbon source, $NaNO_3$, Na_2HPO_4, and KH_2PO_4, while $CaCl_2$ and $MgSO_4$ were autoclaved separately. Following the preparation of stock solutions for the other components, including the trace elements that had been filtered, sterilized, and volume calculated, then, the solutions were added to the autoclaved media. MSM had its pH adjusted to pH7 before being autoclaved.

2.2 Bacterial Strains

B. thailandensis E264 is the strain used in this study. Cultures prepared from the freezer stocks were placed in nutrient agar plates and grew for 24 h at 30 °C. For the starter cultures, single colonies were used and grown overnight in a nutrient broth at 30 °C with shaking at 200 rpm. For the inoculum concentration, 10% (v/v) of 10 ml of starter culture was added to 90 ml of nutrient broth and incubated for 24 h at 30 °C at 200 rpm [2].

2.3 Shake Flask Fermentation

The time profiling experiment started with the shake flask fermentation. *B. thailandensis* E264 was cultivated in a 250 mL Erlenmeyer flask and two distinct carbon sources were used, each with varied concentrations (2%, 4%, and 6% (w/v). This indicates that the concentration is expressed as the weight of the solute (glycerol/oleic acid) per volume of the solvent (fermentation media) (e.g. 2% (w/v) glycerol means that 2 parts out of 100 parts are glycerol).

The shake flask cultures were grown in MSM supplemented with either glycerol or oleic acid as the sole carbon source. Glycerol and oleic acid were chosen for their contrasting physicochemical properties as glycerol is readily soluble (miscible), while oleic acid is less soluble (immiscible) and because they are commonly used in previous rhamnolipid production studies. This selection allows for comparison with existing data and enables testing the hypothesis that carbon source solubility significantly impacts rhamnolipid production and characteristics. The use of these well-established substrates facilitates future studies focusing on optimizing carbon source concentrations and exploring a wider range of substrates.

The cultivation was conducted at a temperature of 30 °C and a rotary shaking speed of 200 rpm for nine days. Subsequently, the media let it reach a state of ambient temperature before the inoculation process. The inoculation process involved using 10 ml of an overnight starting culture to introduce microorganisms into flasks containing 90 ml of nutrient broth. This results in a 10% concentration of the inoculum. Then, the flasks were incubated for 24 h at a temperature of 30 °C and a speed of 200 rpm [2]. An immediate sample was collected to ascertain the initial microbial concentration. Then, the flasks were placed in an incubator at a temperature of 30 °C for a duration of 216 h, which is equivalent to nine days. The shake flask experiments were conducted in triplicates. To ascertain cell proliferation, the samples were collected at regular intervals of time of 24 h. Various analyses were performed to measure bacterial growth, including by OD600 and determining the CDW. The production of biosurfactants was measured indirectly through oil spreading activity and the emulsification index. Then, the FTIR analysis was conducted to characterise the RL produced.

2.4 Cell Dry Weight

To ascertain the concentration of biomass, 1 ml of fermentation broth was transferred to a pre-weighed Eppendorf tube and subjected to centrifugation for 15 min at a force of 13,500 times the acceleration due to gravity. This sample was taken daily, and before centrifuging, optical density (OD600) values were measured. After centrifuging, the liquid portion was removed, and the solid residue was rinsed with phosphate-buffered saline (PBS), subjected to centrifugation again, and the PBS was discarded. The wet pellets were subjected to a drying process at a temperature of 80 °C. Before drying, the pellet resuspended in 1 ml of distilled water and vigorously vortexed for 30 s to ensure thorough mixing. The solution is transferred into a pre-weighed Eppendorf tube and dried in an oven at 80 °C overnight. The pre-weighed Eppendorf tube is inserted in a desiccator and its weight is measured instantly (referred to as the final weight, B). The dry cell weight was determined by weighing the pellet before and after drying it at 80 °C until a constant weight was achieved, following centrifugation of the pellet [2, 16]. It will take 6–24 h to dry the sample completely, depending on the oven temperature and the thickness of the paste. Calculate the difference in weight and express the dry weight in g/L as in Eq. 1.

$$CDW\,(g/L) = \frac{Final\ Weight\ (g) - Initial\ Weight\ of\ Empty\ Tube\ (g)}{0.001\,L} \tag{1}$$

2.5 Oil Displacement Test

1ml of oil was added to 25 ml distilled water in a petri dish, then, 15 μl of supernatant was added and the diameter of the clear zone in the oil was measured as an indicator for RL production [17]. The type of oil used is lubricant engine oil.

2.6 Emulsification Index (EI_{24})

The emulsification index of RL from each carbon source type was evaluated to measure its interfacial activity on lubricant engine oil. The emulsifying activity of RL was assessed by combining 2 ml of lubricant engine oil with an equal volume of RL in a test tube. The tube was vortex for 2 min and then allowed to settle for 24 h. The EI% was determined in three replicates for each triplicate from each distinct carbon concentration of the carbon source category. By referring the Eq. 2, the emulsification index (EI%) was calculated using the Eq. 2 [18, 19].

$$EI24(\%) = \frac{\text{Height of emulsified layer (mm)}}{\text{Height of liquid mixture (mm)}} \times 100 \tag{2}$$

2.7 FTIR Analysis

The molecular structure of RL was identified using Fourier-transform infrared spectroscopy [20]. The RL was extracted from the supernatant fluid (2 ml) using chloroform (2 ml), then dried with Na2SO4 and evaporated using a rotary evaporator. To avoid band saturation, spectra are obtained using the ATR technique. The IR spectra were acquired using the Bruker IFS113vFTIR-spectrometer, which covered the spectral range of 4,000–400 cm^{-1} with a resolution of 2 cm^{-1}. Each spectrum consisted of 100 scans and was produced using a 0.23 mm KBr liquid cell. The pure RL paste was compared to the daily sampling of bacterial growth scans throughout 9 days.

2.8 Statistical Analysis

All statistical analyses were conducted using Microsoft Excel. The difference in physicochemical characteristic data from various carbon sources was analyzed using one-way ANOVA with Tukey's multiple comparisons ($P \geq 0.05$). Each value represents mean ± standard deviation ($n = 3$). Results are statistically significant at ($P < 0.05$).

3 Results

3.1 Effect of Glycerol and Oleic Acid on the Growth Profile of *B. thailandensis* E264

Based on Fig. 1 a, in the media supplemented with 4% (w/v) glycerol, it had the highest cell growth as was still increasing to the highest OD600 value of 2.597 until 216 h. In contrast, glycerol 6% (w/v) had the lowest cell growth as it reached the highest OD600 value of only 2.022 after 168 h and it remained constant until the end of fermentation. Meanwhile, glycerol of 2% (w/v) achieved the highest OD600 value of 2.175 and the cell growth decreased until 192 h and increased back to 1.979 OD600 value after 216 h of fermentation.

In this study, the cell growth pattern was highly similar across different glycerol concentrations (w/v), particularly during each bacterial growth phase (Fig. 1(a)). *B.*

thailandensis E264 reached the stationary phase at 48 h in media supplemented with glycerol. It maintained this phase until 216 h when fermentation stopped and thus the death phase for glycerol of all concentrations is unknown. The cell growth pattern has a high degree of similarity across different glycerol concentrations (w/v), particularly during each bacterial growth phase.

For microbial growth supplemented with oleic acid, in Fig. 1(b) showed that 2% (w/v) of oleic acid observed that the growth of cells decreased after only 96 h of fermentation when obtained OD600 value of 1.556 and decreased until the end of fermentation. In contrast, when the cultures were supplemented with a higher concentration of 4% and 6% (w/v), the cell growth increased or remained constant after 96 h of fermentation. Thus, Fig. 1b shows poor cell growth as low concentrations of oleic acid of 2% and 4% (w/v) were used. The death phase begins at 96 h for 2% (w/v) oleic acid and at 168 h for 4% (w/v) oleic acid.

In addition, by increasing the concentration of oleic acid as a carbon source, it does promote rapid bacterial culture growth development. As the bacterial culture was supplemented with a concentration of 2% (w/v) of oleic acid, the growth of cells decreased after 96 h of fermentation. In contrast, when the cultures were supplemented with a higher concentration of 4% and 6% (w/v), the growth increased or remained constant after 96 h of fermentation. Thus, the poor cell growth shown as low concentrations of oleic acid of 2% and 4% (w/v) were used suggesting that hydrolysis of the oils was the bottleneck step [21].

Another complementary method used to monitor culture growth during fermentation is the cell dry weight measurement. Triplicate extractions were carried out for each carbon source concentration with 24-h time intervals of fermentation to assess the cell dry weight of the biomass produced.

The culture that contained 2%, 4%, and 6% (w/v) glycerol yielded 0.2290 g/L, 0.1740 g/L, and 0.0990 g/L as the highest values for the cell dry weight respectively during the nine-day cultural period. In the meantime, the culture that contains 2%, 4%, and 6% oleic acid yields 0.1460 g/L, 0.1590 g/L, and 0.0650 g/L, as the greatest values for the cell dry weight respectively. Thus, it could be concluded that both media supplemented with different types of carbon sources showed that as the concentration of carbon increases, it promotes a poor cell dry weight.

As compared with the study by Funston et al., 2016 [1], the *B. thailandensis* E264 culture grown generated a larger dry cell biomass throughout the growth period with 7.71 g/l after 264 h supplemented with 4% glycerol. Meanwhile, as compared to cell dry weight produced by *P. aeruginosa* with the same growth condition, *P. aeruginosa* USM-AR2 growth of 9.4 g/l was accompanied by RL production [22]. In addition, the peak value of the biomass multiplication of the cells was attained at the 144 h for glycerol 2% (w/v) with the value of 605 mg produced by *P. aeruginosa* [7]. Thus, by comparing the bacterial growth of *B. thailandensis* E264 with other studies with the same growth conditions, it could be concluded that *B. thailandensis* E264 showed poor cell growth between the two different types of carbon sources used and this may be due to some biomass loss during the drying process or possibly due to substrate inhibition [23]. Each microorganism has an optimal nutrient concentration range; exceeding this leads to reduced growth due to enzyme saturation and potential enzyme inhibition. Therefore,

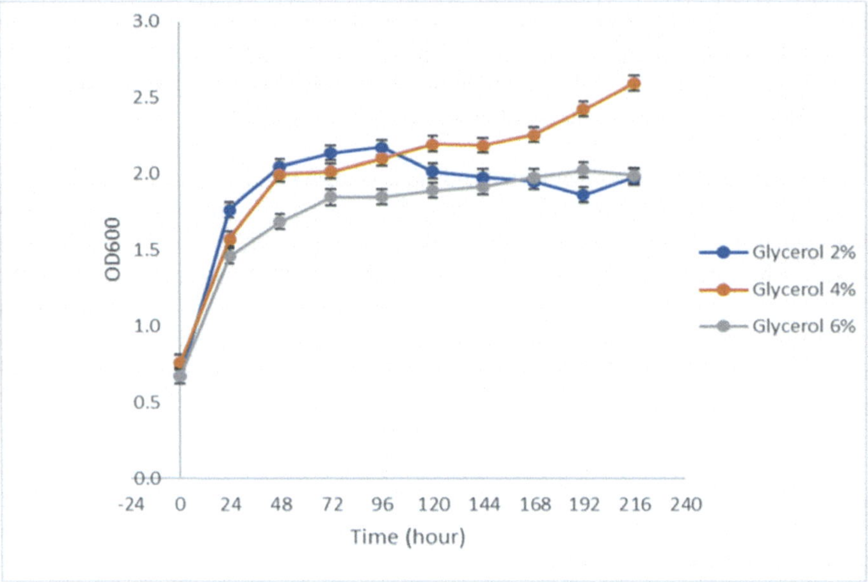

(a)

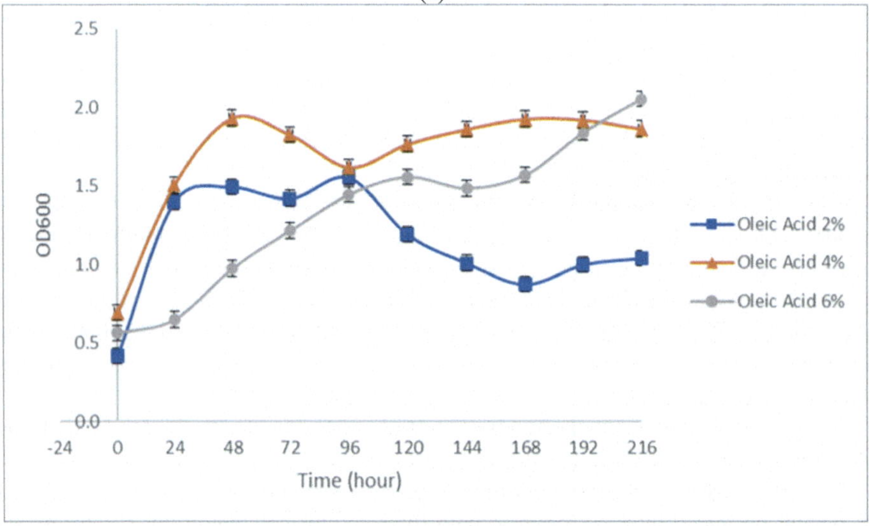

(b)

Fig. 1. The growth curve of *B. thailandensis* E264 (a) with glycerol (2%, 4%, 6% (w/v)) and (b) with oleic acid (2%, 4%, 6% (w/v)).

identifying the optimal nutrient concentration for *B. thailandensis* E264 is crucial for maximizing growth and rhamnolipid production for future studies.

Thus, future research should concentrate on improving cell growth to enhance RL production. Cell proliferation can be influenced by different substrates or nutrients. In addition to carbon, the presence of nitrogen, phosphate, and trace elements such as metals and iron all have an impact on the ultimate output of the product [8]. The production of RLs is mostly influenced by growth circumstances such as by several environmental parameters, including agitation, oxygen levels, pH, and temperature [8]. Given the significant demand and expensive nature of RLs, this study provides a solid basis for researchers to concentrate on improving manufacturing and optimizing conditions to achieve a high yield of RLs.

3.2 Oil Displacement Test

This assay is an easy and rapid method to detect biosurfactant production. If the oil spreads forming a clear zone, then it indicates the presence of biosurfactant in the culture supernatant. The diameter of this clear zone correlates to biosurfactant activity also termed oil displacement activity [8]. The region affected by the biosurfactants' action on the oil exhibited a direct correlation with the quantity of biosurfactants tested [24]. The larger the diameter of the clear zone, the higher the surface activity of the tested biosurfactant solution [8].

In this study, rhamnolipid produced by *B. thailandensis* E264 demonstrated the ability to displace the oil (Fig. 2) at 1.70 ± 0.985 cm from 4% (w/v) of glycerol which was the greatest value observed among the three concentrations. Similarly, *B. thailandensis* E264 displaced the oil at 3.60 ± 1.217 cm from 6% (w/v) of oleic acid, which was the highest value observed among the three concentrations of oleic acid while negative control does not show any displacement (diameter) of the oil (Fig. 2).

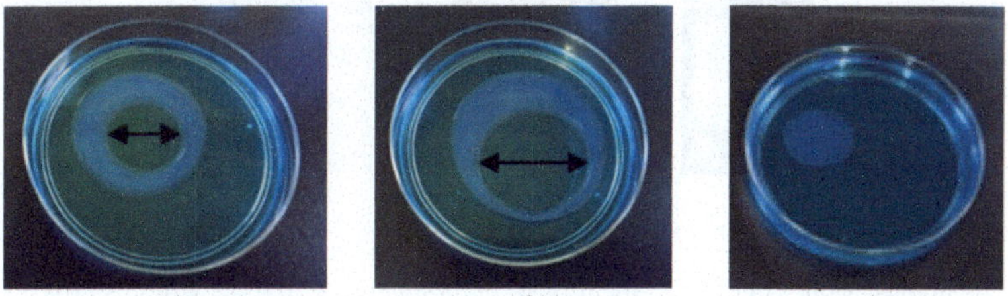

(a) (b) (c)

Fig. 2. The highest value oil displacement of *B. thailandensis* E264 (a) with glycerol 4% (w/v) and (b) with oleic acid 6% (w/v) c) with negative control (distilled water dyed with methylene blue)

As for the oil displacement test for glycerol with different concentrations, the highest value is 1.70 ± 0.99 cm which is for glycerol with 4% (w/v) concentration followed by 0.87 ± 0.49 cm and 0.66 ± 0.29 cm for glycerol 2% (w/v) and 6% (w/v). Meanwhile, for the carbon source oleic acid, the highest value measured is 3.60 ± 1.22 cm with 6% (w/v) concentration followed by 2.53 ± 0.15 cm and 2.20 ± 0.27 cm for 2% and 4% (w/v) concentration respectively. The oil displacement with oleic acid is significantly higher than the sample supplemented with glycerol at 0.066 and 0.045 respectively ($P < 0.05$). As a limited study known by *B. thailandensis* E264, these results could only be compared with *P. aeruginosa* PAO1, *P. aeruginosa* DS10–129 and *P. aeruginosa* KVD-HM52 supplemented with 2% (w/v) molasses with 2.98 ± 0.11 cm, 3.05 ± 0.08

cm and 3.37 ± 0.12 cm of oil displacement respectively [25]. Thus, it could be concluded that the results obtained showed a significant value for the oil spreading capacity.

3.3 Emulsification Index (EI_{24})

Emulsification is an indirect method used to screen biosurfactant production. It was presumed that if the cell-free culture broth contained biosurfactant then it would emulsify the hydrocarbons present [26]. Here, lubricant engine oil was used as a hydrophobic substrate. One of the functions of biosurfactants is to enhance contact between oil and water. The ability of biosurfactants to enhance contact between oil and water can be measured by the emulsification index [27]. If the value of the emulsification index is high, the contact between oil and water is also high.

The highest value of EI_{24} obtained for this study, as in Fig. 3, is 50% for 6% (w/v) of glycerol, in comparison, 30% for 2% (w/v) oleic acid, which is lower than 60% of EI_{24}, indicates that the rhamnolipid produced by *B. thailandensis* E264 exhibits a weak surface activity. A high emulsification index of up to 60–70% demonstrated a strong emulsifying activity and a strong surface activity [28]. A rhamnolipid with a high emulsification index indicates a strong interaction between oil and water [27].

(a) (b) (c)

Fig. 3. The highest value oil displacement of *B. thailandensis* E264 (a) with glycerol 6% (w/v) and (b) with oleic acid 2% (w/v) c) with negative control.

For the media culture supplemented with glycerol, the highest values of the emulsification index for each concentration during 216 h of fermentation are 25.0%, 34.78% and 50.0% for 2%, 4% and 6% (w/v) respectively. Meanwhile, the highest value obtained supplemented with oleic acid is 30.0%, 15.0% and 13.64% for 2, 4 and 6% (w/v) respectively. The emulsification index with glycerol is significantly higher than the sample supplemented with oleic acid at 0.09 and 0.045 respectively ($P < 0.05$). Studies reported that RLs produced from *P. aeruginosa* with glucose-emulsified kerosene, liquid paraffin, and crude oil with emulsion index (EI_{24}) of 59.2%, 57.6% and 65.5%. RLs produced from soybean oil emulsified kerosene, liquid paraffin, and crude oil with EI_{24} of 62.8%, 60.4% and 76.1% [29].

3.4 Statistical Analysis

ANOVA Test of Oil Displacement

ANOVA test is a crucial statistical tool used to determine if there are significant differences in oil displacement and emulsification index among different concentrations of glycerol and oleic acid. The oil displacement test was to identify the presence of biosurfactants [8]. Thus, it also be related to surface tension and as a lower value of surface tension of the liquid, much easier for it to flow and reach every tight gap [30]. The lower value of surface tension was much more desirable for most industries.

In this study, oil displacement was measured at various time points for different glycerol and oleic acid concentrations. The ANOVA test was applied to the oil displacement data for each time point separately. A significant ANOVA result (p-value ≤ 0.05) indicates that at least one of the glycerol or oleic acid concentrations caused a significantly different average oil displacement compared to the others at that specific time point. This helps determine if the concentration of the carbon source has a real effect on the oil displacement capacity of the resulting biosurfactant.

Based on the ANOVA analysis, there are no significant differences in the oil displacement test for crude rhamnolipid grown at different concentrations of oleic acid and glycerol (%). This was concluded by the lower value of the p-value by oleic acid which was 0.2714 while glycerol had a p-value of 0.5512 (Table 1, 2). When the p-value obtained is higher than 0.05, the h-null hypothesis is accepted and justifies that all means are equal and there is no significant difference between the different carbon concentrations. Thus, there is no significant difference between different concentrations of carbon source impact on the oil displacement test. However, 2% (w/v) oleic acid showed consistently high oil displacement values, particularly in the early stages of fermentation.

Table 1. ANOVA test for oil displacement test on different glycerol concentrations.

SUMMARY

Groups	*Count*	*Sum*	*Average*	*Variance*
G_2%	10	3.95	0.395	0.034628
G_4%	10	5.3	0.53	0.179289
G_6%	10	4.7	0.47	0.011489

ANOVA

Source of Variation	*SS*	*df*	*MS*	*F*	*P-value*	*F crit*
Between Groups	0.0915	2	0.04575	0.608902	0.551248	3.354131
Within Groups	2.02865	27	0.075135			
Total	2.12015	29				

Table 2. ANOVA Test for Oil Displacement Test on Different Oleic Acid Concentrations.

SUMMARY						
Groups	*Count*	*Sum*	*Average*	*Variance*		
0_2%	10	12.1	1.21	0.827667		
0_4%	10	8.5	0.85	0.431667		
0_6%	10	16.1	1.61	1.907667		
ANOVA						
Source of Variation	*SS*	*df*	*MS*	*F*	*P-value*	*F crit*
Between Groups	2.890667	2	1.445333	1.369119	0.271428	3.354131
Within Groups	28.503	27	1.055667			
Total	31.39367	29				

ANOVA Test of Emulsification Index (EL_{24})

Similarly, ANOVA test was used to determine if there were statistically significant differences in EI_{24} among the different concentrations of carbon sources and significantly impacts the emulsification properties of the produced rhamnolipid.

Based on the ANOVA test analysis, EI_{24} showed a significant difference between both types of carbon sources. Both obtained p-values lower than 0.05 with 0.0447 for glycerol and 0.0086 for oleic acid (Table 3, 4). The EI_{24} was to determine the ability of the liquid to able to emulsify and remove dirt particles or impurities. Thus, a high value of EI_{24} is the most desirable as it could be beneficial to industries especially those related to detergency, cosmetical skincare or soil bioremediation. As the ANOVA test analysis showed the p-values obtained were lower than 0.05, it could be justified that the concentration of carbon sources does affect the emulsification index EI_{24} of the rhamnolipid as the biosurfactant. By reference to values obtained, oleic acid with 2% (w/v) showed the highest value of EI_{24} compared to all meanwhile, glycerol 4% (w/v) a significant value was the most consistent compared to other glycerol concentrations.

In summary, the ANOVA test for emulsification index with varying oleic acid concentrations yielded a low p-value (<0.05) except in the oil displacement test, which indicates that at least one oleic acid or glycerol concentration significantly differs from the others. By examining the means (average values) of oil displacement and emulsification index for each oleic acid concentration, it is likely found that 2% (w/v) of oleic acid consistently showed either the highest or among the highest values across multiple time points. Similarly, a low p-value from the ANOVA test for glycerol concentrations would indicate significant differences among the groups. Examination of the mean values for oil displacement and emulsification index would then show that 4% (w/v) glycerol consistently performed well, often exhibiting the highest or near-highest values. However, further optimization and consideration of environmental factors are crucial for achieving optimal rhamnolipid yield and desired physicochemical properties.

Table 3. ANOVA Test for Emulsification Index (EI_{24}) on Different Glycerol Concentrations.

SUMMARY

Groups	*Count*	*Sum*	*Average*	*Variance*
G_2%	10	136.5263	13.65263	28.09554
G_4%	10	233.8835	23.38835	130.2023
G_6%	10	306.087	30.6087	463.1779

ANOVA

Source of Variation	*SS*	*df*	*MS*	*F*	*P-value*	*F crit*
Between Groups	1448.086	2	724.0429	3.495114	0.044682	3.354131
Within Groups	5593.282	27	207.1586			
Total	7041.367	29				

Table 4. ANOVA Test for Emulsification Index (EI_{24}) on Different Oleic Acid Concentrations

SUMMARY

Groups	*Count*	*Sum*	*Average*	*Variance*
O_2%	10	152.0813	15.20813	122.9285
O_4%	10	48.63636	4.863636	22.27043
O_6%	10	66.72727	6.672727	15.48246

ANOVA

Source of Variation	*SS*	*df*	*MS*	*F*	*P-value*	*F crit*
Between Groups	610.4487	2	305.2243	5.698688	0.008617	3.354131
Within Groups	1446.132	27	53.56046			
Total	2056.581	29				

3.5 FTIR Analysis

As for the structural characterization, the chemical structure of RL is detected by using Fourier-transform infrared spectroscopy. Irradiation of RL molecules in IR light activates the oscillation of chemical bonds at characteristic frequencies. The transmission is measured over a frequency of 400–4000 cm^{-1} [8, 20, 21].

By using the commercial rhamnolipid (95%) as a standard (Fig. 4), the fraction of the chemical structure of biosurfactant could be compared and identified. For instance, the presence of C-O stretching bands at 1065.39 cm^{-1} shows the existence of bonds between carbon atoms and hydroxyl groups in the chemical structure of the rhamnose ring [31], the detection of a feeble symmetric stretching signal at 1640.66 cm^{-1} suggests the existence of ester carbonyl groups (C = O in COOH) in the biosurfactant [17].

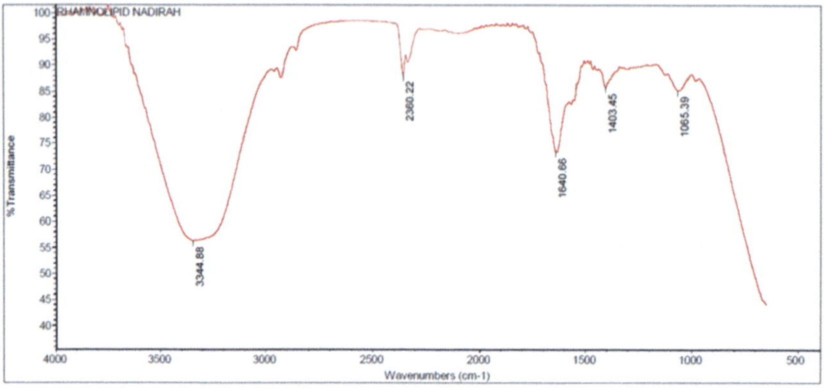

Fig. 4. The IR spectrum of commercial rhamnolipid (95%) as a standard.

At the initial stage of the fermentation process with a glycerol content of 2% (w/v), a wide band was detected at s3285.44 cm^{-1}. This band corresponds to the stretching vibrations of hydroxyl groups in the O-H bonds. The presence of C-O stretching bands in the range of 1300–1000 cm^{-1}, specifically at 1043.55 cm^{-1}, shows the existence of bonds between carbon atoms and hydroxyl groups in the chemical structure of the rhamnose ring the detection of a feeble symmetric stretching signal at 1635.55 cm^{-1} suggests the existence of ester carbonyl groups (C = O in COOH) in the biosurfactant [17].

Biosurfactants can also be demonstrated at the initial fermentation using a glycerol concentration of 4% (w/v). The existence of carbon-oxygen stretching bands at 1043.55 cm^{-1} verifies the formation of bonds between carbon atoms and hydroxyl groups in the chemical structure of the rhamnose ring [31]. Regrettably, only this number indicates the existence of biosurfactant, whereas the remaining days primarily indicate the presence of hydroxyl groups, alkenyl, aliphatic primary amine, and secondary amine.

Furthermore, the presence of RL is seen throughout the fermentation process, except on the 8th day, for the glycerol content of 6% (w/v) (Fig. 5). The presence of bonds formed between carbon atoms and hydroxyl groups in the chemical structure of the rhamnose ring is confirmed by the observation of C-O stretching bands at specific wavenumbers: 1044.66 cm^{-1}, 1044.84 cm^{-1}, 1044.20 cm^{-1}, 1044.49 cm^{-1}, 1044.45 cm^{-1}, 1044.64 cm^{-1}, 1045.10 cm^{-1}, 1045.10 cm^{-1}, and 1045.5 cm^{-1} [31].

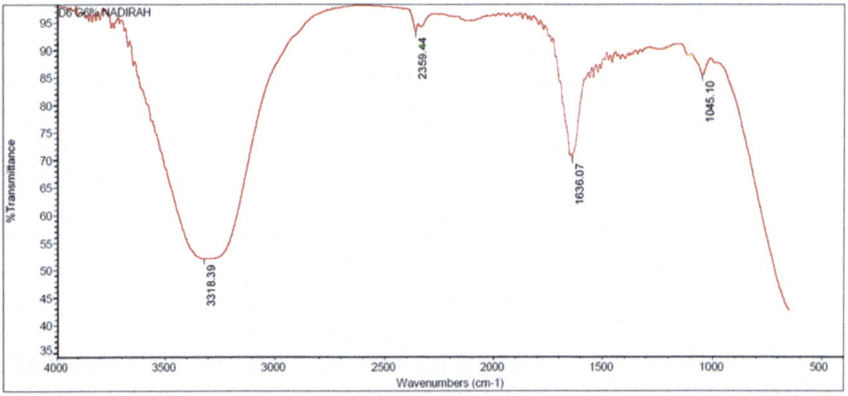

Fig. 5. The IR spectrum of supernatant with glycerol 6% (w/v) on Day 6 of fermentation.

When utilizing a shake flask with oleic acid as the carbon source, the presence of the RL is observed when a concentration of 6% (w/v) is used (Fig. 6). It is visible from the start until the fifth day of fermentation. The presence of links between carbon atoms and hydroxyl groups in the chemical structure of the rhamnose ring was confirmed by the peaks observed at 1044.88 cm^{-1}, 1044.98 cm^{-1}, 1045.02 cm^{-1}, 1045.01 cm^{-1}, 1044.98 cm^{-1}, and 1044.9 cm^{-1} [31].

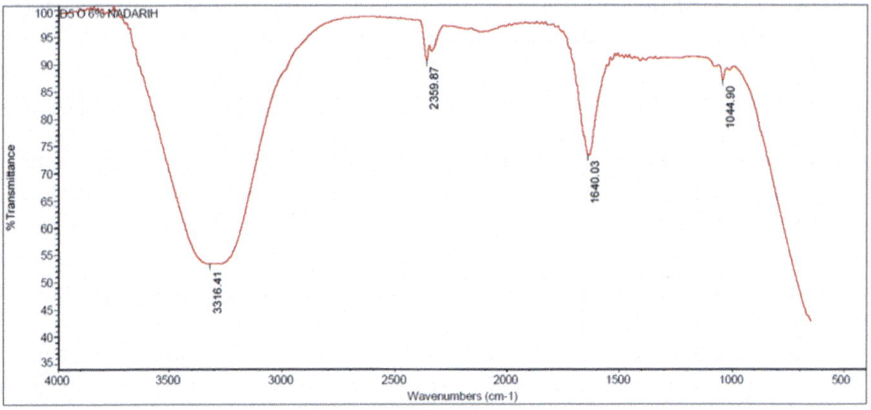

Fig. 6. The IR spectrum of supernatant with oleic acid 6% (w/v) on Day 5 of fermentation.

4 Discussion

The shake flask fermentation process will investigate the impact of various concentrations of glycerol and oleic acids, used as carbon sources, on the production of RLs. Two distinct carbon sources were utilized due to their contrasting possession of hydrophilic and hydrophobic properties. Following a 9-day cultivation period, the bacterial cultures undergo physicochemical and structural characterization analysis. The physicochemical characterization employed the oil spreading method and the emulsification index as an indirect measurement of the generated RL. Regarding the structural characterization, the chemical composition of RL is determined through the application of Fourier-transform infrared spectroscopy. Exposing RL molecules to infrared (IR) light triggers the vibration of chemical bonds at specific frequencies. The transmission is quantified in a frequency range of 400–4000 cm^{-1} [20, 27].

From the growth analysis, it was observed that the growth pattern of *B. thailandensis* E264 supplemented with glycerol is different from oleic. Both cultures had a similar growth pattern, characterized by a rapid increase in population size during the first 24 h (exponential/log phase). This was followed by a period of slower growth (early stationary phase) between 48 and 168 h, where the bacteria continued to grow, but at a slower rate. Starting from 168 h, both cultures reached a late stationary phase characterized by minimal cell growth and steady or slightly increasing OD600 values (Fig. 1). However, it is important to note that the death phase for glycerol is unknown, as the stationary phase remains constant. On the other hand, for oleic acid, the death phase begins at 96 h for a concentration of 2% (w/v) and at 168 h for a concentration of 4% (w/v). This contrasts with the 2% (w/v) concentration, where cell growth is slightly increasing. Therefore, it demonstrated that *B. thailandensis* E264 had limited efficiency in utilizing glycerol and oleic acid at increasing concentrations, leading to a low dry cell weight.

In comparison with other studies, the growth analysis in Fig. 1(a) is similar to Irorere et al., 2018 [2] with 2% (w/v) glycerol and Funston et al., 2016 [1] with 4% (w/v) glycerol that showed *B. thailandensis* E264 culture under 30 °C followed the same growth pattern with the exponential/log phase occurring within the first 24h, and a nearly stationary phase was then observed between 24 and 120 h where the growth of the bacterium continued at a slower rate and OD600 values either remained stable or increased very slowly later on.

Based on growth analysis in Fig. 1(b), the 2% (w/v) oleic acid had similar growth in Irorere et al., 2018 [2], the OD600 values either remained stable or increased very slowly after 24 h until the end of fermentation. This contrasts with the 6% (w/v) concentration, where cell growth is slightly increasing until the end of fermentation. Thus, by increasing the concentration of oleic acid as a carbon source, it does promote rapid bacterial culture growth development.

Another complementary method used to monitor culture growth during fermentation is the cell dry weight measurement. Triplicate extractions were carried out for each carbon source concentration with 24-h time intervals of fermentation to assess the cell dry weight of the biomass produced.

The culture that contained 2%, 4%, and 6% (w/v) glycerol yielded 0.2290 g/L, 0.1740 g/L, and 0.0990 g/L as the highest values for the cell dry weight respectively during the nine-day cultural period. In the meantime, the culture that contains 2%, 4%, and 6% oleic acid yields 0.1460 g/L, 0.1590 g/L, and 0.0650 g/L, as the greatest values for the cell dry weight respectively.

Increasing carbon source concentration, regardless of the specific carbon source, resulted in decreased *B. thailandensis* E264 biomass. This suggests suboptimal growth, possibly due to substrate inhibition. High carbon source concentrations can overwhelm metabolic pathways, interfering with substrate uptake and processing, thus reducing biomass. Thus, this study showed that *B. thailandensis* E264 had responses to different carbon sources that can vary significantly, which might impact their growth.

For the indirect measurement of the rhamnolipid activities, an oil displacement test and emulsification index were used. The presence of a distinct area of clarity on the surface of oil is a significant characteristic of biosurfactant which is the RL. This feature serves as an indicator for assessing the potential ability of bacteria to produce RL [32]. In this study, *B. thailandensis* E264 demonstrated the ability to displace the oil at 1.70 $\pm$ 0.985 cm, which was the greatest value observed among the three concentrations of glycerol. Similarly, *B. thailandensis* E264 displaced the oil at 3.60 $\pm$ 1.217 cm, which was the highest value observed among the three concentrations of oleic acid. The oil displacement tests conducted in this study demonstrate the RL's strong surface activity and confirm the existence of RL, resulting in positive outcomes for the created RL. These results demonstrated that higher production of RL synthesis was observed when the carbon sources consisted of 4% (w/v) glycerol and 6% (w/v) oleic acid, as compared to other carbon concentrations.

In comparison with other research studies related to surface tension assay tests, *P. aeruginosa* had been reported to displace a clear zone diameter up to 8 cm and 11.7 $\pm$ 0.67 cm in oil-spreading tests (*P. aeruginosa* L10; *Pseudomonas* sp. CQ2) [33, 34], and *Burkholderia* species as in this study that had been displaced with a clear zone diameter of 2–4 cm in similar tests. Although certain *Burkholderia* species can produce rhamnolipids, the composition and the effectiveness of these biosurfactants can vary. The *Burkholderia* species might produce rhamnolipids with slightly different molecular structures or produce them in lower quantities compared to *P. aeruginosa.* Consequently, the oil spreading diameter in an assay involving *Burkholderia* species might be smaller, indicating a lower biosurfactant activity or concentration. [35, 36]. Furthermore, the oil displacement test showed the surface tension properties of the RL produced and this specific physicochemical property is vital in bioremediation industries [37] especially as it influences their effectiveness in applications like oil spill cleanup and enhanced oil recovery. Their ability to enhance bioavailability, stimulate microbial activity, and work synergistically with other bioremediation approaches makes them a cost-effective and environmentally friendly option for addressing environmental pollution.

Next, the highest value of EI24 obtained for this study is 50% for 6% (w/v) of glycerol concentration while 30% for 2% (w/v) oleic acid concentration which is lower than 60% indicates that the RL produced by B. thailandensis E264 exhibit a weak surface activity. A high emulsification index of up to 60–70% demonstrated a strong emulsifying activity and a strong surface activity [38]. A RL with a high emulsification index indicates a strong interaction between oil and water [27].

Emulsifications occur when the reduction of surface tension and the decreased interface between oil and water results in significant improvements in surface tension reduction [38]. RL exhibited an emulsion form known as water in oil (w/o). The emulsification index (EI_{24}) is used to quantify the effectiveness of RL in improving the interaction between oil and water. The RL significantly reduced the interfacial tension by over 50% in the emulsion of water with lubricating engine oil. This effect was observed when the culture media was supplemented with a glycerol concentration of 6% (w/v), which was the highest value among all tested concentrations. These distinctive characteristics of RLs make them well-suited for a wide range of industrial and environmental applications that facilitate the interaction between oil and water. Higher emulsification indices, signifying their superior ability to stabilize emulsions (mixtures of immiscible liquids) [9]. Thus, rhamnolipids can stabilize oil-water mixtures, making them valuable in pharmaceuticals and cosmetics [39]. For instance, cosmetics containing rhamnolipids, developed by various companies, were patented and used as an antiaging and anti-wrinkle product [40].

According to the IR spectra data, it can be proven that the RLs produced by *B. thailandensis* E264 are categorized as glycolipids, which are composed of aliphatic acids and esters. The identified adsorption bands correspond to the observations made by Singh et al., 2013 [31], which suggest the presence of rhamnose rings and hydrocarbon chains showed that the type of biosurfactant produced is the RL. The FTIR spectra exhibited a small shoulder, possibly due to the existence of a biosurfactant abundant in di-RLs generated by *P. aeruginosa* TMN [41]. The obtained findings are consistent with the framework described as biosurfactant which consists of aliphatic acid and the glycolipid moiety [21].

In summary, it is crucial to investigate the physicochemical characteristics of the RL produced by *B. thailandensis* E264 to develop a superior RL that can serve as a viable substitute for the primary producer of RL, namely *P. aeruginosa.* It has been documented that even little variations in the structure of the mixture can significantly impact its physiochemical properties. Various physiochemical properties of RL are essential to determine their future suitability for use in various industries applications.

5 Conclusion

This study demonstrates that Burkholderia thailandensis E264, a non-pathogenic bacterium, can produce RLs with varying physicochemical properties depending on the carbon source used. The results showed that *B. thailandensis* E264 was able to produce RL of higher surface activity on 6% (w/v) oleic acid as able to displace oil at 3.60 $\pm$ 1.217 cm compared to 4% (w/v) of glycerol with 1.70 $\pm$ 0.985 cm. Higher oil displacement (larger clear zone) means the biosurfactant is more effectively reducing the surface tension between the oil and water and it is vital for application such as in bioremediation industries. Meanwhile, the RL produced had the highest value of EI_{24} obtained of 50% for 6% (w/v) of glycerol, in comparison with 2% (w/v) oleic acid of 30%. Higher EI24 of RLs can stabilize oil-water mixtures, making them valuable in pharmaceuticals and cosmetics. Thus, using both miscible (glycerol) and immiscible (oleic acid) carbon sources yielded RLs with distinct characteristics, indicating the possibility of tailoring RL properties by manipulating the carbon source. The optimal RL for industrial applications, as indicated by the oil displacement and emulsification index (EI_{24}), might be achieved using specific concentrations of each carbon source. Thus, the ability of promising RLs produced by *B. thailandensis* E264 highlights its great potential as a safer and alternative rhamnolipid producer strain compared to commonly used, *P. aeruginosa.* To fully realize the industrial potential of *B. thailandensis* E264-derived RLs as a potentially more cost-effective alternative in future, several key steps are recommended such as to scale-up the production size by investigating the scalability of *B. thailandensis* E264 RL production in larger bioreactors, optimizing fermentation parameters (e.g., pH, temperature, aeration) for maximum yield and consistent product quality. Next, conduct a comprehensive cost-benefit analysis comparing *B. thailandensis* E264 and *P. aeruginosa* RL production, considering factors such as substrate cost, fermentation conditions, downstream processing, and safety regulations. Lastly, it is suggested to tailor RL production by manipulating carbon sources and fermentation conditions to optimize specific properties for targeted applications (e.g., enhanced bioremediation, specific cosmetic or food applications). By addressing these next steps, the potential of *B. thailandensis* E264 as a sustainable and economically viable alternative for RL production can be fully realized, contributing to the development of safer and more environmentally friendly industrial processes.

References

1. Funston, S.J., et al.: Characterising rhamnolipid production in Burkholderia thailandensis E264, a non-pathogenic producer. Appl. Microbiol. Biotechnol. **100**(18), 7945–7956 (2016)
2. Irorere, V.U., Smyth, T.J., Cobice, D., McClean, S., Marchant, R., Banat, I.M.: Fatty acid synthesis pathway provides lipid precursors for rhamnolipid biosynthesis in Burkholderia thailandensis E264. Appl. Microbiol. Biotechnol. **102**(14), 6163–6174 (2018)
3. Chebbi, A., et al.: Burkholderia thailandensis E264 as a promising safe rhamnolipids' producer towards a sustainable valorization of grape marcs and olive mill pomace (2021)
4. Brett, P.J., DeShazer, D., Woods, D.E.: Burkholderia thailandensis sp. nov., a Burkholderia pseudomallei-like species. Int. J. Syst. Bacteriol. **48 Pt 1**(1), 317–320 (1998)
5. Dubeau, D., Déziel, E., Woods, D.E., Lépine, F.: Burkholderia thailandensis harbors two identical RHL gene clusters responsible for the biosynthesis of rhamnolipids. BMC Microbiol. **9** (2009)
6. Koh, S.F., Tay, S.T., Sermswan, R., Wongratanacheewin, S., Chua, K.H., Puthucheary, S.D.: Development of a multiplex PCR assay for rapid identification of Burkholderia pseudomallei, Burkholderia thailandensis, Burkholderia mallei and Burkholderia cepacia complex. J. Microbiol. Methods **90**(3), 305–308 (2012)
7. Ehinmitola, E.O., Aransiola, E.F., Adeagbo, O.P.: Comparative study of various carbon sources on rhamnolipid production. S. Afr. J. Chem. Eng. **26**, 42–48 (2018)
8. Kumar, R., Das, A.J.: Rhamnolipid Biosurfactant. In: Rhamnolipid Biosurfactant. Springer, Singapore (2018)
9. Abdel-Mawgoud, A.M., Lépine, F., Déziel, E.: Rhamnolipids: diversity of structures, microbial origins and roles. Appl. Microbiol. Biotechnol. **86**(5), 1323–1336 (2010)
10. Ali, I., Jamal, A., Saqib, Z., Ali, M.I., Masood, A.B.: Biosurfactants: introduction and classification. In: Industrial Applications of Biosurfactants and Microorganisms: Green Technology Avenues from Lab to Commercialization, pp. 1–23 (2024)
11. Randhawa, K.K.S., Rahman, P.K.S.M.: Rhamnolipid biosurfactants-past, present, and future scenario of global market. Front. Microbiol. **5**(SEP) (2014)
12. Siddiqui, A., Ceppi, P.: A non-proliferative role of pyrimidine metabolism in cancer. Mol. Metab. **35**, 100962 (2020)
13. Wittgens, A., Rosenau, F.: Heterologous rhamnolipid biosynthesis: advantages, challenges, and the opportunity to produce tailor-made rhamnolipids. In: Frontiers in Bioengineering and Biotechnology, vol. 8. Frontiers Media S.A (2020)

14. Mendes, A.N., et al.: Physicochemical properties of rhamnolipid biosurfactant from Pseudomonas aeruginosa PA1 to applications in microemulsions. J. Biomater. Nanobiotechnol. **6**(1), 64–79 (2015)
15. Abbot, V., Paliwal, D., Sharma, A., Sharma, P.: A review on the physicochemical and biological applications of biosurfactants in biotechnology and pharmaceuticals. Heliyon **8**(8) (2022)
16. Panikov, N.S.: Microbial growth dynamics. In: Comprehensive Biotechnology, 2nd edn., vol. 1, pp. 257–283. Elsevier Inc. (2011)
17. Pornsunthorntawee, O., Wongpanit, P., Chavadej, S., Abe, M., Rujiravanit, R.: Structural and physicochemical characterization of crude biosurfactant produced by Pseudomonas aeruginosa SP4 isolated from petroleum-contaminated soil. Biores. Technol. **99**(6), 1589–1595 (2008)
18. Cooper, D.G., Goldenberg, B.G.: Surface active agents from two Bacillus species. Appl. Environ. Microbiol. **53**, 224–229 (1987)
19. Wei, Y.H., Chou, C.L., Chang, J.S.: Rhamnolipid production byindigenous Pseudomonas aeroginosa J4 originating frompetrochemical waste water. Biochem. Eng. J. **24**, 146–154 (2005)
20. Heyd, M., et al.: Development and trends of biosurfactant analysis and purification using rhamnolipids as an example. Anal. Bioanal. Chem. **391**(5), 1579–1590 (2008)
21. Moussa, T.A.A., Mohamed, M.S., Samak, N.: Production and characterization of di-rhamnolipid produced by Pseudomonas aeruginosa TMN. Braz. J. Chem. Eng. **31**(4), 867–880 (2014)
22. Md Noh, N.A., Abdullah, A.A.-A., Mohamad Ibrahim, M.N., Mohd Yahya, A.R., Md Noh, N.A., et al.: J. Gener. Appl. Microbiol. 58, 153–161 (2012)
23. Wang, X., Xia, K., Yang, X., Tang, C.: Growth strategy of microbes on mixed carbon sources. Nat. Commun. **10**(1) (2019)
24. Morikawa, M., Hirata, Y., Imanaka, T. A study on the structure function relationship of lipopeptide biosurfactants (2000)
25. Deepika, K.V., Raghuram, M., Bramhachari, P.V.: Rhamnolipid biosurfactant production by Pseudomonas aeruginosa strain KVD-HR42 isolated from oil contaminated mangrove sediments. Afr. J. Microbiol. Res. **11**(6), 218–231 (2017)
26. Nayarisseri, A., Singh, P., Singh, S.K.: Screening, isolation and characterization of biosurfactant producing Bacillus subtilis strain ANSKLAB03. Bioinformation **14**(06), 304–314 (2018)
27. Sumiardi, A., Soetarto, E.S., Susilaningsih, D.: Screening and characterization of biosurfactant produced by bacterial consortium in degrading polycyclic aromatic hydrocarbon compound. In: AIP Conference Proceedings, vol. 2002 (2018)
28. Gudiña, E.J., Rodrigues, A.I., Alves, E., Domingues, M.R., Teixeira, J.A., Rodrigues, L.R.: Bioconversion of agro-industrial by-products in rhamnolipids toward applications in enhanced oil recovery and bioremediation. Biores. Technol. **177**, 87–93 (2015)
29. Zhao, F., Han, S., Zhang, Y.: Comparative studies on the structural composition, surface/interface activity and application potential of rhamnolipids produced by Pseudomonas aeruginosa using hydrophobic or hydrophilic substrates. Bioresour. Technol. **295** (2020)
30. Safari, P., Hosseini, M., Lashkarbolooki, M., Ghorbani, M., Najafpour Darzi, G. Evaluation of surface activity of rhamnolipid biosurfactants produced from rice bran oil through dynamic surface tension. J. Pet. Explor. Prod. (2023)
31. Singh, D.N., Tripathi, A.K.: Coal induced production of a rhamnolipid biosurfactant by Pseudomonas stutzeri, isolated from the formation water of Jharia coalbed. Biores. Technol. **128**, 215–221 (2013)
32. Deng, Z., et al.: One biosurfactant-producing bacteria achromobacter sp. A-8 and Its potential use in microbial enhanced oil recovery and bioremediation. Front. Microbiol. **11** (2020)

33. Sun, W., et al.: Isolation and characterization of biosurfactant-producing and diesel oil degrading Pseudomonas sp. CQ2 from Changqing oil field, China. RSC Adv. **8**(69), 39710–39720 (2018)
34. Wu, T., et al.: Pseudomonas aeruginosa L10: a hydrocarbon-degrading, biosurfactant-producing, and plant-growth-promoting endophytic bacterium isolated from a Reed (Phragmites australis). Front. Microbiol. **9**(MAY), 1087 (2018)
35. Ibrahim, H.M.M.: Characterization of biosurfactants produced by novel strains of Ochrobactrum anthropi HM-1 and Citrobacter freundii HM-2 from used engine oil-contaminated soil. Egypt. J. Pet. **27**(1), 21–29 (2018)
36. Youssef, N.H., Duncan, K.E., Nagle, D.P., Savage, K.N., Knapp, R.M., McInerney, M.J.: Comparison of methods to detect biosurfactant production by diverse microorganisms. J. Microbiol. Methods **56**(3), 339–347 (2004)
37. Shao, B., et al.: Effects of rhamnolipids on microorganism characteristics and applications in composting: a review. Microbiol. Res. **200**, 33–44 (2017)
38. Almansoory, A.F., Rozaimah, S., Abdullah, S., Anuar, N. Screening for potential biosurfactant producing bacteria from hydrocarbon-degrading isolates (2016)
39. Tiso, T., et al: Production, performance, and application. Consequences of microbial interactions with hydrocarbons, oils, and lipid: production of fuels and chemicals, pp. 1–37 (2017)
40. Takahashi, M., Morita, T., Fukuoka, T., Imura, T., Kitamoto, D.: Glycolipid biosurfactants, man nosylerythritol lipids, show antioxidant and protective effects against H2O2-induced oxidative stress in cultured human skin fibroblasts. J. Oleo Sci. **61**(8), 457–464 (2012)
41. Rahman, K.S.M., Rahman, T.J., McClean, S., Marchant, R., Banat, I.M.: Rhamnolipid biosurfactant production by strains of Pseudomonas aeruginosa using low-cost raw materials. Biotechnol. Progr. **18**(6), 1277–1281 (2002)

Biology

Assessing Forest Health and Ecosystem Integrity: A Comparative Review of Methodologies, Indicators, and Emerging Technologies

Umi Syafiqah Ahmad Fahmi and Aqilah Awg Abdul Rahman(✉)

Centre of Research for Sustainable Uses of Natural Resources (COR-SUNR), Faculty of Applied Sciences and Technology, Universiti Tun Hussien Onn Malaysia, Pagoh Higher Education Hub, 84600 Pagoh, Johor Darul Takzim, Malaysia
aqilah@uthm.edu.my

Abstract. Forests provide critical ecological, cultural, and economic services but are increasingly threatened by deforestation, climate change, and anthropogenic pressures. Monitoring forest ecosystem integrity is crucial for addressing these challenges. This systematic review employs the PRISMA framework to evaluate current methodologies for assessing forest health, analyzing 13 studies from an initial dataset of over 75,000 articles. The review examines the strengths and limitations of traditional field-based approaches and advanced technologies such as remote sensing, machine learning, and GIS tools. Traditional methods offer detailed site-specific insights, whereas advanced tools provide broader spatial and temporal coverage but lack standardized frameworks and consistent data integration. Key findings highlight the need for clear definitions of forest health, standardized indicators, and improved integration of monitoring approaches across ecosystems. Identified research gaps include limited applications of tools like LiDAR and hyperspectral imaging and insufficient cross-ecosystem comparisons. This study recommends refining methodologies and fostering scalable, robust monitoring systems to support sustainable forest management worldwide.

Keywords: Remote sensing · GIS tools · Machine learning · Biodiversity indicators · Sustainable forest management

1 Introduction

Forests are among the planet's most dynamic and diverse ecosystems, covering approximately 31% of Earth's land area according to the United Nations (UN). They play a pivotal role in maintaining ecological balance by regulating climate, conserving biodiversity, and delivering essential ecosystem services such as carbon sequestration, water purification, and soil stabilization [22]. Beyond their ecological significance, forests also hold immense cultural and economic value, providing recreational spaces, timber, and medicinal resources that sustain livelihoods and economies [7].

P. Chang et al. (Eds.): SCIEMATHIC 2024, SPPHY 427, pp. 181–193, 2025.
https://doi.org/10.1007/978-981-96-7679-8_15

Despite their critical importance, forest ecosystems face escalating threats from a combination of anthropogenic and natural stressors, including habitat fragmentation, pest outbreaks, deforestation, and the intensifying impacts of climate change. These stressors not only compromise the health of forests but also undermine the vital functions and services they provide [16, 19]. Given these challenges, monitoring forest health has become an essential priority to ensure the sustainability of forest ecosystems and the services they offer.

Over the years, forest health monitoring methods have evolved significantly. Traditional approaches, such as ground-based surveys, remain valuable for collecting detailed, site-specific data. However, these methods are often resource-intensive and limited in spatial and temporal coverage. Advances in technology, particularly in remote sensing, unmanned aerial vehicles (UAVs), and machine learning, have enabled large-scale, high-resolution, and cost-effective monitoring of forest ecosystems [15, 17]. While these technologies offer significant advantages, challenges remain in integrating modern and traditional approaches due to issues such as data accuracy, standardization, and applicability across diverse forest types and geographic regions [2, 4].

This systematic literature review aims to evaluate existing methodologies for forest ecosystem assessment, with a focus on their applications, technologies, and key elements. By analyzing current practices, this review seeks to identify strengths, limitations, and research gaps, contributing to a more comprehensive understanding of forest health monitoring. Specifically, this study addresses the following research questions:

RQ1: What are the commonly used methods for assessing forest ecosystem integrity, and how do they differ in terms of accuracy and applicability?
RQ2: How do traditional field-based methods compare to modern technologies in measuring forest ecosystem integrity?
RQ3: What are the challenges in integrating multi-scale and multi-method approaches to assess forest ecosystem integrity?

By addressing these questions, this review not only synthesizes existing knowledge but also provides a foundation for developing robust, scalable, and standardized methodologies that can guide sustainable forest management and policymaking.

2 Material and Method

This study employs the systematic literature review (SLR) methodology, recognized for its impartial and structured approach to reviewing scientific literature. As noted by Lame [13], systematic reviews treat the literature review process as a scientific endeavor, applying the principles of empirical research to enhance transparency, replicability, and reduce the potential for bias. By adopting this method, researchers can more effectively identify knowledge gaps and stay informed about current research trends.

2.1 Protocol and Registration

The selection of studies was guided by clearly defined inclusion and exclusion criteria, as recommended by Xiao and Watson [23]. These criteria ensured a rigorous and consistent approach to the literature search (Table 1).

Table 1. Eligibility criteria for study selection

Inclusion criteria	Exclusion criteria
Articles published from 2020 to 2024	Papers written in a language other than English
Full – text article	Online forum
Empirical (qualitative, quantitative, mixed methods, design science) and conceptual articles related to the assessment of forest health or ecosystem service	–

The systematic search yielded 75,538 papers across the two search engines. Following the application of the eligibility criteria, 33 papers were selected for detailed review. Of these, 13 were comprehensively analyzed due to their strong alignment with the study's aims and their potential to provide clear insights into methodologies that can inform future research and environmental management decisions.

2.2 Identification

Three fundamental stages of the systematic review process were employed. Using dictionaries, encyclopedias, thesaurus, and previous research, the first step involves identifying keywords and searching for related, related terms. Search strings have been created for Science Direct and Scopus after all pertinent terms have been chosen (Table 2). The first stage of the study has able to successfully obtain 68, 960 papers from Scopus and 3, 123 papers from Science Direct.

2.3 Study Selection and Eligibility Criteria

The PRISMA framework guided the screening process (Fig. 1). Titles and abstracts of the collected papers were reviewed for relevance, followed by full-text screening to ensure alignment with the study's objectives. Papers that were unrelated to the research questions or failed to meet quality standards were excluded. The final selection of 13 articles was based on their ability to provide comprehensive insights into the methodologies and frameworks utilized in forest ecosystem integrity assessments.

2.4 Data Extraction

Data extraction was conducted following the guidelines outlined by Afifi, Stryhn, and Sanchez [1]. To facilitate analysis and provide a summary of the selected literature, key data were filtered based on the categories that were created during the data extraction procedure. This study's categories include research gaps, study findings, study issues, background information, and framework types that have been employed. Abstracts of the selected articles were reviewed and evaluated in alignment with the study's research

Table 2. Search query

Database	Search Query
SCOPUS	TITLE-ABS-KEY (terrestrial OR ecosystem OR index) AND PUBYEAR > 2019 AND PUBYEAR < 2025 AND (LIMIT-TO (DOCTYPE, "ar") OR LIMIT-TO (DOCTYPE, "dp")) AND (LIMIT-TO (SUBJAREA, "ENVI") OR LIMIT-TO (SUBJAREA, "EART")) AND (LIMIT-TO (EXACTKEYWORD, "Forestry") OR LIMIT-TO (EXACTKEYWORD, "Environmental Monitoring") OR LIMIT-TO (EXACTKEYWORD, "Ecosystems") OR LIMIT-TO (EXACTKEYWORD, "Forest") OR LIMIT-TO (EXACTKEYWORD, "Forest Ecosystem") OR LIMIT-TO (EXACTKEYWORD, "Environmental Management") OR LIMIT-TO (EXACTKEYWORD, "Ecosystem Service") OR LIMIT-TO (EXACTKEYWORD, "Environmental Protection") OR LIMIT-TO (EXACTKEYWORD, "Environmental Impact") OR LIMIT-TO (EXACTKEYWORD, "Satellite Imagery") OR LIMIT-TO (EXACTKEYWORD, "Satellite Data") OR LIMIT-TO (EXACTKEYWORD, "Spatial Analysis") OR LIMIT-TO (EXACTKEYWORD, "Anthropogenic Effect") OR LIMIT-TO (EXACTKEYWORD, "Ecosystem Restoration"))
Science Direct	TITLE-ABS-KEY (assessment AND forest health OR ecosystem integrity) AND YEARS > 2020 AND YEARS < 2024 AND LIMIT TO (ARTYPE, "Research articles")

questions. The extracted data were then synthesized to provide a comprehensive comparison of the findings and to highlight key trends, gaps, and opportunities for future research.

3 Result and Discussion

Table 3 is following data were extracted from each study: study information (first author and year of publication) and database.

Table 4 shows the selected articles and the methods or frameworks the studies used, with proper citations provided in the References section. Articles that did not meet the inclusion criteria were excluded. The analysis of the articles cover the novel method used to measure forest condition, importance of site productivity, comparison with conventional methods and future applications. Articles coded as (A) are analyzed to answer the research question coded as (RQ). By focusing on innovative techniques, comparisons with conventional methods, and applications of modern technologies, these articles contribute valuable insights to the research questions posed in this review: RQ1 (common methods for assessing forest integrity), RQ2 (comparison of traditional and modern approaches), and RQ3 (challenges in multi-scale and multi-method integration).

The process has resulted in 13 articles focusing on multiple types of methods to assess forest ecosystem integrity which are discussed in the following paragraphs. The study by Yue et al. [24] (A1) aims to develop a periodic site index based on the growth-effective

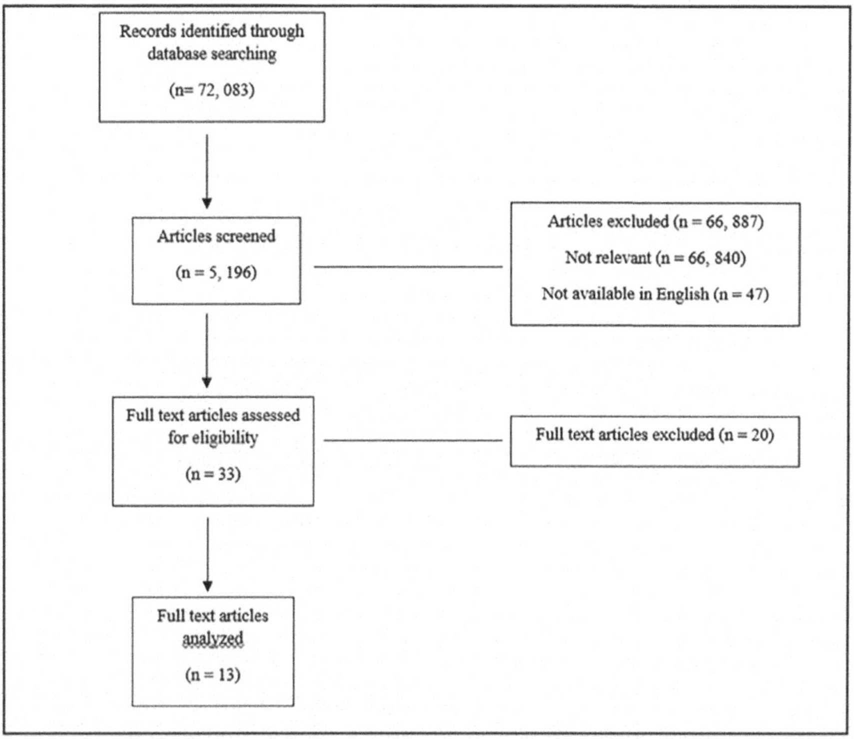

Fig. 1. Screening process of 72, 083 records from two search engines which are Scopus and Science Direct to obtain articles that are most relevant for data analysis.

age of Norway spruce (*Picea abies L.*), emphasizing the importance of precise site productivity forecasts for sustainable forest management under changing environmental conditions. Using measurements of dominant height across various parts of Southwest Germany, the study employs a recursive process to update the site index model with multiple measurements. The findings reveal that the new periodic site index statistically outperforms traditional methods in estimating stand height and volume increments. Historical analyses uncover trends in stand age and productivity, aligning with previous regional studies on forest growth. Furthermore, the method demonstrates adaptability to remote sensing data, enhancing its potential to forecast forest productivity across diverse stand conditions.

The study by Kotikot et al. [12] (A2) developed high-resolution (10m) mangrove cover maps for 2020–2022 using Random Forest machine learning and Sentinel-2 imagery, integrated with the Google Earth Engine as a decision support tool for managers. The findings revealed a 17.87% increase in mangrove cover in Puerto Rico from 2020 to 2021, followed by a 13.12% decline from 2021 to 2022. Degradation primarily occurred in small patches within existing mangrove ecosystems, driven by factors such as defoliation, death, stress, and human activities. The study's novel approach enables differentiation between intact and degraded mangrove vegetation, the mapping of small mangrove

Table 3. Data extraction

ID	Author	Year	Title	Database
A1	Yue et al., 2024	2024	A growth-effective age-based periodic site-index for the estimation of dynamic forest site productivity under environmental changes [24]	Scopus
A2	Kotikot et al., 2024	2024	Building a mangrove ecosystem monitoring tool for managers using Sentinel-2 imagery in Google Earth Engine [12]	Scopus
A3	Goodbody et al., 2024	2024	Integration of Airborne Laser Scanning data into forest ecosystem management in Canada: Current status and future directions [6]	Scopus
A4	Haq et al., 2023	2023	Measuring forest health at stand level: A multi-indicator evaluation for use in adaptive management and policy [8]	Scopus
A5	Baidoo & Obeng, 2023)	2023	Evaluating the impact of land use and land cover changes on forest ecosystem service values using landsat dataset in the Atwima Nwabiagya North, Ghana [3]	Science Direct
A6	Tsafack et al., 2023	2023	Arthropod-based biotic integrity indices: A novel tool for evaluating the ecological condition of native forests in the Azores archipelago [21]	Science Direct
A7	Tanács et al., 2022	2022	Assessing ecosystem condition at the national level in Hungary - indicators, approaches, challenges [20]	Science Direct
A8	Zelený et al., 2021	2021	Towards the evaluation of regional ecosystem integrity using NDVI, brightness temperature and surface heterogeneity [25]	Science Direct

(continued)

Table 3. (*continued*)

ID	Author	Year	Title	Database
A9	Ibarra-Bonilla et al., 2021	2021	Characterizing the impact of Land-Use/Land-Cover changes on a Temperate Forest using the Markov model [10]	Scopus
A10	Jiménez-Carmona et al., 2020	2020	Ants (Hymenoptera: Formicidae) as an indicator group of human environmental impact in the riparian forests of the Guadalquivir river (Andalusia, Spain) [11]	Science Direct
A11	Forsberg et al., 2020	2020	Fruit feeding butterflies as indicator taxon, pitfalls and concerns demonstrated in the Atlantic Forest [5]	Science Direct
A12	Queiroz et al., 2020	2020	The Forest Line Mapper: A Semi-Automated Tool for Mapping Linear Disturbances in Forests [18]	Scopus
A13	He et al., 2020	2020	Evaluation of Health Level of Land-use Ecosystem Based on GIS Grid Model [9]	Scopus

patches previously overlooked in lower-resolution studies, and temporal analysis of mangrove changes. Applications of this method include rapid assessment of ecosystem changes, detection of disturbance impacts, and monitoring of restoration progress. The framework provides critical, accessible data for continuous mangrove ecosystem monitoring and management, particularly in the face of environmental uncertainties and climate change.

The study by Goodbody et al. [6] (A3) explores the integration of Airborne Laser Scanning (ALS) data into forest management and environmental applications. ALS technology has proven particularly effective for stream network mapping and watershed management. Conducted in Canada, a country with vast forest coverage, this study highlights the opportunities and challenges of leveraging ALS for large-scale data management. The technology enhances operational forest planning by providing precise road network mapping, soil compaction assessment, and timber resource evaluation. However, barriers to its implementation include data gaps in certain regions, accessibility challenges, and limited technical understanding among users. While ALS offers valuable information on terrain and vegetation structure, it faces limitations in predicting species composition and forest health indicators.

The study by Haq et al. [8] (A4) aims to evaluate multiple indicators of forest health, focusing on tree composition, diversity, age, structure, regeneration status, and

Table 4. Distribution of methods used by each study in conducting forest health assessment.

ID	Bioindicator	Satellite imagery	Remote sensing	Machine learning	Indices/Multico-mponent indicator	Modelling data
A1	√					
A2		√				
A3			√	√		
A4					√	
A5		√		√		
A6	√					
A7					√	
A8					√	
A9		√				
A10	√					
A11	√					
A12			√	√		
A13		√				√

soil characteristics. Key findings include the identification of 22 tree species, with a notable 41% being native and 59% exotic species. The broadleaved forest type displayed the highest diversity, while scrub forest had the lowest. Exotic species demonstrated better regeneration performance compared to native species, highlighting their influence on forest dynamics. The study's management implications provide a foundation for developing conservation plans to preserve existing forest ecosystems and restore invaded areas.

Baidoo and Obeng [3] (A5) investigate land use and land cover (LULC) transformations in the Atwima Nwabiagya North District over a 20-year period, from 2002 to 2022. Using remote sensing and GIS tools, the study analyzes the extent of LULC changes and their implications for forest ecosystem services. The results show a significant decline in both high-density and low-density forests, with increases in built-up areas and bare ground. Forest areas have been converted to urban and agricultural land, leading to detrimental effects on the local ecosystem. The study also highlights the decline in forest ecosystem service values, emphasizing the need for better management practices to protect natural resources and improve rural livelihoods in the rapidly urbanizing region of Ghana.

Tsafack et al. [21] (A6) focus on developing two biological integrity indices (IBIs) that use arthropod communities as indicators of forest site quality in the Azores. The indices, IBI-SLAM for the understory and IBI-Canopy for the canopy stratum, incorporate multiple parameters sensitive to environmental disturbances, offering a more comprehensive understanding of ecosystem complexity compared to traditional single-measurement methods. The study highlights the need for effective assessment tools to support biodiversity management, particularly in island ecosystems. However, the indices are tailored for specific strata (understory and canopy) and unique arthropod communities in the Azores, limiting their applicability to other regions or ecosystems without further adaptation. Further research and refinement are necessary to enhance the robustness and broader applicability of these indices.

Tanács et al. [20] (A7) discuss the assessment of ecosystem condition at the national level in Hungary, focusing on the development of a comprehensive set of indicators tailored to various ecosystem types, enabling a more nuanced understanding of ecosystem health. The study combines qualitative expert assessments with quantitative machine learning methods to enhance the robustness and accuracy of the evaluations. A simplified scoring system was introduced to facilitate the evaluation of ecosystem conditions, making it easier to communicate findings and inform policy decisions. However, the study shares limitations with conventional methods, such as sampling bias. Biodiversity-based assessments, especially those relying on bird observations, were influenced by sampling efforts that varied by location and observer availability, leading to uneven spatial representation. Additionally, the study provides a baseline but lacks repeated assessments to track temporal changes effectively. Future studies could address these limitations by improving data collection (e.g., using remote sensing), integrating additional indicators, and refining methods and frameworks iteratively.

The study by Zeleny et al. [25] (A8) integrates three key indicators—Normalized Difference Vegetation Index (NDVI), Brightness Temperature (BT), and Vegetation Surface Heterogeneity (HG)—to assess ecosystem integrity. These indicators correspond to different ecosystem functions: energy capture (NDVI), water flow (BT), and structural diversity (HG). The method was tested in two regions, Třeboňsko (Czech Republic) and Schleswig-Holstein (Germany), where it revealed differences in ecosystem integrity due to varying land-use practices. The findings indicate that dense forests and wetlands exhibited the highest levels of integrity, while urban areas and croplands had lower integrity scores. Additionally, ecosystem integrity was found to be higher in more natural landscapes like Třeboňsko. The approach is simple, scalable, and uses open-access satellite data and GIS tools, making it accessible to a wide range of users. However, there are limitations, particularly in the insufficient representation of biodiversity and ecosystem maturity in low-productivity areas such as sand dunes. Despite these constraints, this study offers a practical framework for assessing ecosystem integrity at a regional scale, which can serve as a valuable decision-support tool for environmental management and policy development. The study emphasizes the need for future research to integrate more comprehensive data and refine methodologies to address these gaps.

Ibarra-Bonilla et al. [10] (A9) investigate the impact of Land-Use/Land-Cover (LULC) changes on a temperate forest ecosystem using the Markov model. The study analyzes the transition probabilities of land-cover types, providing a valuable framework to predict future changes in forest ecosystems under varying land-use scenarios. The Markov model offers insights into the dynamic nature of land-cover changes and their ecological implications. The findings highlight the potential for managing forest landscapes by understanding how past land-use changes influence future trends. This study's modeling approach can be applied to other regions to predict the consequences of land-use changes on biodiversity and forest health.

Jiménez-Carmona et al. [11] (A10) explore the role of ants (Hymenoptera: Formicidae) as bioindicators of human environmental impacts in riparian forests along the Guadalquivir River in Andalusia, Spain. By assessing the ant communities, the study evaluates their potential as indicators of habitat quality and disturbance in these ecosystems. The findings suggest that ants can serve as reliable bioindicators for detecting environmental degradation, with different species responding to varying levels of disturbance. This research contributes to improving biodiversity monitoring techniques in riparian forests and emphasizes the importance of ants in assessing ecosystem health.

Forsberg et al. [5] (A11) examine the use of fruit-feeding butterflies as an indicator taxon to assess biodiversity in the Atlantic Forest of Brazil. The research discusses the advantages and challenges of using butterflies as bioindicators, particularly in tropical ecosystems. The findings underscore the role of butterflies in reflecting environmental changes and their potential as a tool for monitoring ecosystem health. However, the study also identifies several concerns, such as the variability in the ecological roles of different butterfly species, which could complicate their use in biodiversity assessments. The article provides valuable insights into the application of butterflies for ecological monitoring.

Queiroz et al. [18] (A12) introduce the Forest Line Mapper, a semi-automated tool designed to map linear disturbances in forests. This tool provides a robust methodology for detecting and mapping disturbances such as roads, powerlines, and other linear features that can fragment forest ecosystems. The Forest Line Mapper combines remote sensing data and machine learning techniques to improve the accuracy of disturbance mapping. The tool's application can enhance forest management and conservation strategies by offering an efficient way to monitor and evaluate disturbances over large forested areas.

He et al. [9] (A13) evaluate the health level of land-use ecosystems using a GIS grid model. The model integrates spatial data to assess the ecological health of ecosystems under various land-use conditions. By analyzing factors such as vegetation cover, water availability, and human impact, the study provides a comprehensive approach to land-use health evaluation. The GIS grid model allows for fine-scale analysis, making it applicable to large-scale land-use assessments. The findings highlight the importance of land-use planning in maintaining ecosystem health and the potential of GIS-based tools for environmental management.

The 13 studies collectively highlight the diverse methods and indicators employed to assess and monitor ecosystem health and biodiversity in varying landscapes. Advances in remote sensing technologies, such as Random Forest machine learning and Airborne Laser Scanning (ALS), have provided high-resolution, scalable approaches to detect ecosystem changes, including mangrove cover (A2) and forest health (A3). The integration of multiple indicators, such as vegetation indices, biodiversity measures, and land-use models, allows for a nuanced understanding of environmental transformations, particularly in temperate forests (A9) and riparian ecosystems (A10). Studies on arthropod communities, such as ants and butterflies (A10, A11), demonstrate their utility as bioindicators of ecological disturbances, offering valuable insights into biodiversity health. The application of GIS and machine learning tools (A12, A13) offers practical frameworks for mapping disturbances and evaluating ecosystem integrity. However, challenges persist, such as the need for improved temporal data (A7) and the limitations of species-specific indices (A6). Overall, these studies contribute to the development of sophisticated monitoring tools, emphasizing the importance of integrating multiple data sources and refining methodologies to address critical environmental issues like climate change and land-use change. These findings underscore the necessity for ongoing innovation in ecosystem management to preserve biodiversity and ecosystem services.

4 Conclusion

In summary, this review aims to critically assess and synthesize the methodologies employed to evaluate forest health, focusing on their effectiveness in measuring ecosystem integrity. Key indicators, such as biodiversity, conservation status, and resilience to environmental stressors, are examined in depth. The review places particular emphasis on comparing conventional field-based methods with advanced technologies, such as remote sensing, GIS, and modeling approaches, to assess forest productivity and soil quality. The findings reveal a broad spectrum of opportunities for further exploration, particularly in the integration of diverse tools and metrics. This analysis provides valuable insights into current practices and highlights the importance of refining standardized indicators to enhance comparability across ecosystems. It also emphasizes the challenges related to data availability, accuracy, and the need for consistent methodologies. The review outlines clear guidelines for future research, calling for the development of more comprehensive, robust, and adaptive approaches to ecosystem monitoring that can effectively address emerging environmental threats, including climate change, land-use change, and biodiversity loss.

Acknowledgements. This work was supported in part by the Ministry of Higher Education (MOHE) Malaysia through Fundamental Research Grant Scheme (FRGS/1/2023/WAB 11/UTHM/03/2).

References

1. Afifi, M., Stryhn, H., Sanchez, J.: Data extraction and comparison for complex systematic reviews: a step-by-step guideline and an implementation example using open-source software. Syst. Rev. **12**(1), 226 (2023). https://doi.org/10.1186/s13643-023-02322-1
2. Apostol, B., et al.: Data collection methods for forest inventory: a comparison between an integrated conventional equipment and terrestrial laser scanning. Ann. For. Res. **61**(2), 189 (2018). https://doi.org/10.15287/afr.2018.1189
3. Baidoo, R., Obeng, K.: Evaluating the impact of land use and land cover changes on forest ecosystem service values using landsat dataset in the Atwima Nwabiagya North, Ghana. Heliyon **9**(11), e21736 (2023). https://doi.org/10.1016/j.heliyon.2023.e21736
4. Blanco, J.A., Lo, Y.-H.: Latest trends in modelling forest ecosystems: new approaches or just new methods? Curr. For. Rep.s **9**(4), 219–229 (2023). https://doi.org/10.1007/s40725-023-00189-y
5. Forsberg, F., Barfod, A.S., Francisco, A.J., Ribeiro, M.C.: Fruit feeding butterflies as indicator taxon, pitfalls and concerns demonstrated in the Atlantic Forest. Ecol. Ind. **111**, 105986 (2020). https://doi.org/10.1016/j.ecolind.2019.105986

6. Goodbody, T.R.H., et al.: Integration of airborne laser scanning data into forest ecosystem management in Canada: current status and future directions. For. Chron. **100**(2), 238–258 (2024). https://doi.org/10.5558/tfc2024-014
7. Grammatikopoulou, I., Vačkářová, D.: The value of forest ecosystem services: a meta-analysis at the European scale and application to national ecosystem accounting. Ecosyst. Serv. **48**, 101262 (2021). https://doi.org/10.1016/j.ecoser.2021.101262
8. Haq, S.M., et al.: Measuring forest health at stand level: a multi-indicator evaluation for use in adaptive management and policy. Ecol. Ind. **150**, 110225 (2023). https://doi.org/10.1016/j.ecolind.2023.110225
9. He, W., Li, Z., Jing, D.: Evaluation of health level of land-use ecosystem based on GIS grid model. Nat. Environ. Pollut. Technol. **19**(4), 1475–1482 (2020). https://doi.org/10.46488/NEPT.2020.v19i04.013
10. Ibarra-Bonilla, J.S., Villarreal-Guerrero, F., Prieto-Amparán, J.A., Santellano-Estrada, E., Pinedo-Alvarez, A.: Characterizing the impact of land-use/land-cover changes on a temperate forest using the Markov model. Egypt. J. Remote Sens. Space Sci. **24**(3), 1013–1022 (2021). https://doi.org/10.1016/j.ejrs.2021.11.004
11. Jiménez-Carmona, F., Heredia-Arévalo, A.M., Reyes-López, J.L.: Ants (Hymenoptera: Formicidae) as an indicator group of human environmental impact in the riparian forests of the Guadalquivir river (Andalusia, Spain). Ecol. Ind. **118**, 106762 (2020). https://doi.org/10.1016/j.ecolind.2020.106762
12. Kotikot, S.M., et al.: Building a mangrove ecosystem monitoring tool for managers using Sentinel-2 imagery in Google earth engine. Ocean Coast. Manag. **256**, 107307 (2024). https://doi.org/10.1016/j.ocecoaman.2024.107307
13. Lame, G.: Systematic literature reviews: an introduction. In: Proceedings of the Design Society: International Conference on Engineering Design, vol. 1, no. 1, pp. 1633–1642 (2019). https://doi.org/10.1017/dsi.2019.169
14. Moher, D., et al.: Preferred reporting items for systematic review and meta-analysis protocols (PRISMA-P) 2015 statement. Syst. Rev. **4**(1), 1 (2015). https://doi.org/10.1186/2046-4053-4-1
15. Paheding, S., Saleem, A., Siddiqui, M.F.H., Rawashdeh, N., Essa, A., Reyes, A.A.: Advancing horizons in remote sensing: a comprehensive survey of deep learning models and applications in image classification and beyond. Neural Comput. Appl. **36**(27), 16727–16767 (2024). https://doi.org/10.1007/s00521-024-10165-7
16. Prăvălie, R.: Major perturbations in the Earth's forest ecosystems. Possible implications for global warming. Earth-Sci. Rev. **185**, 544–571 (2018). https://doi.org/10.1016/j.earscirev.2018.06.010
17. Quamar, M.M., Al-Ramadan, B., Khan, K., Shafiullah, M., El Ferik, S.: Advancements and applications of drone-integrated geographic information system technology—a review. Remote Sens. **15**(20), 5039 (2023). https://doi.org/10.3390/rs15205039
18. Queiroz, G.L., McDermid, G.J., Rahman, M.M., Linke, J.: The forest line mapper: a semi-automated tool for mapping linear disturbances in forests. Remote Sens. **12**(24), 4176 (2020). https://doi.org/10.3390/rs12244176
19. Randhir, T.O., Erol, A.: Emerging threats to forests: resilience and strategies at system scale. Am. J. Plant Sci. **04**(03), 739–748 (2013). https://doi.org/10.4236/ajps.2013.43A093
20. Tanács, E., et al. Assessing ecosystem condition at the national level in Hungary - indicators, approaches, challenges. One Ecosyst. **7** (2022). https://doi.org/10.3897/oneeco.7.e81543
21. Tsafack, N., et al.: Arthropod-based biotic integrity indices: a novel tool for evaluating the ecological condition of native forests in the Azores archipelago. Ecol. Ind. **154**, 110592 (2023). https://doi.org/10.1016/j.ecolind.2023.110592

22. Weiskopf, S.R., et al.: Climate change effects on biodiversity, ecosystems, ecosystem services, and natural resource management in the United States. Sci. Total. Environ. **733**, 137782 (2020). https://doi.org/10.1016/j.scitotenv.2020.137782
23. Xiao, Y., Watson, M.: Guidance on conducting a systematic literature review. J. Plan. Educ. Res. **39**(1), 93–112 (2019). https://doi.org/10.1177/0739456X17723971
24. Yue, C., Kahle, H.-P., Klädtke, J., Kohnle, U.: A growth-effective age-based periodic site-index for the estimation of dynamic forest site productivity under environmental changes. Ann. For. Sci. **81**(1), 33 (2024). https://doi.org/10.1186/s13595-024-01253-8
25. Zelený, J., Mercado-Bettín, D., Müller, F.: Towards the evaluation of regional ecosystem integrity using NDVI, brightness temperature and surface heterogeneity. Sci. Total. Environ. **796**, 148994 (2021). https://doi.org/10.1016/j.scitotenv.2021.148994

Exploratory Analysis of Pollen Morphology in Selected *Zingiber* Species (Zingiberaceae) from Peninsular Malaysia

Aimi Syazana Sedek[1], Salasiah Mohamad[1(✉)], and Sam Yen Yen[2]

[1] Department of Technology and Natural Resources, Faculty of Applied Sciences and Technology, Universiti Tun Hussein Onn Malaysia (Pagoh Campus), KM 1, Jalan Panchor, 84600 Muar, Johor, Malaysia
salasiah@uthm.edu.my

[2] Flora Biodiversity Program, Forest Biodiversity Department, Forest Research Institute Malaysia, 52109 Kepong, Selangor, Malaysia

Abstract. The genus *Zingiber* is traditionally categorized into four sections based on the position of the inflorescence: sect. *Zingiber* (radical, with an erect peduncle), sect. *Cryptanthium* Horan. (radical, with a procumbent peduncle), sect. *Pleuranthesis* Benth. (with erect inflorescences emerging from the side of the leafy stem), and sect. *Dymczewiczia* (Horan.) Benth. (with inflorescences borne apically on the leafy stem). This classification has been supported by studies on the pollen morphology of *Zingiber* in various regions, which suggest that sect. *Zingiber* (cerebroid exine sculpturing), sect. *Cryptanthium* (striate exine sculpturing) while sect. *Dymczewiczia* should be merged with sect. *Zingiber* due to their similar pollen characteristics. However, an initial exploratory study on the pollen morphology of seven *Zingiber* species in Peninsular Malaysia revealed that the pollen morphology not supported the traditional classifications into sect. *Zingiber* and sect. *Cryptanthium*, regardless of whether the inflorescence is erect or procumbent. This finding suggested to challenge the previous classification based on inflorescence peduncle habit. Further research is ongoing, with more samples being analysed to include a broader range of *Zingiber* species found in Peninsular Malaysia.

Keywords: Palynology · taxonomy · SEM

1 Introduction

1.1 Palynology of *Zingiber*

Zingiberaceae is the largest family within the Zingiberales order, comprising 57 genera and 1900 species (Banaticla-Hilario & Altamirano, 2023; POWO, 2024). It is widely distributed in tropical regions, and Southeast Asia is a hotspot diversity for this family. These perennial herbs are usually terrestrial, but some are seen as epiphytic (POWO, 2024). Pollen morphological studies of Zingiberaceae began in the early 1900s by several prominent researchers (Walker & Doyle, 1975: Liang, 1988; Mangaly & Nayar,

P. Chang et al. (Eds.): SCIEMATHIC 2024, SPPHY 427, pp. 194–208, 2025.
https://doi.org/10.1007/978-981-96-7679-8_16

1990) that have continued to progress in recent years, including those by Theilade et al. (1993), Acma et al. (2018) on tribe Alpinieae from the Philippines, and by Kajornjit et al. (2018) on *Globba* in Thailand, Zou et al. (2022) on 33 genera of the family, and recently by Mohamad et al. (2023) on tribe Alpinieae species from Malaysia that provide insights to taxonomic significance. In the Zingiberaceae family, pollen grains are characterised as apolar and may exhibit various shapes, including spheroidal, subspheroidal, ovoid, or ellipsoidal. These grains are inaperturate, with a thin exine typically measuring 1–2 μm, which can be continuous or discontinuous across different species. The intine layer is notably thick, averaging 3–5 μm others can range from 1.5 to 24 μm. Pollen ornamentation within the family is diverse, encompassing psilate, echinate, microrugulate, striate, fossulate, cerebroid, areolate, micro-granulate, and verrucate patterns. Pollen size also shows significant variation within species, generally falling into the categories of medium (35–50 μm), large (51–100 μm), or very large (100–139 μm), with the majority classified as large (51–100 μm) (Zou et al. 2022; Mohamad et al. 2023). Walker and Doyle (1975) revealed pollen grains of the genus zingiber are notably rare for having a single aperture, unlike other genera in the zingiberaceae family. They also noted that Zingiberaceae pollen grains can reach sizes of 200 μm or more. However, Liang (1988) reported that Chinese *Zingiber* pollen is inaperturate and exhibits striate sculpturing, particularly from the section *Cryptanthium*, based on light and scanning electron microscope (SEM) examinations. Mangaly and Nayar (1990) added complexity to these findings in their study of South Indian pollen, noting that the genera *Curcuma* and *Zingiber*, particularly *Z. roseum*, displayed faint sulci when viewed under a light microscope. In 1993, Theilade et al. conducted the most comprehensive pollen study on the genus *Zingiber*, examining 18 species from several regions including Malaysia, Thailand, and Burma using light microscopy and SEM. Their findings indicated that *Zingiber* pollen is inaperturate and can be categorized into two groups based on pollen characteristics: sect. *Zingiber* and sect. *Dymczewiczia*, features cerebroid or reticulate sculpturing, while sect. *Cryptanthium* possessed elliptic pollen with spiro-striate sculpturing (Theilade et al., 1993). They also proposed merging sects. *Dymczewiczia* into sect. *Zingiber* based on cerebroid exine sculpturing is further supported by phylogenetic analysis done by Theerakulpisut et al. (2012). However, subsequent studies on the pollen morphology of *Zingiber* only focused on newly discovered species to verify the sections within *Zingiber* traditionally (Triboun et al., 2002; Ye et al., 2015; Ardiyani et al., 2017; Odyuo et al., 2019; Jayakrishnan et al., 2021; Zou et al., 2022). The latest pollen study by Zou et al. (2022) provided a comprehensive overview of Zingiberaceae pollen but included only 11 *Zingiber* species, with just one native species from Peninsular Malaysia, *Z. longibracteatum*. To date, pollen from only eight *Zingiber* species in Peninsular Malaysia has been examined (Theilade et al., 1993; Zou et al., 2022). This paper presented and discussed a preliminary analysis of selected *Zingiber* pollen from Peninsular Malaysia, marking an exploratory investigation into the pollen of *Zingiber* species in the region.

2 Materials and Methods

2.1 Study Sites

Fertile plants of seven *Zingiber* species were collected from various locations in Peninsular Malaysia, including their type localities, including Royal Belum State Park, Fraser's Hill, Taka Melor Eco Forest, Panti Forest Reserve, and Teluk Bahang Forest Reserve. The latitude and longitude of each collection site were recorded using a Global Positioning System (GPS) for precise documentation. Field notes were also taken, including information on the type of habitat and environmental conditions.

2.2 Species Identification

Morphological examinations were conducted on selected *Zingiber* species, specifically *Z. griffithii*, *Z. raja*, *Z. spectabile*, *Z. flaviflorum*, *Z.* aff. *belumense*, *Zingiber* sp. (bahangense 130), and *Zingiber* sp. 2. Detailed observations and descriptions of all plant parts, including leaves, inflorescences, flowers, and fruits (when available), were made to aid in species identification. Specimens were compared with herbarium specimens at KEP and digital images from SING and K to validate their identification (Royal Botanic Gardens, 2024; Singapore Herbarium Online, 2024). Additionally, protologues and online databases (POWO, 2024; IPNI, 2024) were consulted. Photographs of the plants and their habitats were taken for visual documentation. Herbarium specimens were deposited at both the Forest Research Institute Malaysia (FRIM)(KEP) and Universiti Tun Hussein Onn Malaysia (UTHM).

2.3 Pollen Examination

Scanning Electron Microscope (SEM). The freshly collected pollen grains were immediately fixed by immersing them in a 70% ethanol solution. The pollen samples were then dehydrated sequentially using 80%, 90%, and 100% ethanol. The grains were oven-dried at 50 °C for 24 h on a carbon tape. After drying, the pollen grains were coated with gold at a 40 mA sputter current using a Quorum (Q150R S). The prepared and coated samples were examined with a Scanning Electron Microscope (SEM) (COXEM EM30AX), a technique that provides detailed information on the specimens' three-dimensional(3D) structure. SEM enabled high-resolution imaging of the morphology and analysing parameters including polarity, symmetry, grain size, shape, apertures, and exine sculpturing. The pollen terminology followed Erdtman, (1952), Punt et al. (2007), Agashe & Coulton (2009), Halbritter et al. (2018), and Zou et al. (2022).

Light Microscope (LM). Fresh pollen grains were collected from the anther thecae and immediately preserved in a 70% ethanol solution for further analysis. A small amount of pollen was evenly spread onto a glass microscope slide and carefully covered with a coverslip. The general shape and surface features of the pollen were then observed in two-dimensional (2D) images at 40 × magnifications using an AmScope 40 × -2500 × LED Digital Binocular Compound Microscope.

3 Results

3.1 Plant Morphology

The earlier classification of the genus *Zingiber* into sections was primarily based on the habit of the inflorescence peduncle, which is sect. *Zingiber* (radical, erect peduncle), sect. *Cryptanthium* (radical, procumbent peduncle), sect. *Pleuranthesis* (erect from the side of the leafy stem) and sect. *Dymczewiczia* (terminal, born apically on the leafy stem). In this exploratory study, seven selected *Zingiber* species, namely *Z. flaviflorum, Z. griffithii, Z. raja, Z. spectabile, Zingiber* aff. *belumense*, *Zingiber* sp. (bahangense 130), and *Zingiber* sp. 2 were examined morphologically. The majority of the species had a decumbent inflorescence with a procumbent peduncle, or called scape, namely *Z. flaviflorum, Z.* aff. *belumense*, *Zingiber* sp. (bahangense 130), and *Zingiber* sp. 2 (Fig. 1A–B, L–R). In contrast, the rest of the species were erect (Fig. 1C–K). *Zingiber* sp. (bahangense 130) is potentially a new species, as described in the previous paper by Sedek et al. (2024), and currently under thorough morphological and molecular investigation. The description of each selected species is presented herein.

1. *Zingiber flaviflorum* C.K. Lim & Meekiong. – Type: Fraser's Hill. Locality: Fraser's Hill, montane forest, altitude 1300 m, 29 July 2023. Figures 1A–B

Rhizome creamish yellow, aromatic. Leafy shoots 2.97–3.0 m tall, bending about 45°, 8 pairs of leaf, pseudostem green yellowish, hairy or puberulous, margin obvious yellowish puberulous. Ligule 0.4 cm long, emarginate, densely pubescent. Petiole very short to sessile, densely pubescent. Lamina ovate to lanceolate, 12.0–39.0 cm × 3.5–5.0 cm, cuneate at base, acuminate at apex, adaxial green, glabrous; abaxial pale green and sparsely pubescent at midrib, leathery. Inflorescence procumbent to decumbent, 16.0–41.0 cm long. Scape procumbent, 8.0 cm to 31.0 cm long, densely pubescent, cream-green colour. Spike ovoid with pointed apex, 8.0–10.0 cm long, closely imbricate. Bract ovate, apex acute, pink gradually dark pink with green at the base, light pink towards the apices, glabrous. Bracteole oblong, 2.5 cm long, apex truncate, glabrous. Floral tube 6.0 cm long. Calyx boat-shaped, 2.5 cm long, apex retuse, translucent white. Corolla yellow, semi translucent; dorsal lobe 3.0 cm long, oblong to ovate, apex acuminate; lateral lobe 2.5 cm long, oblong to obovate with deeply bifid. Labellum 2.7 cm long, midlobe ca 1.5 cm × 1.4 cm, semi-translucent, prominent veined, apex acute; lateral staminodes connate to labellum, well developed, ca 0.2 cm × 0.3 cm, yellow semi-transluscent, apices truncate.

Anther ca 1.4 cm long, anther crest bifid apex, yellow; stigma arching downward, hairy.

Ovary densely villous; epigynous gland two. Fruit unknown.

Distribution: Endemic to Fraser's Hill, Pahang, Malaysia.

Ecology: Lower montane forest, open ground, along the trail, alt. 1300 m.

Phenology: Flowering period initiated middle of May and flowering by the end of July. Fruiting may start on August to October.

2. *Zingiber griffithii* Baker. – Type: Melaka. Locality: Kota Tinggi, Johor, lowland dipterocarp forest, altitude 30–50 m, 25 September 2023. Figures 1C–E.

Rhizome creamy yellow. Leafy shoots 0.37–0.5 m tall, erect, 6 pairs of leaf, pseudostem dark purple to maroon at the lower part, green upper part, pubescent. Ligule 0.3 cm long, deeply bilobed, entire, glabrous. Petiole prominent pulvinus, very short to sessile. Lamina broadly lanceolate to ovate, 16–23.5 cm × 3.4–9.5 cm, cuneate at base, mucronate at apex, adaxial finely raised veins, glabrous, green; abaxial sparsely pubescent, pale green. Inflorescence radical, erect, 12.0–15.0 cm long. Scape erect, 10.0 cm long, pubescent, reddish-brown pink colour. Spike fusiform, with acute apex, 12.0 cm long. Bract pinkish orange to pink, apex acute, pubescent, pulpy, less tough, not closely overlapped. Floral tube 5.2 cm long, cream. Calyx 1.5 cm long, boat-shaped, apex bifid, translucent white. Corolla white-yellow; dorsal lobe 2.4 cm long, oblong to ovate, apex acuminate, semi-translucent white yellowish; lateral lobe 2.0 cm long, oblong to obovate with deeply bifid or split, apex acuminate, semi translucent, yellow to pale cream. Labellum 1.5 cm long, midlobe ca 0.8 cm × 1.6 cm, deeply bifid, semi-translucent, apex acute to semi – truncate; lateral staminodes connate to labellum, well developed, ca 0.5–0.6 cm × 0.2–0.3 cm, yellowish pale to creamish white, semi-transluscent, apex truncate to bluntly truncate. Anther ca 0.4 cm long, brownish – purple colour, anther crest horn-shaped, apex acute, yellow; stigma arching downward, cilia. Ovary densely villous. Fruit unknown.

Distribution: Johor, Selangor, Perak, Borneo.

Ecology: Lowland dipterocarp forest, near small streams/creeks, humid rich soils, alt. 30–50 m.

Phenology: Flowering period initiated from August to late September.

3. *Zingiber raja* C.K. Lim & Kharukanant. – Type: Royal Belum. Locality: Royal Belum, Perak, lowland dipterocarp forest, altitude 250 m, 9 July 2023. Figures 1F–H.

Leafy shoots ca. 1.3 m tall, pseudostem green. Ligule ca. 0.2 cm long, broadly bilobed, truncate apex, glabrous, green. Petiole prominent pulvinus, 0.5 cm to short. Lamina broadly lanceolate, ca. 23–38 cm × 8.5–10 cm, cuneate at base, shortly caudate at apex, adaxially green, abaxially green, glabrous for both. Inflorescence arise from the rhizome, erect, up to 97.0 cm long. Scape erect, up to 60.0 cm long, dark red. Spike fusiform with pointed apex, up to 35.0 cm long. Bract 6.3 cm, ovate, reddish orange – yellow, apex acute, sparsely pubescent. Bracteole 3.8 cm long, oblong, apex acute. Floral tube 6.0 cm long. Calyx 3.0 cm long, elongate navicular, apex acuminate, translucent white. Corolla cream; dorsal lobe 3.0 cm long, oblong to ovate, apex acute, cream semi-translucent; lateral lobe narrowly boat-shaped 3.0 cm long, apex acute. Labellum ca. 2.1 cm, midlobe 1.0 cm × 0.8 cm, lateral staminodes connate to labellum, well developed, ca 0.4 cm × 0.4 cm, apex bilobed, purple mottled yellow, margin undulate. Anther ca 2.0 cm long, purple, anther crest acuminate apex; stigma hairy. Ovary glabrous, three locucidal; epigynous gland two. Fruit unknown.

Distribution: Peninsular Thailand, Royal Belum.

Ecology: Primary and lowland dipterocarp forest, riverine, freshwater swamp, forested hill slope, alt. 250 m.

Phenology: Flowering period initiated in early July to August. Fruiting may start on early September to October

4. *Zingiber spectabile* Griffith. – Type: Melaka. Locality: Taka Melor Eco Forest, Johor, lowland dipterocarp forest, altitude 30–50 m, 8 June 2023 (Sedek et al. 2024a). Figures 1I–K.

Leafy shoots up to 3.14 m tall, bending downward, with flushed pink at the base, 21 pairs of leaf, pseudostem green, glabrous. Ligule 1.5 cm long, deeply bilobed, thin scarious, acute to bluntly acute apices, rotten. Petiole prominent pulvinus, very short to sessile. Lamina lanceolate, ca 40 cm × 6–7 cm, rounded at base, acuminate at apex, adaxially green, abaxially dull green, sparsely pubescent and compactly pubescent at midrib, margin undulate. Inflorescence arise from the rhizome, erect, ca. 55 cm long. Scape erect, 40 cm, overlapping bract, green. Spike 15 cm long, cylindrical with incurved bracts, apex entire. Bract obovate, orange-pink colour. Bracteole 4.5 cm long, tubular, apex acuminate, deeply bifid, pink at the apex, yellow at the bottom of the bracteole. Floral tube 7.5 cm long. Calyx 2.2 cm long, boat-shaped, apex acute, semi-translucent white. Corolla yellow; dorsal lobe 3.2 cm long, boat-shaped, apex acuminate; lateral lobe 3.2 cm long, boat-shaped with deeply bifid, apex acuminate. Labellum ca 3.0 cm, midlobe 1.5 cm × 0.5 cm, apex bifid, yellow with mottled purple patched; lateral staminodes connate to labellum, well developed, ca 0.5 cm × 0.5 cm. Anther ca 1.5 cm long, yellow, anther crest purple; stigma arching downward, hairy. Ovary densely villous, three locules. Fruit ca. 2.0–3.0 cm, capsule, ovoid.

Distribution: Kedah, Perak, Selangor, Melaka, Johor, Pahang

Ecology: Lowland dipterocarp forest, along the trails, observed at the roadsides, alt. 30–50 m.

Phenology: Flowering initiated in early May to June.

5. *Zingiber* aff. *Belumense* Griffith. – Type: Belum Titiwangsa "Highpoint", Perak. Locality: Royal Belum, Perak, lowland dipterocarp forest, altitude < 100 m, 9 July 2023. Figures 1L–M.

Leafy shoot ca. 1.2–1.3 m tall, erect, pseudostem green, pubescent, margin densely pubescent. Ligule 0.8–1.0 cm long, bilobed, obtuse apices, pubescent at the edge of the apices, green. Petiole prominent pulvinus, 0.5 cm long, densely pubescent. Lamina lanceolate, 41.5–43 cm × 9.5–10.3 cm, attenuate at base, aristate to caudate at apex, adaxially green, glabrous, abaxially green, pubescent on the blade and midrib. Inflorescence arise from the rhizome, procumbent, 9.8–10.0 cm long. Scape procumbent, ca. 3.0–4.8 cm long, brownish maroon, pubescent, acute apices. Spike ovoid with blunt or rounded apex, 5.2–7.0 cm long. Bract ovate, brownish maroon with tinged green, apex acute, pubescent. Floral tube 4.5 cm long. Calyx 2.5 cm long, oblong, apex truncate, translucent white. Corolla pale yellow; dorsal lobe 2.3 cm long, boat shape, apex acuminate, semi-translucent; lateral lobe 2.3 cm long, boat shaped, apex acute, semi-translucent. Labellum 1.6 cm long, midlobe ca 1.0 cm x 0.7 cm, semi-translucent, apex retuse; lateral staminodes connate to labellum, well developed, ca 0.5 cm × 0.4 cm, pale yellow semi-transluscent, apices acuminate. Stamen ca 1.9 cm long; anther ca 1.0 cm long; stigma hairy. Fruit unknown.

Distribution: Royal Belum

Ecology: Lowland dipterocarp forest, alt. <100 m.

Phenology: The flowering period initiated early July to August. Fruiting may start on early September to October.

6. *Zingiber* sp. (bahangense 130) Aimi Syazana & Salasiah. – Type: Teluk Bahang, Penang. Locality: Teluk Bahang, hill dipterocarp forest, altitude ca. < 300 m, 9 July 2023. Figures 1N–P.

Leafy shoot 2.8–3.0 m tall, long trailing arching downward, pseudostem green, pubescent. Ligule bifid and short, acute apices, sparsely pubescent at the edge of the apices, green. Petiole prominent pulvinus, 0.5 cm to short. Lamina elliptic, 19.5–43 cm × 3.5–9.5 cm, attenuate at base, aristate to caudate at apex, adaxially shiny green, abaxially shiny pale green, and compactly pubescent at midrib. Inflorescence decumbent. Scape procumbent, subterranean in the ground, creamish white with tinged brownish pink at the edge of the margin, glabrous. Spike ovoid with pointed apex, 15.0–16.0 cm long. Bract 6.0 cm, ovate, brownish maroon-dark purple with green, apex acute, pink margin with prominent line, densely pubescent. Bracteole 2.9 cm long, oblong, apex obtuse. Floral tube 6.0 cm long. Calyx 2.8 cm long, ovate, bifid, apex acute, translucent white. Corolla yellowish cream; dorsal lobe 2.5 cm long, oblong to ovate, apex acuminate, semi-translucent; lateral lobe 2.5 cm long, oblong to obovate with deeply bifid, apex acuminate, semi-translucent. Labellum yellowish cream semi-translucent, apices bluntly undulate deflexed. Fruit unknown.

Distribution: Penang Hill

Ecology: Lowland and hill dipterocarp forest, alt. <300 m.

Phenology: Flowering period initiated at the end of July to August. Fruiting may start on early September to October.

7. *Zingiber* sp. 2 (Fraser's Hill). – Type: Fraser's Hill, Pahang. Locality: Fraser's Hill, lower montane forest, altitude 1300 m, 25 September 2023. Figures 1Q-R.

Leafy shoot up to 2 m tall, erect. Ligule truncate apex. Petiole prominent pulvinus, very short to sessile. Lamina elliptic to broadly lanceolate, 34.0 cm × 11.0 cm, cuneate to oblique at base, caudate at apex, adaxially glabrous and glossy, green colour; abaxially glabrous and glossy, pale green. Inflorescence decumbent. Scape procumbent, 20.0 cm long, whitish to pink colour. Spike elongated ovoid with pointed apex, 16.0 cm long. Bract pale pink with green tinged, apex acute. Floral tube 5.0 cm long, yellowish cream. Calyx 2.0 cm long, boat-shaped, semi-translucent. Corolla white-yellow; dorsal lobe 2.7 cm long, boat-shaped, apex acute, semi-translucent; lateral lobe 2.5 cm long, boat-shaped with deeply bifid or split, apex acute, semi translucent. Labellum 2.5 cm long, pale yellow. Fruit capsule, ovoid, 2.5 cm long, glabrous, cream.

Distribution: Fraser's Hill

Ecology: Lower montane forest, alt. 1300 m.

Phenology: The flowering period may initiate from August to late September.

Fig. 1. Seven selected *Zingiber* species collected from Peninsular Malaysia. A–B. *Z. flaviflorum*, C–E. *Z. griffithii*, F–H. *Z. raja*, I–K. *Z. spectabile*, L–M. *Z.* aff. *Belumense*, N–O. *Zingiber* sp. (bahangense 130), P. *Zingiber* sp. 2.

Fig. 1. (*continued*)

3.2 Pollen Micromorphology

The pollen of seven selected native *Zingiber* species from Peninsular Malaysia was identified as monad, consisting of single, inaperturate grains (Figs. 2 and 4). Preliminary results from SEM indicate that the majority of *Zingiber* species have large-sized grains, ranging from 50 μm to 66 μm. The size classification follows the method outlined by Walker and Doyle (1975), which is based on the measurement of the longest axis of the pollen grains. All of the selected species have spheroidal shapes, categorized into three sub-groups; prolate- spheroidal, oblate-spheroidal and spherical, following established classifications (Erdtman, 1952). The cerebroid sculpturing is prevalent in *Zingiber*, but exceptions include *Z. flaviflorum* and *Zingiber* sp. (bahangense 130), which exhibit areolate sculpturing, and *Z. griffithii*, which is distinctly verrucate (Table 1, Fig. 3).

Table 1. Characteristics of selected Zingiber pollen grains from Peninsular Malaysia

Species	Characteristics		P/E ratio	Size	Shape	Exine sculpturing
	Polar Axis (um)	Equatorial Axis (um)				
Z. flaviflorum	54.87	54.84	1.00	Large	Spherical	Areolate
Z. griffithii	55.72	52.94	1.05	Large	Prolate-spheroidal	Verrucate
Z. raja	65.32	62.75	0.97	Large	Oblate-spheroidal	Cerebroid
Z. spectabile	63.92	61.10	1.05	Large	Prolate-spheroidal	Cerebroid
Zingiber aff. *belumense*	63.83	61.13	1.04	Large	Prolate-spheroidal	Cerebroid
Zingiber sp. (bahangense 130)	60.98	65.79	0.93	Large	Oblate- spheroidal	Areolate
Zingiber sp. 2	65.94	65.94	1.00	Large	Prolate-spheroidal	Areolate

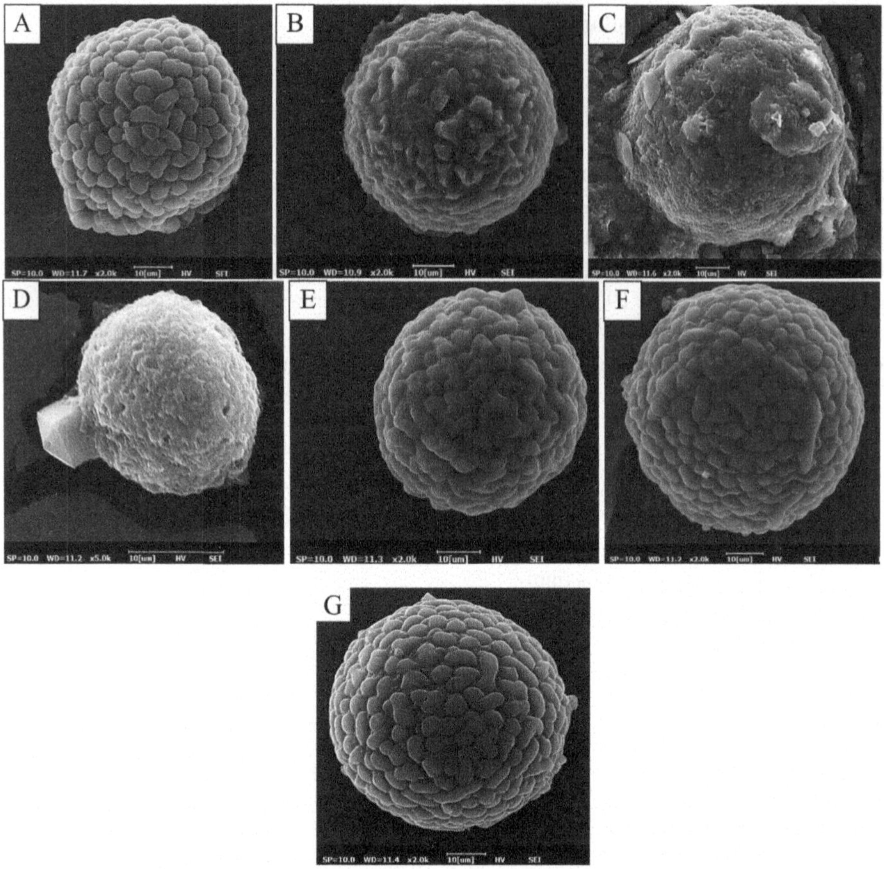

Fig. 2. Micrographs of pollen of selected *Zingiber* in Peninsular Malaysia using SEM. A. *Z. flaviflorum*, B. *Z. griffithii*, C. *Z. raja*, D. *Z. spectabile*, E. *Z.* aff. *belumense*, F. Zingiber sp. (*bahangense*130), G. *Zingiber* sp. 2.

4 Discussion

4.1 Pollen Micromorphology in Relation to Taxonomic Classification

Previous research on *Zingiber* pollen grains has revealed that pollen morphology as a crucial element in the taxonomic classification of the genus (Theilade et al. 1993; Triboun et al. 2002; Ye et al. 2015; Ardiyani et al. 2017; Odyuo et al. 2019; Jayakrishnan et al. 2021; Zou et al. 2022).

Based on the previous pollen morphological classification, *Zingiber* species can be categorized into two sections: sect. *Zingiber* and sect. *Cryptanthium*. Sect. *Zingiber* is well known for possessing spheroidal and cerebroid exine sculpturing, similar to sect. *Dymczewiczia*, as suggested by Theilade et al. (1993), and this is also supported by phylogenetic analysis conducted by Theerakulpisut et al. (2012). In contrast, sect.

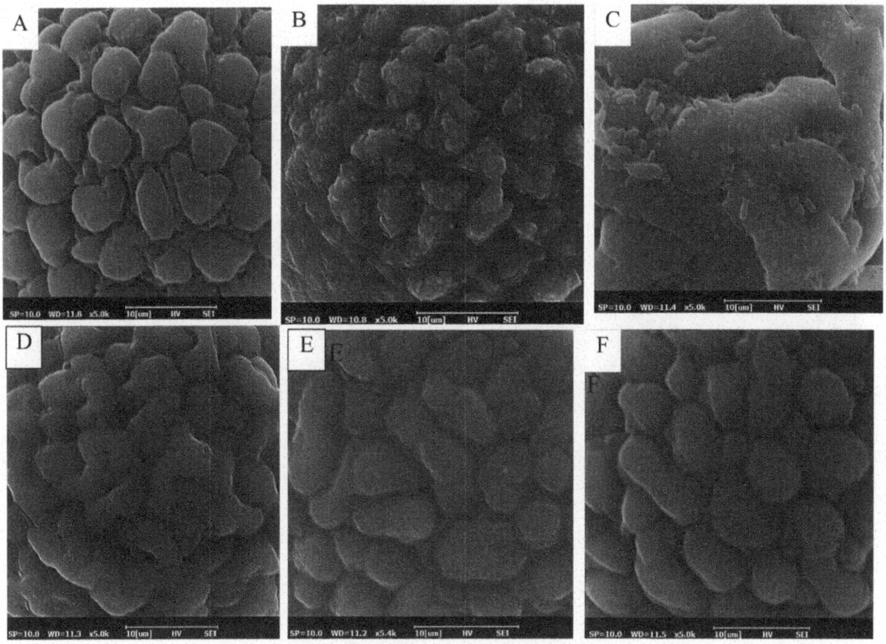

Fig. 3. Closed-up of exine sculpturing pollen of selected *Zingiber* in Peninsular Malaysia using SEM. Areolate as in *Z. flaviflorum* (A), *Zingiber* sp. (*bahangense*130) (E), *Zingiber* sp. 2 (F), Verrucate as in *Z. griffithii* (B), and Cerebroid as in *Z. raja* (C) and *Z.* aff. *belumense* (D)

Cryptanthium is distinguished by its unique pollen morphology, characterized by ellipsoidal and spiro-striate exine sculpturing. However, recent studies by Ye et al. (2015) and Jayakrishnan et al. (2021) revealed that the pollen of sect. *Cryptanthium* is not consistently ellipsoidal; it may vary from sub-spherical to ellipsoidal to ovate-ellipsoidal, but it consistently exhibits striate exine sculpturing.

The results from seven *Zingiber* species of Peninsular Malaysia revealed to challenge the previous classification. Regardless of whether the inflorescence has an erect or procumbent peduncle, the shape of the grains and the exine sculpture, does not correlate with or classified the species into separate sections. Although the exine sculptures vary slightly, ranging from areolate to verrucate to cerebroid, these differences are insufficient to support a distinct separation into subsections.

The primary discoveries suggest that the taxonomic classifications of the genus *Zingiber* should not rely solely on the type of the inflorescence peduncle. Instead, pollen morphology, supported by key morphological and molecular data, provides a more reliable basis for categorization. This hypothesis aligns with the findings of Odyuo et al. (2019), who observed that despite reporting procumbent inflorescences, the pollen of *Zingiber* species consistently displays an ellipsoidal shape with striate sculpturing, even when the inflorescence is notably erect. Similarly, Theerakulpisut et al. (2012) highlighted that some species within the *Dymczewiczia* section exhibit inconsistent inflorescence habits, producing both apical and erect inflorescences. These inconsistencies suggest that environmental factors may influence the habit, alongside genetic variations

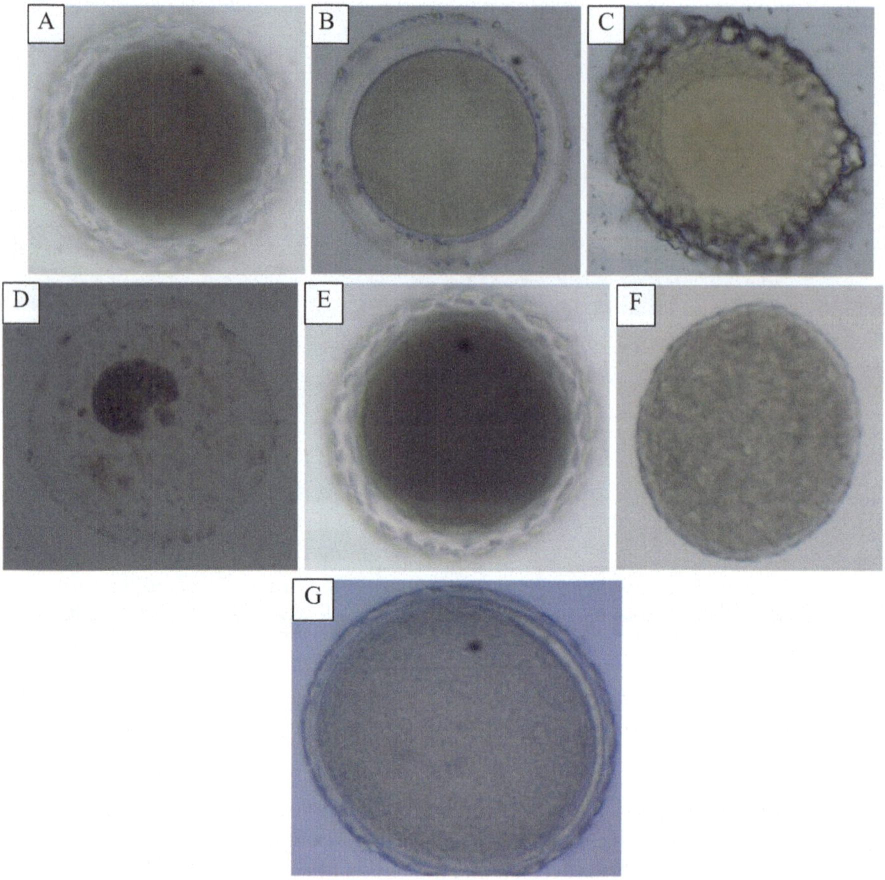

Fig. 4. Pollen images of the studied *Zingiber* using LM. A. *Z. flaviflorum*, B. *Z. griffithii,* C. *Z. raja*, D. *Z. spectabile*, E. *Z.* aff. *belumense*, F. *Zingiber* sp. (*bahangense*130), G. *Zingiber* sp. 2. Magnification A–E. 40x, F. 10x, G. 40

governing inflorescence expression (Theilade, 1999). Therefore, it is inappropriate to classify the genus *Zingiber* into distinct sections based solely on peduncle habit. Further research is ongoing, incorporating pollen samples from various localities across Peninsular Malaysia to better understand the taxonomic classification of the genus *Zingiber* in relation to pollen morphology.

5 Conclusion

In conclusion, preliminary investigation on the pollen grain shape and exine sculpture in the studied *Zingiber* from Peninsular Malaysia does not align with the traditional classification based on inflorescence peduncle habit. Further research, incorporating a broader range of species and localities, is essential to refine our understanding of pollen morphology to taxonomic significance.

Acknowledgement. This research was supported by Universiti Tun Hussein Malaysia (UTHM) through Research Enhancement-Graduate Grant (RE-GG) Vot No. Q201. The authors like to extend our appreciation to Dr. Qammil Muzzamill @ Meekiong Kalu and Datuk Seri Lim Chong Keat for their insights on *Zingiber* species. Additionally, we are grateful to the Jabatan Perhutanan Semenanjung Malaysia for granting the research permit and to the rangers for their assistance in the field.

References

Acma, F., Mendez, N.P.: Pollen morphology and pollen elemental composition of selected Philippine native gingers in tribe Alpinieae (Alpinioideae: Zingiberaceae). Biol. Forum **10**(1), 1–10 (2018)

Agashe, S.N., Coulton, E.: Pollen and Spores: Applications with Special Emphasis on Aerobiology and Allergy, 400 p. Science Publisher, Enfield (2009)

Ardiyani, M., Newman, M.F., Poulsen, A.D.: A new species of *Zingiber* (Zingiberaceae) east of Wallace's line. Gardens' Bull. Singapore **69**(2), 189–199 (2017)

Banaticla-Hilario, M.C.N., Altamirano, M.R.B.: Conservation of IUCN threatened Zingiberaceae species in tropical Asia: challenges, gaps, and opportunities. In: Ramamoorthy, S., Buot Jr., I.E., Rajasekaran, C. (eds.). Plant Diversity in Biocultural Landscapes, pp. 673–724. Springer (2023)

Erdtman, G.: Pollen morphology and plant taxonomy, angiosperm, p. 553. Almquist and Wiksell, Stockholm (1952)

Halbritter, H.: Palynology: history and systematic aspects. In: Illustrated Pollen Terminology. Springer, Cham (2018)

IPNI. International Plant Names Index., http://www.ipni.org. Accessed 11 Sept 2024

Jayakrishnan, T., Joe, A., Hareesh, V.S., Sabu, M.: Two new *Zingiber* (Zingiberaceae) species from Arunachal Pradesh, Northeastern India. Taiwania **66**(1), 101–112 (2021)

Kajornjit, P., Saensouk, S., Saensouk, P.: Pollen morphology and leaf anatomy of genus Globba in Thailand. Sci. Asia **44**, 146–161 (2018)

Liang, Y.H.: Pollen morphology of the family Zingiberaceae in China – Pollen types and their significance in the taxonomy. Acta Phytotaxonomica Sinica **26**, 265–281 (1988)

Mangaly, J.K., Nayar, J.: Palynology of south Indian Zingibearaceae. Bot. J. Linn. Soc. **103**, 351–366 (1990)

Mohamad, S., Kalu, M., Sedek, A.S.: Preliminary study of the pollen morphology of Malaysian Zingiberaceae (Tribe Alpinieae) and the taxonomic relationship. Reinwardtia **22**(2), 91–102 (2023)

Odyuo, N., Roy, D.K., Lyngwa, C., Mao, A.A.: *Zingiber perenense*, a new species in *Zingiber* section *Cryptanthium* (Zingiberaceae) from Nagaland, India. Thailand Nat. Hist. Mus. J. **13**(1), 1–10 (2019)

POWO. http://www.plantsoftheworldonline.org/. Accessed 11 Sept 2024

Punt, W., Hoen, P.P., Blackmore, S., Nilsson, S., Le Thomas, A.: Glossary of pollen and spore terminology. Rev. Palaeobot. Palynol. **143**, 1–81 (2007)

Royal Botanic Gardens. https://data.kew.org/?lang=en-US. Accessed 11 Sept 2024

Singapore Herbarium Online. https://herbaria.plants.ox.ac.uk/bol/sing. Accessed 11 Sept 2024

Sedek, A.S., Mohamad, S., Sam, Y.Y..: Diversity of *Zingiber* Mill. (Zingiberaceae) in peninsular malaysia including short remarks of an undescribed taxon. J. Trop. Biodivers. Biotechnol. **9**(3), jtbb93187 (2024)

Sedek, A.S., Mohamad, S., Abd-Aziz, N., Sarjuni, N.H., Hakim, A.M.A.: Diversity of wild gingers (Zingiberaceae) in southern Peninsular Malaysia: Panti forest reserve and Labis forest reserve. Pertanika J. Trop. Agric. Sci. **47**(1), 11–24 (2024a)

Theerakulpisut, P., Triboun, P., Mahakham, W., Maensiri, D., Khampila, J., Chantaranothai, P.: Phylogeny of the genus *Zingiber* (Zingiberaceae) based on nuclear ITS sequence data. Kew Bull. **67**, 389–395 (2012)

Theilade, I.: A synopsis of the genus *Zingiber* (Zingiberaceae) in Thailand. Nord. J. Bot. **19**, 389–410 (1999)

Theilade, I., Mærsk-Møller, M.L., Theilade, J., Larsen, K.: Pollen morphology and structure of Zingiber (Zingiberaceae). Grana **32**(6), 338–342 (1993)

Triboun, P., Chantaranothai, P., Larsen, K.: *Zingiber idae* a new name for *Z. villosum*, (Theilade) (Zingiberaceae). Thai For. Bull. (Bot.) **30**, 62–68 (2002)

Walker, J.W., Doyle, J.A.: The bases of angiosperm phylogeny: palynology. Ann. Mo. Bot. Gard. **62**, 664–723 (1975)

Ye, Y.S., Bai, L., Cia, N.H.: *Zingiber hainanense* (Zingiberaceae), a new species from Hainan, China. Phytotaxa **217**(1), 073–079 (2015)

Zou, P.U., Newman, M., Liao, J.: Systematics of Zingiberaceae. Grana **61**(6), 448–470 (2022)

Effect of Egg Shells Waste as Bio Fertilizer on Bird's Eye Chilli (*Capsicum Frutescens*) Growth and Yield Improvements

Pang Hui Shan[1] and Furzani Pa'ee[2(✉)]

[1] Faculty of Applied Sciences and Technology, Universiti Tun Hussein Onn Malaysia Pagoh, KM1, Jalan Panchor, 84600 Muar, Johor, Malaysia

[2] Advanced Herbal and Ethnomedical Research (AdHerb) Focus Group, Universiti Tun Hussein Onn Malaysia, KM1, Jln Panchor, 84600 Panchor, Johor, Malaysia

furzani@uthm.edu.my

Abstract. Eggshell wastes are one of the main solid wastes due to the consumption of eggs in daily diets and the general way of disposal for eggshell is in landfills that would lead to pollution and public health. To reduce negative impacts on the environment, eggshell wastes have been found to be used as a bio fertilizer in agriculture. In Malaysia, the local production of bird's eye chili, or called cili padi (*Capsicum frutescens*) in large scale is still relatively low even though it is frequently used as spice in daily cooking. In order to determine the best fertilization formula for eggshell wastes in order to improve plant growth and yield, this study examined the effects of varying amounts of eggshell powder, also known as ESP (0 g, 45 g, 60 g, 75 g, and 90 g), on bird's eye chilli using the soil's pH value, soil nutrients, morphological analysis, and metabolite analysis. Following ESP application, the pH of the soil remained acidic in all treatments. 90 g with 60.43% was the optimal ESP formulation in terms of growth rate for plant height. When it came to increasing the number of leaves, the 45 g ESP formulation yielded the best results (382.36%), while the 60 g ESP formulation was the most effective in growing the number of branches (379.17%) and nodes (80.14%). For flowering and fruit yield, the highest number was formulated by 75 g ESP, that was 8 and 6, respectively. The total chlorophyll content (30.723 mg/mL) and total phenolic content (0.414 ± 0.01 mg GAE/g extract) had the highest value when applying 75 g ESP. The optimum formulation of eggshell wastes as bio fertilizer differs due to its differential effects on individual plant growth parameters. The formulation of 90g ESP was the best formulation that mostly highlighted the improvement in most parameters such as the plant height and metabolite (chlorophyll and total phenolic content).

Keywords: Eggshell wastes · bio fertilizer · bird's eye chili · growth rate · formulation

P. Chang et al. (Eds.): SCIEMATHIC 2024, SPPHY 427, pp. 209–218, 2025.
https://doi.org/10.1007/978-981-96-7679-8_17

1 Introduction

In Malaysia, the eggshell wastes from household, food sector or growing egg cracking industries is estimated to generate 70,686 tonnes of waste per year based on the production rate from the total of egg production in 2012. [1] In general, eggshell wastes are always disposed of in landfills directly due to no economic value [2], however, unpleasant odours that would draw pests also raise the potential of environmental contamination and harm to the general public. [3] Eggshell has a significant calcium and nitrogen content, however its toxic chemical component is less abundant than natural calcium. [4]. Therefore, developing eggshell bio fertilizer [5] that can reduce the negative impact of chemical fertilizer use is very useful for plant growth and yield.

Based on the data statistics from Department of Statistics Malaysia, self-sufficiency ratio (SSR) described that the chili species obtained 30.9% only in 2020 while import dependency ratio (IDR) showed chili gained the highest percentage that was 72.4%. This data also indicated that there was not enough production to meet the domestic need for chili [6] and on the other hand, the import of chili is still higher than the local supply in Malaysia even though chili species are used as ingredients in daily cooking.

The chemical fertilizer is well-known as one of the agricultural ways to provide fertilization for all leafy vegetables to affect plant growth development, fruit yield, and quality. However, the excessive use of chemical fertilizer causes damage to the environment [7] and affects the accumulation of nitrate in the plant. Excessive usage of nitrogen fertilisers damages the surrounding ecology by increasing nitrate discharge [26]. The alternative strategy to replace the chemical fertilizer is applying bio fertilizer which is made up of renewable resources. Bio fertilizer enhances nitrogen fixation, phosphorus solubilization, and stimulating and synthesizing plant growth [8].

The objectives of this study were to evaluate and study the effect of different concentration of eggshell powder on pH, and morphological growth parameters (morphological analysis), and biochemical parameter (metabolite analysis) of bird's eye chili (cili padi). The optimum fertilization formula of eggshell wastes in plant growth and yield improvement was determined from the result of morphological analysis and metabolite analysis.

2 Material and Methods

2.1 Sample Collection

The eggshell wastes were collected from home or restaurant nearby residential area or hostel. The washing, and boiling process within 10 min was conducted to remove bacterial and viral contact. [9] After drying and grinding the egg shell into a fine powder that was $\leq$250 μm in size [10], 1 kg of eggshell powder (ESP) was obtained for this investigation.

2.2 Preparation of Plant Treatments and Experimental Design

The experimental design was a Completely Randomized Design (CRD) with three replications. All the bird eye's chili seedlings treatments were located at the Herbs Garden

UTHM Pagoh Campus. The plant treatments were cultivated using the same type of soil (biochar) and a daily water application of 500 mL was provided to each plant. The plant seedlings treatments were divided into 5 groups with 3 replications followed by different concentration of ESP: 0 g (control), 45 g, 60 g, 75 g, and 90 g. [11].

2.3 Determination of Percentage of Calcium Carbonate (CaCO3)

By referring to Eq. 1 and Eq. 2, The percentage of calcium carbonate (CaCO3) was calculated and determined from the eggshells by back titration with 0.050 M sodium hydroxide (NaOH).

$$CaCO_3(s) + 2HCl(aq) \rightarrow CaCl_2(aq) + H_2O(l) + CO_2(g) \tag{1}$$

$$HCl(aq) + NaOH\,(aq) \rightarrow NaCl_2(aq) + H_2O(l) \tag{2}$$

2.4 pH Measurement on Soil for Different Concentration of Eggshell Powder (ESP)

Before applying ESP, pH value of soil was measured by using 4-in-1 digital soil analyzer (Model L1E14-L2D10). During week 1 after ESP application, pH value was measured again and continuously until week 8.

2.5 Morphological Analysis in Plant Growth and Yield

Similar to pH, morphological analysis involved weekly manual measurements of growth characteristics, such as plant height, number of leaves, number of branches, number of nodes, number of flowers, and number of fruits, from week 0 (the first week) to week 8.

Each group treatment's outcome data was entered into the table in triplicate, and the average was determined. In the final week (week 8), the fruits were gathered. The growth rate [12], which is often stated as a percentage that characterises the change over time by subtracting the initial value from the final value and dividing by the beginning value as indicated in Eq. 3, was used to determine all plant growth metrics (apart from flowers and fruits) [13].

$$The\,percentage\,of\,growth\,rate = \frac{Final\,value - Initial\,value}{Initial\,value} \times 100 \tag{3}$$

Plant Height. The plant height was measured from its base to its highest point by the ruler [14].

Number of Leaves. The leaves per plant which were visible and involved the newly emerged leaves on the tips of the plant was counted thoroughly [14].

Number of Branches. The branches per plant was counted radically [14].

Number of Nodes. The nodes that occurred at all of branches on whole plant were counted [14].

Number of Flowers. The flowers per plant were observed and counted virtually [14].

Number of Fruits. The fruits on each plant were counted by sight [14].

2.6 Biochemical Analysis on Plant Leaves and Fruit Yield

The chlorophyll content of leaves and total phenolic content of fruits were parameters used in this study for metabolite analysis.

Determination of Chlorophyll Content in Leaves. The chlorophyll content of leaves for each treatment was determined by acetone method [15] which was traditionally extracting the leaves sample by using 80% acetone and measuring the absorbance at 663 nm and 645 nm respectively using T60 UV/Vis Spectrophotometer. The absorbance was substituted into Eq. 4, Eq. 5, and Eq. 6. [16].

$$Chl'a' = 12.7(A_{663}) - 2.69(A_{645}) \tag{4}$$

$$Chl'b' = 22.9(A_{645}) - 4.68(A_{663}) \tag{5}$$

$$Total\ Chlorophyll\ (a + b) = 20.2(A_{645}) + 8.02(A_{663}) \tag{6}$$

where, A = Absorbance at wavelength.

Determination of Total Phenolic Content in Fruits. The fruit yield calculated from Eq. 7 in this study was used in Folin-Ciocalteu reagent method [17]. The gallic acid standard curve with 0.01, 0.02, 0.03, 0.04 and 0.05 mg/mL concentration [18] was used. The mixture for fruit ethanolic extracts followed by Folin-Ciocalteu reagent and sodium carbonate was measured by the wavelength 725 nm through T60 UV/Vis Spectrophotometer [19] The total phenolic content was calculated by substituting the regression equation from standard curve of gallic acid (Eq. 8) [18] into Eq. 8.

$$Extraction\, yield\,(\%) = \frac{W1}{W2} \times 100 \tag{7}$$

where, W1 = The mass of crude extract
W2 = The mass of the sample

$$y = 7.895x + 0.008, R^2 = 0.997 \tag{8}$$

$$Total\, phenolic\, content = c \times \frac{V}{m} \tag{9}$$

where, c (x-axis value) = The concentration of gallic acid established from the calibration curve in mg/mL
V = The volume of extract in mL
m = The weight of plant extract in g.

3 Results and Discussion

3.1 Percentage of Calcium Carbonate (CaCO3) in Egg Shell Powder (ESP)

The calcium carbonate level in ESP was calculated to be 79.65%, which is lower than the average range of 94% to 97% [3]. This mismatch could be related to incomplete decontamination during preparation, such as insufficient drying before grinding, resulting in partial dissolving in hydrochloric acid [3]. As illustrated in Table 1, the presence of calcium carbonate is critical for its role in nutrient supply, plant growth, and soil improvement.

3.2 Soil pH Value Before and After Applying Egg Shell Powder (ESP)

Weekly measurements of the pH of the soil were made from week 0 to week 8. The trends in soil pH for each treatment are shown in Table 2. It is noteworthy that after eight weeks, the administration of 90 g ESP (T4) produced the greatest pH value of 5.9, while the control group (T0) had a pH value of 5.2. Given that chillies thrive in a pH range of 5.5 to 7.0, this suggests that ESP has a moderate acid-neutralizing impact that is advantageous for chilli growth [20].

Table 1. Soil pH Value Before and After Eggshell Powder (ESP) Application

Treatment	pH value	
	Before	After
T0	5.5	5.2
T1	5.4	5.5
T2	5.4	5.5
T3	5.7	5.7
T4	5.6	5.9

3.3 Morphological Plant Growth and Fruit Yield

According to Table 3, 90 g ESP (T4) had the fastest growth rate for plant height, increasing by 60.43% over the course of eight weeks. This shows the beneficial effects of increased calcium and nutrition availability in ESP and is a notable improvement above the control (28.17%).

Table 2. Plant Height Growth Rate for 8 Weeks

Treatment	Week 0	Change (cm)	Growth rate (%)
T0	24.67	6.95	28.17
T1	19.53	11.73	60.06
T2	20.17	7.35	36.44
T3	17.90	7.33	40.95
T4	16.20	9.79	60.43

With the maximum leaf growth rate of 382.36%, as shown in Table 3, the 45 g ESP treatment (T1) suggested that lower concentrations would maximise vegetative growth by preventing nutrient overload.

The most effective way to improve both metrics of number of branches and nodes was with 60 g ESP (T2), which increased the number of branches by 379.17% and the

Table 3. Number of Leaves on Week 0 and Leaves Growth Rate for 8 Weeks

Treatment	Week 0	Change (cm)	Growth rate (%)
T0	30.33	42.46	139.99
T1	19.67	75.21	382.36
T2	31.00	56.66	182.77
T3	20.00	46.81	234.05
T4	30.00	55.13	183.77

number of nodes by 80.14% (Tables 4 and 5). This implies that plant development and nutrient delivery are balanced by intermediate ESP concentrations [21].

Table 4. Number of Branches on Week 0 and Leaves Growth Rate for 8 Weeks

Treatment	Week 0	Change (cm)	Growth rate (%)
T0	14.33	32.21	224.77
T1	15.00	43.88	292.53
T2	12.00	45.50	379.17
T3	13.00	36.13	277.92
T4	13.00	46.00	353.85

Table 5. Number of Nodes on Week 0 and Leaves Growth Rate for 8 Weeks

Treatment	Week 0	Change (cm)	Growth rate (%)
T0	8.00	26.13	69.38
T1	8.00	30.75	73.98
T2	7.00	35.25	80.14
T3	8.00	27.63	71.05
T4	7.00	30.63	77.14

The number of the flowers seen in 8 weeks was displayed in Table 6. When compared to other treatments within 8 weeks, T3, which was treated with 75 g of ESP, produced the most flowers making it the best formulation for flowering.

The number of fruits observed within 8 weeks was recorded in Table 7. For number of fruits, T3 that applied with 75 g ESP was the best formulation compared to other treatments within 8 weeks, with 6 fruits.

Table 6. Number of Flower for 8 Weeks

Treatment	T0	T1	T2	T3	T4
Week 0	0	0	0	0	0
Week 1	0	0	0	0	0
Week 2	0	0	0	0	0
Week 3	0	0	0	0	0
Week 4	0	0	0	0	0
Week 5	0	3	1	8	0
Week 6	0	1	1	7	1
Week 7	2	2	1	1	2
Week 8	4	7	0	0	0

Table 7. Number of Fruits for 8 Weeks

Treatment	T0	T1	T2	T3	T4
Week 0	0	0	0	0	0
Week 1	0	0	0	0	0
Week 2	0	0	0	0	0
Week 3	0	0	0	0	0
Week 4	0	0	0	0	0
Week 5	0	0	0	0	0
Week 6	0	0	0	4	0
Week 7	0	0	0	6	1
Week 8	0	0	0	6	4

3.4 Chlorophyll Content in Leaves

90g ESP (T4) produced the highest levels of chlorophyll "a" (19.121 μg/mL), chlorophyll "b" (16.962 μg/mL), and total chlorophyll content (8.200 μg/mL), according to the chlorophyll content analysis (Table 8). Because calcium and phosphorus stabilize chlorophyll-protein complexes and improve photosynthetic efficiency, this increase can be explained [22].

Calcium from ESP promotes photosynthesis by stabilizing chlorophyll-protein complexes and activating photosystem enzymes. Furthermore, phosphorus in ESP helps ATP production, which meets the energy requirements of chlorophyll synthesis [22, 24].

Table 8. Chlorophyll a, Chlorophyll b And Total Chlorophyll (a + b) of Leaves with Different Concentration of Eggshell Powder (ESP)

Treatment	Chl a (μg/mL)	Chl b (μg/mL)	Total chlorophyll (a + b) (μg/mL)
T0	16.030	11.761	4.798
T1	20.692	17.414	8.078
T2	15.187	11.305	4.682
T3	31.046	19.484	6.509
T4	19.121	16.962	8.200

3.5 Total Phenolic Content in Fruits

90 g ESP (T4) produced the highest total phenolic content (0.414 mg GAE/g extract), according to the phenolic content analysis (Table 9). The 75 g ESP treatment (T3), on the other hand, resulted in a decreased phenolic content but a higher fruit production. This implies that nutrient allocation within the plant drives a trade-off between yield amount and phenolic component concentration.

ESP promotes phenolic biosynthesis by supplying calcium and other micronutrients that activate key enzymes like phenylalanine ammonia-lyase (PAL), which is involved in the phenylpropanoid pathway leading to phenolic compound production [17, 25].

Table 9. Fruit Yield and Total Phenolic Content of Fruit Extract

Sample extract	Percentage of yield (%)	Total phenolic content (mg GAE/g extract)
T3	2.89	0.291 ± 0.01
T4	2.46	0.414 ± 0.01

From the results, the best eggshell powder (ESP) fertilization formulation for plant height improvement was treated with 90 g ESP, while 45 g ESP formulation would improve the growth rate of leaves. Both the branches and nodes would be increased by 60 g ESP. The stimulation of flowering rate and fruit yield were formulated by 75 g ESP. The formulation of 90 g ESP was performed the best in determining the chlorophyll content of leaves and the total phenolic content of chili fruits. The level of concentration for ESP had different effects on different parts of chili plant. The concentration of ESP may affect the level of nutrients in the soil. A little amount of fertilizer will react to plant growing poorly. The plant will grow faster when applying more fertilizer. However, excessive use of fertilizer will slow down the plant development and increase the risk of suffering root damage or even dying. [23] Therefore, the optimum fertilization formula is important to improve and maintain the development of plant growth. From the data, it is suggested that the 90 g eggshell powder (ESP) formulation was the most effective, promoting significant improvements in plant height, chlorophyll content in leaves, and the total phenolic content of chili fruits. This formulation also led to enhanced flowering rate and fruit yield, making it the best choice for overall plant growth and productivity.

Aside from its nutrient value, using eggshell powder (ESP) as a biofertilizer has a number of environmental advantages. First, ESP helps with waste management by repurposing biowaste that would otherwise end up in landfills, hence lowering waste accumulation and greenhouse gas emissions [7]. This corresponds with the circular economy concepts by converting waste into a resource.

Second, ESP reduces the environmental dangers associated with chemical fertilisers, such as nitrate leaking into groundwater and soil acidification, which are key concerns in intensive agriculture [22]. These effects are especially harmful to aquatic habitats, where nutrient discharge can induce eutrophication. Furthermore, ESP improves soil health by increasing microbial activity and organic matter content, resulting in long-term soil fertility. Unlike chemical fertilisers, ESP is less likely to degrade soil or cause salinisation, assuring the long-term use of agricultural land.

These environmental benefits highlight ESP's potential as a sustainable option in agriculture, lowering reliance on synthetic inputs while addressing environmental issues such as pollution and resource waste.

4 Conclusion

The expected outcome in this study could be described that the plant growth and yield would be improved when the concentration of eggshell powder (ESP) increases. By using different plant growth parameters (morphological analysis), and metabolite analysis for leaves and fruit yield, different concentration of eggshell powder (ESP) as a type of bio fertilizer had different effect on different part of the bird's eye chili plant. The optimum fertilization formulation for eggshell wastes was determined for each plant part by investigating the growth rate of the plant within the time. This study can be a reference for further studies which are related to eggshell wastes development.

References

1. Rohim, R., et al.: Characterization of calcium oxide catalyst from eggshell waste (2014). Adv. Environ. Biol. **2014**(8), 35–38 (2014)
2. Razali, N., et al.: Thermal decomposition of calcium carbonate in chicken eggshells: study on temperature and contact time. Malays. J. Anal. Sci. **26**(2), 347–359 (2022)
3. Ahmed, T.A.E., et al.: Biotechnological applications of eggshell: recent advances. Front. Bioeng. Biotechnol. **9** (2021). https://doi.org/10.3389/fbioe.2021.675364
4. Oliveira, D., et al.: A literature review on adding value to solid residues: egg shells. J. Clean. Prod. **46**, 42–47 (2013). https://doi.org/10.1016/j.jclepro.2012.09.045
5. Laohavisuti, N., et al.: Simple recycling of biowaste eggshells to various calcium phosphates for specific industries. Sci. Rep. **11**(1) (2021). https://doi.org/10.1038/s41598-021-94643-1
6. Razak, M.Y.A.: Supply and utilization accounts selected agricultural commodities, Malaysia 2016–2020. Department of Statistics Malaysia Office Portal (2021). https://www.dosm.gov.my/v1/index.php?r=column/cthemeByCat. Accessed 30 June 2022
7. Al-fehaid, S.A.S., et al.: Effect of chemical, organic and bio fertilizers on growth, yield and quality of cucumber plant (Cucumis sativus L.) grown under greenhouse conditions. J. Sohag Agriscience (JSAS) **7**(1), 28–40 (2022)

8. Sabier Sae, K., et al.: Effect of bio-fertilizer and chemical fertilizer on growth and yield in cucumber (Cucumis sativus) in green house condition. Pak. J. Biol. Sci. **18**(3), 129–134 (2015). https://doi.org/10.3923/pjbs.2015.129.134
9. Khairnar, M.D., Nair, S.S.: Study on eggshell and fruit peels as a fertilizer. Int. J. Innov. Eng. Res. Technol. 1–3 (2021). https://repo.ijiert.org/index.php/ijiert/article/view/2032. Accessed 02 July 2022
10. Islam, M.K., et al.: Human uptake of eggshell powder as an alternate source of calcium. Dhaka Univ. J. Pharmac. Sci. **18**(2), 249–255 (2019). https://doi.org/10.3329/dujps.v18i2.44465
11. Anugrah, R.D., et al.: The effect of eggshell organic fertilizer on vegetative growth of cayenne pepper (Capsicum frutescens L). Earth Environ. Sci. **755**(1), 012001 (2021). https://doi.org/10.1088/1755-1315/755/1/012001
12. Dambreville, A., et al.: Analysing growth and development of plants jointly using developmental growth stages. Ann. Bot. **115**(1), 93–105 (2014). https://doi.org/10.1093/aob/mcu227
13. Ali, M.A., et al.: Source-sink relationship between photosynthetic organs and grain yield attributes during grain filling stage in spring wheat (triticum aestivum). Int. J. Agric. Biol. **12**, 509–515 (2010)
14. Heady, H.F.: The measurement and value of plant height in the study of herbaceous vegetation. Ecology **38**(2), 313–320 (1957)
15. Rajalakshmi, K., Banu, N.: Extraction and estimation of chlorophyll from medicinalplants. Int. J. Sci. Res. **4**(11), 209–212 (2015)
16. Pérez-Patricio, M., et al.: Optical method for estimating the chlorophyll contents in plant leaves. Sens. (Basel Switz.) **18**(2), 650 (2018). https://doi.org/10.3390/s18020650
17. Ngamkhae, N., et al.: Optimization of extraction method for Kleeb Bua Daeng formula and comparison between ultrasound-assisted and microwave-assisted extraction. J. Appl. Res. Med. Aromat. Plants **28**, 100369 (2022). https://doi.org/10.1016/j.jarmap.2022.100369
18. Azlim Almey, A., et al.: Total phenolic content and primary antioxidant activity of methanolic and ethanolic extracts of aromatic plants' leaves. Int. Food Res. J. **17**(4) (2010)
19. Johari, M.A., Khong, H.Y.: Total phenolic content and antioxidant and antibacterial activities of Pereskia bleo. Adv. Pharmacol. Sci. **2019**, 1–4 (2019). https://doi.org/10.1155/2019/7428593
20. Chatterjee, R., et al.: Quality bird's eye chili production: a retrospective. Int. J. Bio-Resour. Stress Manage. **3**(3), 412–414 (2012)
21. Zanden, A.M.V.D.: Vegetative plant parts (2008). https://extension.oregonstate.edu/node/115961/printable/print. Accessed 10 Jan 2023
22. Siedliska, A., et al.: Identification of plant leaf phosphorus content at different growth stages based on hyperspectral reflectance. BMC Plant Biol. **21**(1) (2021). https://doi.org/10.1186/s12870-020-02807-4
23. Buechel, T.: Relationship between fertilizer & plant stretching | PRO-MIX (2022). https://www.pthorticulture.com/en/training-center/relationship-between-fertilizer-plant-stretching/. Accessed 13 Jan 2022
24. Caires, E.F., et al.: Soil acidity management and crop yields in a no-till system. Agron. J. **113**(5), 4196–4209 (2021). https://doi.org/10.1002/agj2.20847
25. Medeiros, J.R.A., et al.: Application of calcium-rich biowaste as an amendment for agricultural soil: effects on soil properties and plant growth. J. Environ. Manage. **276**, 111349 (2020). https://doi.org/10.1016/j.jenvman.2020.111349
26. Zhang, X., Davidson, E., Mauzerall, D., et al.: Managing nitrogen for sustainable development. Nature **528**, 51–59 (2015). https://doi.org/10.1038/nature15743

Conservation Initiatives and Strategies for the Milky Stork: Case Studies from Southeast Asia

Maizatul Leeqa Razalee[1], Nor Atiqah Norazlimi[1,2(✉)], and Kamarul Rahim Kamarudin[1,3]

[1] Department of Technology and Natural Resources, Faculty of Applied Sciences and Technology, Universiti Tun Hussein Onn Malaysia (Pagoh Campus), Pagoh Higher Education Hub, KM 1, Jalan Panchor, 84600 Muar, Johor, Malaysia
atiqah@uthm.edu.my
[2] Environmental Management and Conservation Research Unit, Faculty of Applied Science and Technology, Universiti Tun Hussein Onn Malaysia (Pagoh Campus), Pagoh Higher Education Hub, KM 1, Jalan Panchor, 84600 Muar, Johor, Malaysia
[3] Centre of Research for Sustainable Uses of Natural Resources (SUNR), Faculty of Applied Sciences and Technology (FAST), Universiti Tun Hussein Onn Malaysia (Pagoh Campus), Pagoh Higher Education Hub, KM 1, Jalan Panchor, 84600 Muar, Johor, Malaysia

Abstract. Native to Southeast Asia, the Milky Stork (*Mycteria cinerea*) is an endangered species experiencing severe losses due to habitat loss, poaching, and environmental degradation. This review examines the ongoing conservation initiatives to preserve and increase Milky Stork populations in Southeast Asian nations. Essential tactics include community engagement projects, captive breeding and reintroduction programs, anti-poaching efforts, and habitat protection and restoration. Recent research has shed important light on population patterns, habitat utilization, and species' ecological needs, emphasizing crucial areas that must be addressed. However, many questions still need to be answered, especially regarding long-term monitoring and including socioeconomic factors in conservation planning. Addressing these gaps is crucial for thriving effective, sustainable conservation programs. This study emphasizes the significance of comprehensive strategies that combine economic approaches with ecological research, international collaboration, and local community involvement to ensure the survival of the Milky Stork. To conclude, this review fills the gaps by acknowledging the conservation effort to increase the population of Milky Stork.

Keywords: Conservation Initiatives · Habitat Protection · Milky Stork · Southeast Asia Countries · Threats

1 Introduction

This paper reviews the conservation efforts that Southeast Asia countries have made in conserving Milky Stork (*Mycteria cinerea*). This study examined 20 selected papers from 1987 to 2023, focusing on Milky Stork conservation efforts in Southeast Asia Countries.

P. Chang et al. (Eds.): SCIEMATHIC 2024, SPPHY 427, pp. 219–236, 2025.
https://doi.org/10.1007/978-981-96-7679-8_18

Conserving the Milky Stork is crucial as it is classified as endangered according to the International Union for Conservation of Nature (IUCN) Red List of Threatened Species, with an estimated global population of 1,500 individuals.

Milky Stork is one of the larger waterbirds in the Ciconiidae family which can be found throughout Southeast Asia's tropical wetland environments, including countries such as Cambodia, Indonesia, and Malaysia (Kaminsin *et al.*, 2023). Despite its wide distribution, currently, this species is thought to be highly at risk of extinction in the wild.

According to Jeyarajasingam and Pearson (2012), stork birds are long-legged, wading birds with long necks, large wings, and long pointed beaks. Approximately, 20 species of stork (Fig. 1) can be found worldwide, most of which occupy temperate or tropical habitats (Gula *et al.*, 2023). In terms of morphology, Milky Stork appeared almost white plumage exteriorly (Fig. 2) (Li *et al.*, 2006; Shepherd & Giyanto, 2009; Ismail & Rahman, 2012).

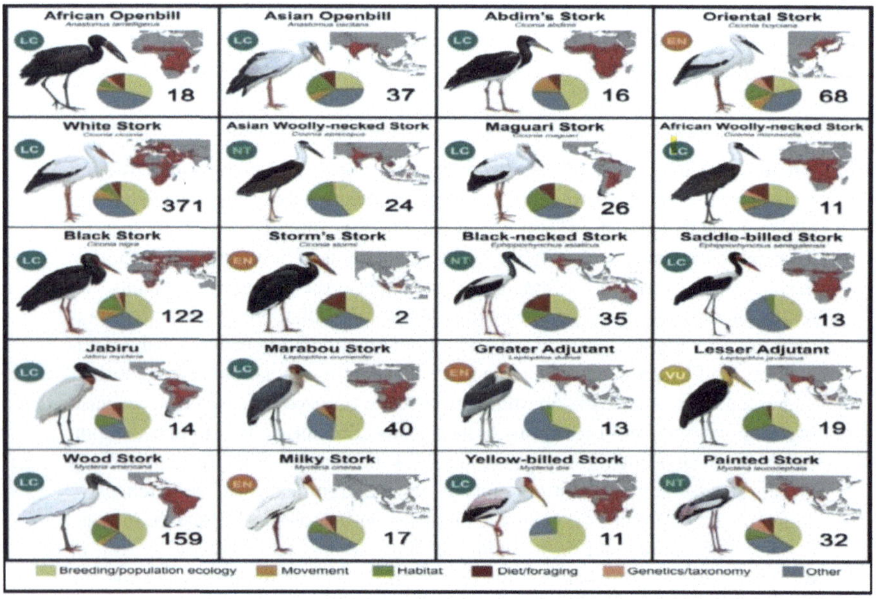

Fig. 1. Shows 20 species of Stork Birds in worldwide (Gula *et al.*, 2023)

Fig. 2. Shows one individual of Milky Stork

Storks are considered waterbirds and play an essential role in this ecosystem. Aquatic areas such as mangroves, benthic zones, and fish communities are primarily consumed by top predators, like waterbirds (Rahman *et al.*, 2019). Waterbirds are an essential bio-indicator of ecosystem change, as they demonstrate noticeable and meaningful responses to their surroundings (Rahman & Ismail, 2018). Besides, their sensitivity to environmental changes may be reflected in their movement or migration activities, making them an excellent bio-indicator of ecosystem health (Rahman & Ismail, 2018).

This stork was classified as an endangered species because their numbers are fewer than those classified as threatened species (Neel *et al.*, 2012). It exemplifies a species whose population has become endangered due to coastal development and deforestation (Ismail & Rahman, 2012; BirdLife International, 2019). Some conservation initiatives have been implemented to address this decline, such as habitat restoration, captive breeding, and releasing the storks into appropriate environments. The International Union for Conservation of Nature/Species Survival Commission (IUCN/SSC 2013) defines reintroduction as the deliberate relocation and release of an organism within its native range that has vanished due to human activity or natural disasters.

Due to the decreasing number of individuals, some conservation measures have already been taken, such as a reintroduction program at Kuala Gula, Malaysia. This site is considered one of the significant birding destinations that serve as a stopover site for migrating shorebirds traveling along the East Asian Australian corridor. As a result, it is essential to the survival of Milky Stork and birds migrating along the northern half of the peninsula (Rahman *et al.*, 2017). Most individuals seen in Malaysia today were likely released during earlier reintroduction initiatives at the Kuala Gula Bird Sanctuary and the Kuala Selangor Nature Park (Rahman *et al.*, 2020).

1.1 Historical Population Trend

The historical population trend of the Milky Stork has shown declining in this species; since the 1980s, Malaysia's wild population has rapidly decreased (Li *et al.*, 2006). However, the release program for Milky Stork from captive breeding has helped boost their numbers in the wild (Faiq *et al.*, 2016). The captive population of Milky Storks can be expanded through captive breeding if appropriate planning is carried out based on their breeding activity in captivity. However, insufficient information remains regarding the successful reintroduction of the species into the wild (Ismail *et al.*, 2011).

According to reports from Li *et al.* (2006), Ismail *et al.* (2011), and Ismail and Rahman (2016), the number of milky storks traveling in the wild in southern peninsular Malaysia and Singapore may be as low as 100–150. This represents roughly 7–10% of the total milky stork population worldwide. With approximately 1500 individuals of this species remaining in the wild, the high prevalence of hybrids is a cause for concern (Birdlife International, 2018).

Over its entire range, the species is declining due to habitat destruction, food hunting, and capture for trade. Its current state in Indonesia is uncertain, although reports suggest that while healthy numbers are still found in north and south Sumatra parts, populations have significantly decreased (Iqbal & Hasudungan, 2008; Shepherd & Giyanto, 2009). In Sumatra, 497 Milky Storks were counted during the 2008–2009 survey. According to Iqbal (2012), during the study period, South Sumatra recorded the highest count of 322 individuals, while Aceh had the lowest population of Milky storks, with just 11 birds. Additionally, with around 100–150 birds remaining in Cambodia and Peninsular Malaysia, and no breeding colonies reported in Thailand or Vietnam, the mainland population of the Milky Stork is currently small.

1.2 Previous Conservation Efforts by Southeast Asia Countries

The conservation efforts in Malaysia began with captive breeding programs initiated in 1987 at Zoo Negara (Ismail & Rahman, 2012). These efforts eventually saw success, as noted by Ismail and Rahman (2016). The first reintroduction program was launched at Kuala Selangor Nature Park (KSNP) in 1998. However, the program at KSNP was unsuccessful, as the stork population made limited nesting attempts, none of which resulted in success. Additionally, financial constraints led to the termination of the program in 2005. Despite this, the reintroduction efforts resumed in 2007 at Kuala Gula, Perak, where the program proved successful, with storks successfully breeding in the area.

Many species of waterbirds, including Milky Stork, have used Pulau Rambut, a small island in Jakarta Bay, Indonesia, as a nesting and roosting area (Firdausy, 2021). Research conducted by Firdausy *et al.*, (2021) on the abundance of waterbirds and the distribution of trees nesting in Pulau Rambut Wildlife Sanctuary found that 78 individuals of Milky Stork inhabit this small island.

In 1998, Milky Stork was also bred in captivity at Bangkok's Dusit Zoo (DZ), Thailand. Nineteen founders were raised alongside its sister taxon, the near-threatened Painted Stork (*Mycteria leucocephala*), for which captivity efforts had already begun in 1997 with an initial 70 founding individuals (Kaminsin *et al.*, 2023).

In Cambodia, according to Van Zalinge *et al.*, (2008), small populations of Milky Storks are limited to the coastal wetlands and floodplain of the Tonle Sap. In northern Siem Reap province, solitary birds are frequently seen in Ang Trapeang Thmor, often in the company of Painted Storks. Most coastal records come from Prek Teuk Sap mudflats in Ream National Park. Additional reports of single or paired Milky Storks have been made in coastal locations such as Sre Ambel (Koh Kong) and Kompong Smach (Sihanoukville). The Tonle Sap habitat is significant for a diverse range of birds due to its productivity as a wetland and the enhanced availability of food during the dry season, which make it an abundant feeding ground.

1.3 Threats to Milky Stork

Numerous factors threaten waterbird populations, including pollution, habitat loss, human activity, climate change, and environmental changes. As urbanization continues to increase, many ecosystems are being destroyed for development to match human demand and the rising population. The Southeast Asian coast, mainly, has experienced intense growth and urbanization (Wong, 1998; Ramesh *et al.*, 2011). Human activities, such as industrial and agricultural development, also pose significant threats to waterbirds, as changes in air temperature, total daily precipitation, and water levels all impact the quality of surface waterways in a broad sense; all of these variables are the outcome of climate change and global warming (Puchlik *et al.*, 2021).

The mangrove ecosystem, a key habitat for waterbirds, is one of the most affected ecosystems. Many wildlife species, including the Milky Stork, are declining rapidly and becoming endangered due to the loss of mudflats and mangroves along the coast (Rahman *et al.*, 2020). According to Abd Rahman and Nazri (2022), the global Milky Stork population has been rapidly declining, with an estimated number of 5500 individuals remaining in Matang Mangrove, Malaysia. Population growth in the area has intensified the hunting pressures, as locals seek Milky Storks' eggs, chicks, and adults for food, trading them in the marketplace. In Indonesia, the Milky Stork population at Kumpai Lake is believed to have declined by 80%, dropping from 500 birds in 2005 to an estimated 100 birds in 2008 due to the hunting of chicks from breeding colonies (Iqbal *et al.*, 2012).

Ismail and Rahman's (2016) case study on the species' reproductive activity, revealed that nest desertion happened in up to 60% of cases due to human disruptions and involvement, such as noise, hunting, deforestation in the neighboring mangroves, etc. Other variables, such as poor nest building, predation, severe weather, etc., accounted for the remaining 40%.

Other than manufactured problems and nature problems mentioned above, the nesting behaviour of the Milky Stork is also disrupted by the hybridization mating between the Milky Stork and the Painted Stork. A recent study recorded multiple instances of the Painted Stork mating with the Milky Stork (Yee *et al.*, 2013; Baveja *et al.*, 2019). The hybrid species may face several challenges because behavioural, anatomical, and genetic distinctions characterize species, and these traits frequently make it difficult for different species to mate or reproduce successfully.

When hybridization does occur, resulting hybrids may experience genetic incompatibilities, variations in reproductive habits, and other issues. This process, known as genetic swamping, is a major threat to highly vulnerable species with limited populations, as it can diminish or eventually erase the genetic heritage of pure species (Kamisin *et al.*, 2023). As shown in Fig. 3(a), the Painted Stork can be distinguished from the Milky Stork based on its morphology, including a black pectoral band and a pink flush on the inner wing coverts and tertials. According to Li *et al.*, (2006), some of these physical traits are mixed in reported hybrid storks, as depicted in Fig. 3(b).

Fig. 3. Shows Painted Stork (*M. leucocephala*) and Hybrid Stork

2 Methodology

The Google Scholar databases were utilized to search the available literature by topic from 1987 to 2023 for this review study. We used the set of terms to look at the idea for this review paper, which is (i) Conservation of the Milky Stork and (ii) Population trends of the Milky Stork. The keywords usually used to review this paper are "Milky Stork" And "Conservation efforts". 20 selected publications were then subjected to an analysis utilizing the Critical Thinking approach, which resulted in a table outlining the key findings, sample region, methodology, strengths and weaknesses of the studies. (Table 1).

Table 1. The list of concepts found in the selected papers.

Paper/year	Sampling areas	Methods	Main findings	Strength	Weakness
Baveja *et al.*, (2019)	Jurong Bird Park, Singapore	Tissue samples	Despite the majority of sampled individuals carrying the signature of varying degrees of introgression from Painted Storks, there are storks with a relatively high proportion of Milky ancestry in the community, which may be pure—furthermore, a group of morphological characteristics closely matched hybridization levels found by genetic methods	Tissue samples are provided. The high-quality DNA, RNA, and protein make comprehensive genetic, genomic, and proteomic investigations possible. This makes it possible for researchers to investigate gene expression, molecular pathways, and genetic diversity	Taking tissue samples from endangered animals presents ethical questions. It is imperative to follow moral rules and obtain informed permission
Faiq *et al.*, (2016)	Kuala Gula Bird Sanctuary, Perak, Malaysia	Monitoring programme	According to estimations, female *M. cinerea* had a greater chance of surviving (30.0%) than males (16.7%). The survival rate of individuals less than a year was roughly 50%; in one to two years, it was 25%; in two to three years, it was 9.1%; and in more than three years, it was 0%	To sustain biodiversity, habitat management, and restoration initiatives might be guided by data on bird populations	Since accessible places are frequently the focus of monitoring, data from distant or challenging-to-reach habitats may be absent
Iqbal *et al.*, (2008)	Kumpai lake, Indonesia	Survey	Discovered 31 nests of Milky Storks and 65–75 adult birds in the accessible areas. There are thought to be between 100 and 115 nests at the location Exploiting eggs and chicks was recognized as a significant threat to the continued existence of Milky Stork at Kumpai Lake	Standardized survey procedures guarantee study replication, enabling trend analysis and long-term observation	Usually, surveys are carried out at particular locations and times, meaning that birds present outside of those times or in unmonitored areas may not be included

(continued)

Table 1. (*continued*)

Paper/year	Sampling areas	Methods	Main findings	Strength	Weakness
Iqbal *et al.*, (2012)	East Coast of South Sumatra, Indonesia	Field Visits	These findings indicate that the number of Milky Storks in South Sumatra has dropped significantly, roughly 70% over 22 years, from 1986 to 2008	By seeing birds up close in their native habitats, researchers can gain first-hand knowledge of their social interactions, behaviours, and ecological roles through trips to the field	There are time and location constraints on field visits. Because researchers can only stay in one location at a time and for short periods, researchers risk missing significant behaviours or events outside of the observation period
Iqbal & Hasudungan, (2008)	South Sumatra, Indonesia	Observations	On August 2, 2005, 500 individuals in Timbul Jaya village had the highest concentration of Milky Storks, followed by 324 birds on June 17, 2002, on the Banyuasin Peninsula	Individual observation methods facilitate long-term monitoring of individual birds or populations. Researchers can see patterns in behaviour, migration, reproduction, and other aspects of the lives of individuals by following them over time	Researchers must devote much time and energy to studying and recording individual birds when using individual observation methods, which can be laborious. This may restrict the variety or size of research, especially for elusive or low-density species
Iqbal, (2012)	East coastal of Sumatra, Indonesia	Interview and surveys	The biggest threats to the species' existence are habitat loss, hunting, and human disturbance. It was determined that the main threat to the breeding colonies' continuous survival was the exploitation of eggs and chicks	Through in-depth data collecting through interviews, a thorough knowledge of the responder's thoughts, feelings, and experienes becomes possible	Analyzing qualitative data from interviews can be difficult and time-consuming; correctly interpreting the respondents' answers demands specialized research skills
Ismail *et al.*, (2012)	Zoo Negara, Malaysia	Bio-logging techniques	Using the bio-logging technique, the study's precise regular activities of the endangered Milky Stork were recognized, along with several ecological characteristics. This approach can be repeated in the wild with minor modifications to observe the behaviour of the birds and get ecological data on their environment	This method can give real-time monitoring. In certain circumstances, bio-loggers can provide researchers with access to real-time data, which helps study the effects of the environment or take prompt action to conserve birds by enabling researchers to track their activity and movements	It is easier to attach equipment to birds correctly by hurting or distressing them. It is essential to consider the attachment method (e.g., adhesive, harness) to avoid interfering with the bird's survival or behaviour

(*continued*)

Table 1. *(continued)*

Paper/year	Sampling areas	Methods	Main findings	Strength	Weakness
Ismail & Rahman, (2011)	Malaysian National Zoo	Historical Data Review	1987 is the first successful nesting of the program. The number of species kept in captivity has increased significantly since then because of improved breeding and handling practices. The zoo's Milky Stork population grew from 10 to over 100 individuals by 2005	It makes it possible to find long-term cycles, patterns, and trends not seen in research conducted over shorter periods	Older gathered using antiquated or imprecise techniques and technologies
Ismail & Rahman, (2012)	Matang Mangrove Forest, Perak, Malaysia	The researchers' involvement and the limited amount of literature review available	Numerous writers have emphasized that, in addition to ecological (such as habitat changes, the presence of natural predators, etc.) and human factors (such as forestry and fishing activities and illegal hunting), one of the most significant problems in the conservation of wildlife, including coastal birds, is the pollution of hazardous chemicals	Reducing bias through limited exposure to previous studies can enable researchers to generate new ideas and techniques	Researchers risk missing crucial background knowledge, hypotheses, and prior discoveries pertinent to their work if they do not do a thorough literature review
Kaminsin *et al.*, (2023)	Dusit Zoo and Nakhon Ratchasima Zoo, Thailand	Sampling and DNA extraction	The morphology of most hybrids can be categorized by the following characteristics: certain pinkish tertial feathers, ambiguous black and white patterns on the wings, and a lack of a clear black breast band	Genetic data can be integrated with other data, such as morphological and ecological, to gain a more thorough understanding of avian biology and ecology	There are ethical concerns with DNA extraction and sampling, especially when working with endangered animals. Getting the required licenses and ensuring minor damage are necessary, which can be difficult
Mardiastuti *et al.*, (2023)	Pulau Rambut, Indonesia	Syma X25 pro drone	The research on Milky Stork nesting behaviour using a small, manually operated drone did not cause any disturbance to the bird. There was no indication of any disturbance when the drone approached the parents directly above them at a distance of up to 2 m	Drones may explore inaccessible or challenging locations like islands, marshes, and wetlands frequently inhabited by waterbirds	Drones are less intrusive than humans, but they might still annoy waterbirds because of their loudness and presence in the air, which may change their natural behaviour

(continued)

Table 1. (*continued*)

Paper/year	Sampling areas	Methods	Main findings	Strength	Weakness
Noor *et al.*, (2020)	Pulau Dua, Banten, In donesia	Irregularly monitored, using binoculars, observations were performed at a distance and photographed	There was just one chick visible on the active nest, and no other individuals or pairs had constructed any additional nests and there were less than ten Milky Storks seen on the island	By providing magnification, binoculars enable users to observe birds up close without disturbing them	Using binoculars at comparatively near ranges is effective. More robust optics may be needed to observe birds in large areas or far away
Ong *et al.*, (2011)	Zoo Negara, Malaysia	Logistic regression	In the examined variables, male storks were considerably larger than females, exhibiting typical sexual size dimorphism. With a total anticipated accuracy of 94.8 and 90.0%, respectively, Bil length accurately categorized the sex of selected individuals from Painted and Milky Stork based on logistic regression analysis	Numerous statistical software programs (such as R, Python, SPSS, and SAS) offer logistic regression, and researchers with various levels of statistical proficiency can use it	High-quality data are necessary for logistic regression. Significant difficulties arise from missing values, necessitating imputation techniques, which can add bias or uncertainty
Rahman *et al.*, (2020)	Kuala Gula Bird Sanctuary, Perak, Malaysia	Vanishing bearing method	The reintroduced Milky Storks are found to be well-suited in Kuala Gula. The results, however, also imply that landscape modification and human activity have impacted Milky Stork foraging activities in both good and negative ways	Because of its high accessibility, monitoring and research in the field are easier	Determining the distance to a bird might take much work, mainly if it is far away. Errors in measuring distance can cause serious problems while gathering data

(*continued*)

Table 1. (*continued*)

Paper/year	Sampling areas	Methods	Main findings	Strength	Weakness
Rahman *et al.*, (2013)	Kuala Gula, Perak, Malaysia	Sample collections of sediment using Ekman grab	The study found elevated levels of certain metals (Cd, Cu, Zn, and Pb) in the sediments of Kuala Gula. These metals are primarily sourced from anthropogenic activities such as aquaculture, fisheries, and oil palm plantations Copper concentrations were highest at station 3 (57 mg/kg), likely due to direct dredging activities. Zinc and lead concentrations were elevated near jetties, which were linked to corrosion and runoff from Zn-coated materials	It works exceptionally well for gathering samples from fine-grained, soft sediments like mud and silt, which are prevalent in various aquatic environments	Since little silt can be grabbed in a single grab, several samples can be taken to gather enough material for examination
Swennen & Martejin (1987)	Sungai Burung, Perak, Malaysia	40x60 telescope	When comparing the weight of the fish consumed in 39 min by the Milky Stork that was seen foraging on an exposed flat, the estimated weight of the fish was 225 g. A Milky Stork's estimated weight of 3 kg means that 630g of fish should be consumed daily	Perfect for birdwatching in wide-open spaces, wetlands, and coastal regions where birds are frequently distant from the observer	Environmental factors that affect telescopes include wind, rain, and heat waves (mirage), which can distort images and make viewing challenging
Tolohah *et al.*, (2023) (9)	All states of Peninsular Malaysia	Citizen science (eBird) data	In Peninsular Malaysia, between 2019 and 2021, the eBird database recorded around 86,881 Asian Openbill occurrences, 2,391 Painted Stork occurrences, and 242 Milky Stork occurrences. These numbers indicate significant variations in the occurrence sizes of the three species	Citizen scientists can contribute observations from various locations and times, including times and places not usually covered by professional scientists, resulting in comprehensive geographical and temporal coverage	Comprehensive validation and verification procedures are necessary to ensure the accuracy and dependability of citizen science data, and these procedures can be resource-intensive

(*continued*)

Table 1. (*continued*)

Paper/year	Sampling areas	Methods	Main findings	Strength	Weakness
Verheugt, (1987)	Sumatra, Indonesia	Through reading and the body of research, paying close attention to current reports—both released and pending—and speaking and corresponding with numerous ornithologists	Recent ornithological research conducted in Indonesia and Malaysia under the auspices of ICBP and Interwader has demonstrated the critical importance of Indonesia's coastal lowlands to the survival of the Milky Stork	Reviews of the literature can be used to pinpoint knowledge gaps or areas in need of additional investigation. This can direct the conversation towards investigating new ideas for observational methods or research inquiries	The validity or application of the results to actual bird observation may be impacted by biases or limitations in the techniques of earlier studies found in the literature
Li *et al.*, (2006)	Matang Mangrove Forest, Perak, Malaysia	Field Survey	The extensive survey verified the severe reduction of Malaysia's Milky Stork population	Frequent field trips can aid in the long-term monitoring of avian populations by providing information on survival rates, breeding success, and population trends, which are essential to conservation efforts	Researchers can disrupt birds, changing their natural behaviour and possibly producing data that is not accurate. Birds may alter their behaviour, conceal, or escape in response to human presence
Yee *et al.*, (2013)	Malaysian national zoo	FTA card-collected blood, Molecular markers, and morphologies	Some hybrid individuals showed alleles unique to either the painted or milky species, and others showed unique alleles of both species. Those birds with intermediate morphological phenotypes were proven to be hybrids when the fingerprinting findings were compared to the morphological characteristics of each hybrid bird	FTA Card-Collected Blood provides stability and convenience Molecular markers offer extensive uses and accurate genetic information Morphologies offer modest genetic resolution and are readily observable, making them valuable for field investigations. However, they are also prone to subjectivity and environmental influences	FTA card-collected Blood sample volume is restricted, and DNA extraction requires particular procedures/ Molecular markers can be expensive and need specialized knowledge and lab space Morphologies are prone to subjectivity and environmental influences

3 Conservation Management and its Conservation Implications

Preserving and replenishing vital wetland habitats, including mudflats, riverine ecosystems, and mangroves, is essential. For Milky Storks, these habitats serve as critical breeding and feeding grounds. For example, Important Bird Areas (IBAs) play a significant role; according to a study by Yong *et al.*, (2022), at least one vulnerable migratory species is supported by more than .4% of Southeast Asia's coastline IBAs.

Preventing poaching is also crucial. As highlighted earlier in this review, hunting and poaching of Milky Stork chicks remain significant threats in Southeast Asia. Therefore, implementing anti-poaching measures to prevent the unlawful taking and trading of Milky Storks is vital. This entails enhanced law enforcement, patrols, and surveillance across their entire range, as the species continues to decline due to habitat destruction, food hunting, and capture for trade (Iqbal *et al.*, 2012).

Other than that, community awareness plays a vital role in the conservation management of the Milky Stork. Involving local communities in conservation initiatives through educational programs can help highlight the importance of the Milky Stork and the need to preserve its habitat. Community-based conservation programs can effectively raise awareness about this endangered waterbird species and foster local stewardship.

Furthermore, captive breeding and reintroduction programs are critical for building a safety-net population and enhancing wild populations. To ensure the success of these efforts, captive-bred birds must be effectively released back into the wild, as emphasized by Ismail and Rahman (2016). These programs are vital for population augmentation and the long-term sustainability of the species.

3.1 Recent Studies, Gaps, and Needs on the Conservation of Milky Stork in Southeast Asia Countries

Recent research has focused on monitoring Southeast Asia's Milky Stork population trends and dispersal. More precise information on their movement and habitat use has been made available by methods like drone surveys and satellite tracking. For waterbird species, drones are particularly valuable for studying habitats in wetland areas, which are often challenging for humans to access, such as deep-water zones or high tree nesting sites.

Drones are beneficial for closely monitoring large wildlife groups, such as waterbird colonies (Sardà-Palomera *et al.*, 2017; Chabot *et al.*, 2015; Borrelle & Fletcher, 2017; Barr *et al.*, 2018; Lyons *et al.*, 2019). They enable the observation of inaccessible regions and the surveying of larger areas with advanced cameras, making it possible to detect more individuals, species, and environments. As a result, employing drones for quantitative waterbird population surveys is increasingly essential (Han *et al.*, 2017). To date, research at Pulau Rambut, Indonesia, is the only study using drones to observe Milky Stork breeding behavior.

Unmanned aerial vehicles (UAVs, often known as drones) have become much more common in recent decades. This new technique overcome the inability to detect specific species using old methods, especially for species that live in isolated places (Shahbazi *et al.*, 2014). Aerial survey imagery has proven beneficial for wildlife surveys, such as reducing human disturbance when tracking endangered raptors (Gallego *et al.*, 2021).

Research on genetic diversity in Milky Stork populations has revealed the extent of inbreeding and hybridization with Painted Stork. Understanding genetic variation is critical for developing effective captive breeding and reintroduction programs. Therefore, it is imperative to detect introgressive hybridization to protect the genetic integrity of the remaining pure species, Milky Stork (Kaminsin *et al.*, 2023). Owever, most reintroduction studies have been conducted in Kuala Gula, Malaysia, with limited research in other regions to evaluate the success of these programs.

Studies have examined how human activities like farming, fishing, and urbanization affect the habitats of Milky Storks. These findings emphasize the need for sustainable land-use strategies to mitigate the adverse effects. Wetlands are crucial for waterbirds in maintaining biodiversity and protecting their habitat (Mas *et al.*, 2021). Among their ecosystem functions, waterbird diversity exemplifies the importance of wetlands for supporting ecological health (Qiu *et al.*, 2024).

More comprehensive studies across the Milky Stork whole range are required to capture an overview of the population to fill the knowledge gap. The data available now could be more consistent and complete; most review papers have only focused on Malaysia and Indonesia, and there needs to be more information about Milky Stork in Thailand and Cambodia. Monitoring populations over time and evaluating the success of conservation efforts require long-term monitoring systems. Studies conducted for brief periods only partially reflected population dynamics; doing research over a long time will give better information and precise knowledge about the species.

Moreover, more research is necessary to fully comprehend the ecological needs of Milky Storks, including their food, foraging habits, and preferred habitats at different phases of life. Understanding the ecology can give a better preview in conserving this species. Conservationists can identify and save vital habitats essential to a species' existence and reproduction by researching the ecological requirements of various species. Understanding ecological principles makes it possible to create conservation plans that are more successful and suited to the unique requirements of different species and habitats.

Furthermore, it is essential to comprehend the socioeconomic elements influencing residents' perceptions of Milky Storks and their habitats. Studies evaluating the efficacy of community-based conservation initiatives are required to understand better a community conserving Milky Stork afterwards, decreasing the hunting and trading of this species at marketplaces.

In order to save the Milky Stork, it is very important to preserve and restore critical habitats, including wetlands, mangroves, and mudflats. The species primarily inhabits tidal coastal areas, where it actively forages in habitats including marshes, swamps, rice paddies, and mangrove habitats (Collar *et al.*, 2001; Shepherd & Giyanto, 2009; Syamal & Harianto, 2018). Conservation efforts should include legal protection, habitat restoration, and promotion of sustainable land-use practices. Adequate funding and resources must be provided for such activities to prevent backsliding, as occurred at Kuala Selangor Nature Park due to financial shortfalls.

The improvement of genetic diversity and sustaining wild populations via enhanced captive breeding and reintroduction programs is equally critical. There needs to be more cooperation among zoos, wildlife reserves, and other conservation organizations involved, as well as international collaboration between Southeast Asian countries to learn from each other's best practices, resources, and knowledge. Joint research projects or cross-border activities aimed at protecting the Milky Stork are all examples of such collaborative efforts.

4 Conclusions

In conclusion, Southeast Asia's Milky Stork conservation is a complex issue that calls for a coordinated strategy incorporating socioeconomic and ecological tactics. Comprehending the Milky Stork's ecology is imperative to recognize and safeguard crucial habitats, guaranteeing the species' access to critical supplies for sustenance, procreation, and endurance. While many issues remain regarding ecological requirements and long-term monitoring, new studies have provided crucial insights into population dynamics, habitat preferences, and the effects of human activity. More stringent anti-poaching measures, community engagement programs to promote local stewardship, and improved preservation and restoration of wetland habitats are all necessary components of effective conservation management. Captive breeding and reintroduction programs support wild populations and boost genetic diversity.

Acknowledgment. This research was supported by Universiti Tun Hussein Onn Malaysia (UTHM) through Tier 1 Grant Research (Vot Q509) and Postgraduate Research Grant (GPPS) (Vot Q598).

References

Abd Rahman, A., Mohd Nazri, A.S.: Land use effect on milky stork habitat selection in Matang forest reserve, Perak. J. Techno-Soc. **13**(2) (2022). https://doi.org/10.30880/jts.2021.13.02.006

Barr, J.R., Green, M.C., DeMaso, S.J., Hardy, T.B.: Detectability and visibility biases associated with using a consumer-grade unmanned aircraft to survey nesting colonial waterbirds. J. Field Ornithol. **89**(3), 242–257 (2018). https://doi.org/10.1111/jofo.12258

Baveja, P., Tang, Q., Lee, J.G.H., Rheindt, F.E.: Impact of genomic leakage on the conservation of the endangered Milky Stork. Biol. Cons. **229**, 59–66 (2019). https://doi.org/10.1016/j.biocon.2018.11.009

BirdLife International, 2018. Species factsheet: Mycteria cinerea. http://www.birdlife.org. Accessed June 2024

BirdLife International, 2019. Species factsheet: Mycteria cinerea. http://www.birdlife.org, Accessed June 2024

Borrelle, S.B., Fletcher, A.T.: Will drones reduce investigator disturbance to surface-nesting seabirds? Mar. Ornithol. **45**, 89–94 (2017)

Chabot, D., Craik, S.R., Bird, D.M.: Population census of a large common tern colony with a small unmanned aircraft. PLoS ONE **10**(4), e0122588 (2015). https://doi.org/10.1371/journal.pone.0122588

Collar, N.J., Andreev, A.V., Chan, S., Crosby, M.J., Subramanya, S., Tobias, J.A.: Threatened birds of Asia: The BirdLife International Red Data Book. BirldLife International, Cambridge (2001)

Faiq, H. I., Safie, M. Y., Shukor, M.N.: Assessment of survival rates and reproductive success of captive bred milky stork released at Kuala Gula Bird Sanctuary, Perak. In: AIP Conference Proceedings, vol. 1784, no. 1 (2016). https://doi.org/10.1063/1.4966848

Firdausy, M.S., Mardiastuti, A., Mulyani, Y.A.: Abundance waterbirds and the distribution of trees nesting in Pulau Rambut (Rambut Island) wildlife sanctuary, Jakarta Bay, Indonesia. In: IOP Conference Series: Earth and Environmental Science, vol. 771, no. 1, p. 012028 (2021). https://doi.org/10.1088/1755-1315/771/1/012028

Gallego, D., Sarasola, J.H.: Using drones to reduce human disturbance while monitoring breeding status of an endangered raptor. Remote Sens. Ecology Conserv. **7**(3), 550–561 (2021). https://doi.org/10.1002/rse2.206

Gula, J., Sundar, K.G., Willows-Munro, S., Downs, C.T.: The state of stork research globally: a systematic review. Biol. Cons. **280**, 109969 (2023)

Han, Y.-G., Jung, S.-H., Kwon, O.: How to utilize vegetation survey using drone image and image analysis software. J. Ecol. Environ. **41**(1) (2017). https://doi.org/10.1186/s41610-017-0035-2

International Union for Consevation of Nature/Species Survival Commission (IUCN/SSC) Guidelines for reintroductions and other conservation translocation. Gland, Switzerland: IUCN/SSC; Version 1.0 (2013)

Iqbal, M., Hasudungan, F.: Observations of Milky Stork Mycteria cinerea during 2001–2007 in South Sumatra province, Indonesia. Birding Asia **9**, 97–99 (2008)

Iqbal, M., Ridwan, A., Takari, F., Mulyono, H.: Rediscovery of a Milky Stork Mycteria cinerea breeding colony in South Sumatra province, Indonesia. Birding Asia **10**, 62–66 (2008)

Iqbal, M., Mulyono, H., Ridwan, A., Takari, F.: An alarming decrease in the Milky Stork Mycteria cinerea population on the east coast of South Sumatra province, Indonesia. Birding Asia **18**, 68–70 (2012)

Iqbal, M.: Survei and Conservation of Milky Stork Mycteria cinerea in Sumatra, Indonesia. Rufford Small Grant (2012)

Ismail, A., Rahman, F., Kin, D.K.S., Ramli, M.N.H., Ngah, M.: Current status of the Milky Stork captive breeding program in Zoo Negara and its importance to the stork population in Malaysia. Trop. Nat. Hist. **11**(1), 75–80 (2011)

Ismail, A., Rahman, F.: An urgent need for Milky Stork study in Malaysia. Pertanika J. Trop. Agric. Sci. **35**(3), 407–412 (2012)

Ismail, A., Rahman, F., Miyazaki, N., Naito, Y.: Initial application of bio-logging techniques on captive Milky Stork (Mycteria cinerea) in Malaysia. Trop. Ecol. **53**(2), 177–181 (2012)

Ismail, A., Rahman, F.: Current status of the milky stork re-introduction programme in Malaysia and its challenges. Tropical Life Sciences Research **27**(2), 13–24 (2016). https://doi.org/10.21315/tlsr2016.27.2.2

Jeyarajasingam, A., Pearson, A.: A Field Guide to the Birds of Peninsular Malaysia and Singapore. Oxford University Press, Oxford (2012)

Kaminsin, D., et al.: Detecting introgressive hybridization to maintain genetic integrity in endangered large waterbird: a case study in milky stork. Sci. Rep. **13**(1), 8892 (2023). https://doi.org/10.1038/s41598-023-35566-x

Li, Z.W.D., Yatim, S.H., Howes, J., Ilias, R.: Status overview and recommendations for the conservation of Milky Stork Mycteria cinerea in Malaysia: Final report of the 2004/2006 Milky Stork field surveys in the Matang Mangrove forest, Perak, Malaysia. Wetlands International and the Department of Wildlife and National Parks, Peninsular Malaysia, Kuala Lumpur, Malaysia (2006)

Lyons, M.B., et al.: Monitoring large and complex wildlife aggregations with drones. Methods Ecol. Evol. **10**(7), 1024–1035 (2019). https://doi.org/10.1111/2041-210x.13194

Mardiastuti, A., et al.: The use of drone to study the nesting behavior of milky stork: some preliminary observations. Biodivers. J. Biol. Divers. **24**(8) (2023). https://doi.org/10.13057/biodiv/d240836

Mas, M., Flaquer, C., Rebelo, H., López-Baucells, A.: Bats and wetlands: synthesising gaps in current knowledge and future opportunities for conservation. Mamm Rev. **51**, 369–384 (2021). https://doi.org/10.1111/mam.12243

Neel, M.C., Leidner, A.K., Haines, A., Goble, D.D., Scott, J.M.: By the numbers: how is recovery defined by the us endangered species act? Bioscience **62**(7), 646–657 (2012). https://doi.org/10.1525/bio.2012.62.7.7

Noor, Y.R., Umar, Mardiastuti A., Gumilang R.S, Iqbal M.: Observation of Milky Stork Mycteria cinerea breeding on Pulau Dua, Banten, Indonesia, after 45 years. SIS Conserv. **2**, 29–32 (2020)

Ong, H.K.A., et al.: Morphometric sex determination of Milky and Painted storks in captivity. Zoo Biol. **31**(2), 219–228 (2011). https://doi.org/10.1002/zoo.20387

Puchlik, M., Piekutin, J., Dyczewska, K.: Analysis of the impact of climate change on surface water quality in North-Eastern Poland. Energies **15**(1), 164 (2021)

Qiu, J., Zhang, Y., Ma, J.: Wetland habitats supporting waterbird diversity: Conservation perspective on biodiversity-ecosystem functioning relationship. J. Environ. Manage. **357**, 120663 (2024). https://doi.org/10.1016/j.jenvman.2024.120663

Rahman, F., Ismail, A., Yusof, S.: Metals contamination in the foraging area of Milky Stork: evidence of anthropogenic inputs in the aquatic environment of Kuala Gula, Malaysia. Toxicol. Environ. Chem. **95**(9), 1499–1505 (2013). https://doi.org/10.1080/02772248.2014.892941

Rahman, F., Ismail, A., Omar, H., Hussin, M.Z.: Exposure of the endangered Milky stork population to cadmium and lead via food and water intake in Kuala Gula Bird Sanctuary, Perak, Malaysia. Toxicol. Rep. **4**, 502–506 (2017). https://doi.org/10.1016/j.toxrep.2017.09.003

Rahman, F., Ismail, A.: Waterbirds: an important bio-indicator of ecosystem. Pertanika J. Sch. Res. Rev. **4**(1) (2018)

Rahman, F., Ismail, A., Yusof, S., Mazlan, N., Engku-Ahmad-Khairi, E.-A.: Evaluation of glyphosate levels in sediments of milky stork foraging areas in Kuala Gula Bird Sanctuary, Perak, Malaysia. Pertanika J. Trop. Agric. Sci. **42**(3), 995–1007 (2019)

Rahman, F., Ismail, A., Shohaimi, S., Omar, H., Zakaria, M.: Foraging activity of the re-introduced milky storks (Mycteria cinerea) in Kuala Gula Bird Sanctuary, Perak, Malaysia. Pertanika J. Trop. Agric. Sci. **43**(2), 217–229 (2020)

Ramesh, R., Lakshmi, A., Purvaja, R.: Integrated coastal and estuarine management in South and Southeast Asia, reference module in earth systems and environmental sciences. Treatise on estuarine and coastal science. Manage. Estuaries Coasts **11**, 227–263 (2011)

Sardà-Palomera, F., Bota, G., Padilla, N., Brotons, L., Sardà, F.: Unmanned aircraft systems to unravel spatial and temporal factors affecting dynamics of colony formation and nesting success in birds. J. Avian Biol. **48**(9), 1273–1280 (2017). https://doi.org/10.1111/jav.01535

Shahbazi, M., Théau, J., Ménard, P.: Recent applications of unmanned aerial imagery in natural resource management. GISci. Remote Sens. **51**(4), 339–365 (2014). https://doi.org/10.1080/15481603.2014.926650

Shepherd, C.R., Giyanto: Observations of milky storks Mycteria cinerea in Percut, North Sumatra, Indonesia. Birding Asia, **11**, 70–72 (2009)

Swennen, C., Marteijn, E.C.L.: Notes on the feeding behaviour of the Milky Stork Mycteria cinerea. Forktail **3**, 63–66 (1987)

Syamal, F.M., Harianto. S.P.: Study of the Milky Stork Bird (Mycteria cinerea) population in Rawa Pacing, Kibang Pacing Village, Menggala Timur Sub-District, Menggala Timur Regency, Lampung Province. J. Sylva Lestari **6**(2), 1–6 (2018). https://doi.org/10.23960/jsl261-6

Tolohah, A.A., Arazmi, F.N., Nor, S.M., Mansor, M.S.: Spatial distribution of Malaysian storks determined based on citizen science (eBird) data. Pertanika J. Sci. Technol. **31**(5), 2329–2338 (2023). https://doi.org/10.47836/pjst.31.5.14

Van Zalinge, R., Evans, T., Sun, V.: A review of the status and distribution of large waterbirds in the Tonle Sap Biosphere Reserve. Wildlife Conservation Society: Cambodia Program (2008)

Verheugt, W.J.M.: Conservation status and action program for the Milky Stork (Mycteria cinerea). Colon. Waterbirds **10**(2), 211 (1987). https://doi.org/10.2307/1521260

Wong, P.P.: Coastal tourism development in Southeast Asia: relevance and lessons for coastal zone management. Ocean Coast. Manag. **38**(2), 89–109 (1998)

Yee, E.Y.S., Zainuddin, Z.Z., Ismail, A., Yap, C.K., Tan, S.G.: Identification of hybrids of painted and milky storks using FTA card-collected blood, molecular markers, and morphologies. Biochem. Genet. **51**(9–10), 789–799 (2013). https://doi.org/10.1007/s10528-013-9607-8

Yong, D.L., et al.: Conserving migratory waterbirds and the coastal zone: the future of South-east Asia's intertidal wetlands. Oryx **56**(2), 176–183 (2022). https://doi.org/10.1017/S00306053200013

Analysing the Gas Exchange Properties of Selected Plant Species at Universiti Tun Hussein Onn Malaysia

Yunusu Audu[1](✉), Alona Cuevas Linatoc[2], Aisha Idris[3](✉), and Alhassan Gabi Usman[1]

[1] Department of Biology, Faculty of Natural Science, Ibrahim Badamasi Babangida University Lapai, PMB 11 Tudun Wada, Niger State, Nigeria
abduldaudu@yahoo.co.uk

[2] Department of Technology and Natural Resources, Faculty of Applied Science and Technology, Universiti Tun Hussein Onn Malaysia (UTHM), Parit Raja, Malaysia

[3] Department of Biological Science, Faculty of Sciences, Federal University Dutse, PMB 7176 Dutse, Jigawa State, Nigeria
idrisaisha24@gmail.com

Abstract. Plant gas exchange properties, including maximum carboxylation rate of Rubisco (Vcmax), maximum electron transport rate (Jmax), light compensation point (LCP), light saturation point (LSP), maximum photosynthetic rate (Amax), and apparent quantum yield (Aqy), are key determinants of photosynthetic efficiency. This study evaluates these gas exchange characteristics in selected plant species at UTHM using a portable photosynthesis system (LICOR). The A/Ci and A/Q response curves demonstrated that photosynthetic assimilation increased with higher CO_2 concentrations and light intensity until saturation was reached. Among the studied species, Spathodea campanulata exhibited the highest Vcmax (52.14 ± 0.005 μmol CO_2 m⁻2 s⁻1) and Jmax (104.46 ± 0.011 μmol CO_2 m⁻2 s⁻1), whereas Rhapis excelsa recorded the highest LCP (69.60 ± 0.067 μmol photons m⁻2 s⁻1) and Costus spicatus displayed the highest LSP (1576.69 ± 0.173 μmol photons m⁻2 s⁻1). These findings suggest that different plant species exhibit varying photosynthetic responses to environmental factors, with higher light and CO_2 concentrations enhancing their carbon sequestration potential. The results underscore the importance of selecting species with higher photosynthetic assimilation rates for optimizing carbon absorption and supporting sustainable environmental managements.

Keywords: Photosynthetic capacity · Gas exchange characteristics · Light compensation point (LCP) · CO_2 assimilation · Carbon sequestration

1 Introduction

Light is a fundamental environmental factor that influences nearly each phase of a plant's life cycle. It is a crucial regulatory signal that influences plant physiology and growth in addition to providing energy for the fixation of carbon dioxide (CO_2) during photosynthesis. Both the quality and intensity of incident light influence the growth and productivity

P. Chang et al. (Eds.): SCIEMATHIC 2024, SPPHY 427, pp. 237–249, 2025.
https://doi.org/10.1007/978-981-96-7679-8_19

of plants, influencing their photosynthetic efficiency and total biomass accumulation [1, 2].

Previous researches have demonstrated that light availability plays a critical role in optimizing photosynthesis. For instance, *Cyanea acuminata* seedlings grown under 75% irradiance exhibited the highest photosynthetic rates, suggesting that this light intensity is optimal for their growth. Conversely, insufficient light can limit photosynthesis, leading to plant stress, reduced net carbon gain, and stunted growth. Furthermore, light quality also influences a plant's ability to assimilate CO_2, as observed in *Arabidopsis thaliana*, where both photosynthetic rates and the light saturation point (LSP) were considerably greater in plants exposed to intense light than in plants maintained in low light. [5, 6].

Variations in light intensity can also affect other physiological traits, such as growth patterns and biochemical composition. For example, in *Artemisia annua*, changes in light levels influence photosynthetic characteristics and the accumulation of secondary metabolites, such as artemisinin. Additionally, increasing CO_2 concentrations can modulate photosynthetic responses, leading to interspecific and intraspecific variations in gas exchange parameters. Given these dynamics, this study aims to evaluate the gas exchange physiognomies of designated species of plants at Universiti Tun Hussein Onn Malaysia (UTHM) under varying light and CO_2 conditions.

2 Methodology

2.1 Location of Study

The main campus of UTHM is situated in Johor's Parit Raja, Batu Pahat, which is located in Peninsular Malaysia's southern area. The investigation was conducted at 1.8531° N, 103.0864° on the main campus. The trees, shrubs, herbaceous, palm plants and bamboos within the UTHM main campus were surveyed. The plants were selected based on the previous knowledge of their photosynthetic rates [5].

2.2 Measurements of the Photosynthetic Carbon Dioxide (A\Ci) Response Curve

To measure ***A\Ci*** curves, a portable photosynthesis system (LI-6400) was used. an infrared gas analyzer designed for precise gas exchange measurements. The youngest fully matured leaves of the selected plant species were used for this study. During *A/Ci* curve measurements, the photosynthetic photon flux density was maintained at a constant level of 1000 µmol m^{-2} s^{-1}, while the flow rate and leaf temperature were set at 500 µmol/s and 30 °C, respectively, to match ambient air temperature.

CO_2 concentrations were initially varied at 450, 400, 300, 250, 200, 150, 100, 50, and 0 µmol/mol, followed by additional concentrations of 450, 550, 650, 800, and 1100 µmol/mol [12]. The *A/Ci* curve was fitted following the methodology described in [10], based on a non-rectangular hyperbola model, as represented in Eq. 1.

$$A_c = \frac{(ci - ci^*)V_{cmax}}{ci + km(Co_2)i} - R_d \quad (1)$$

where A_c is the rate of Rubisco-saturated CO_2 assimilation.

ci Stands for intercellular CO_2 content.
ci * is the photocompensation point for intercellular CO_2.
V_{cmax} is the maximum rate of carboxylation.
R_d is mitochondrial respiration in the dark.
km $(CO_2)i$ is calculated at ci as the apparent Michaelis-Menten constant for CO_2.

2.3 Measurements of the Response Curve for Photosynthetic Light (A\Q)

The portable photosynthesis system LI-6400 was used to measure ***A\Q*** curves. To maintain uniformity throughout the measurements, the reference CO_2 concentration was kept at 400 µmol/mol, the flow rate was kept at 500 µmol/s, and the leaf temperature was fixed at 30 °C. Stepwise decreasing order adjustments were made to the photosynthetic photon flux density (PPFD) from 1500, 1250, 1000, 750, 500, 250, 100, 50, and 25 to 0 µmol m^{-2} s^{-1}.

After gathering data, a non-rectangular hyperbola model was used to fit the A\Q curve, as shown in Eq. 2. According to Eqs. 3 and 4, the light compensation point (LCP) and light saturation point (LSP) were ascertained.

$$A_{net} = \frac{\emptyset PPFD\sqrt{(\emptyset PPFD + A_{max})^2} - \sqrt{4\theta\emptyset PPFD + A_{max}}}{2\theta} - R_d \tag{2}$$

$$LCP = \frac{\theta x(R_d)^2 - R_d x A_{max}}{(R_d x \emptyset) - (\emptyset x A_{max})} \tag{3}$$

$$LSP = \frac{R_d + (0.9xA_{max})xA_{max} - \theta x(R_d + 0.9xA_{max})}{\emptyset(A_{max} - R_d} \tag{4}$$

where

A_{net} is Net Photosynthetic Rates (µmol CO_2 m^{-2} s^{-1}).
A_{max} is the Highest Photosynthetic Rate (µmol CO_2 m^{-2} s^{-1}).
LCP stands for Light Compensation Point. (µmol CO_2 m^{-2} s^{-1}).
LSP stands for Light Saturation Point. (µmol CO_2 m^{-2} s^{-1}).
ϕ is a quantum yield that appears (mol CO_2 mol photons $^{-1}$).
R_d is the rate of dark (day) respiration (at no light; µmol CO_2 m^{-2} s^{-1}).
θ is the convexity of the curve (has no unit).

2.4 Statistical Analysis

Descriptive statistics, including mean and standard deviation, were used to compare different variables. A one-way ANOVA was performed to assess variations among the studied parameters. All statistical analyses were conducted using Excel and IBM SPSS Statistics 22.

3 Results and Discussion

3.1 Identified Plants

The selected plants were identified as *Spathodea campanulata, Pongamia pinnata, Casuarina equisetifolia, Musa sp, Calathea lutea, Costus spicatus, Sanchezia speciose, Eugenia oleina, Terminalia mantaly Tricolar, Roystonea regia, Livistona chinensis, Rhaphis excelsa, Bambusa vulgaris,* and *Phyllostachys aurea.*

3.2 Carbon Dioxide (A\Ci) Reaction Curve for Photosynthetic Processes

From the selected trees (Fig. 1), *S. campunalata* have higher ***A\Ci*** curve compared to *P. pinata and C. equisetifolia.* This is because *S. campunalata* was having higher net photosynthesis while *P. pinata and C. equisetifolia* were having medium and lowest net photosynthesis among the studied trees.

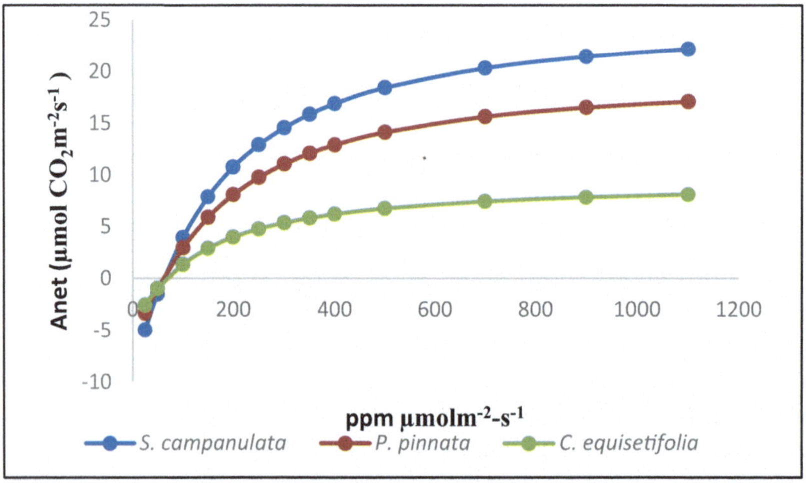

Fig. 1. The ***A\Ci*** curve of a few chosen trees

Figure 2, displays the ***A\Ci*** curve for a few of the herbaceous species under study. The herbaceous plant having the maximum net photosynthesis (*Musa* sp.) was having higher curve compared to the herbaceous plant with medium net photosynthesis (*C. lutea*) and the herbaceous plant with the least photosynthesis (*C. spicatus*).

For shrubs (Fig. 3) and palms (Fig. 4). *S. speciosa* and *R. regia* were having a higher curve. *E. oleina* and *L. chinensis* were having medium curve while *T. mantaly Tricolar* and *R. excelsa* were having the lowest curve.

Among the studied bamboo species, *B. vulgaris* had the highest net photosynthesis, thereby having the highest ***A\Ci*** curve while *P. aurea* had the medium net photosynthesis as shown in Fig. 5.

The fitted curves depict the plants' photosynthetic capacity in response to different CO2 levels. Based on the A/Ci curve, the gas exchange characteristics (Table 1) are the

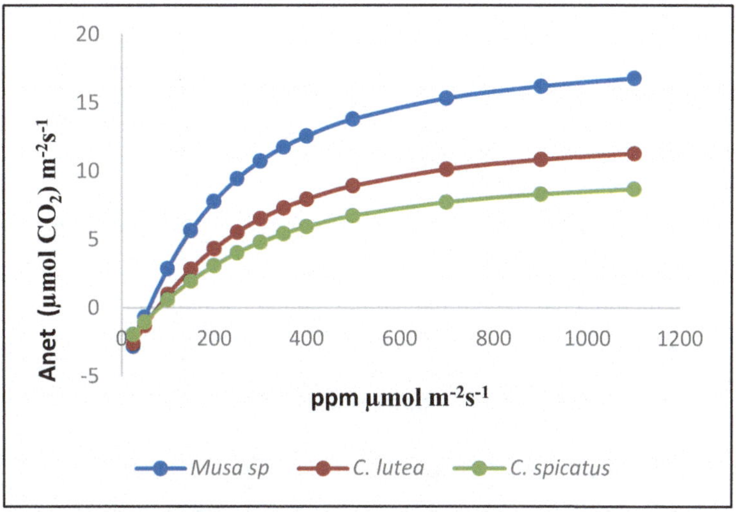

Fig. 2. The *A\Ci* curve for a few of the herbaceous plants under study

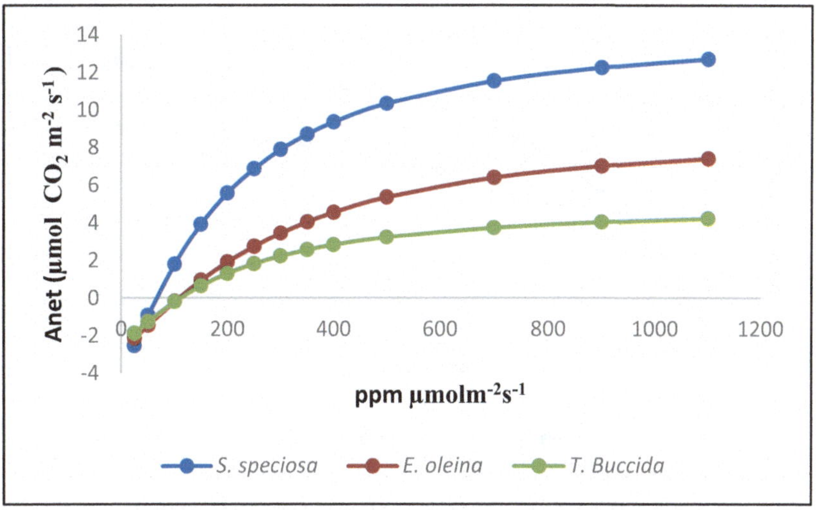

Fig. 3. The *A\Ci* curve for a few of the shrubs under study

maximum electron transport rate (Jmax) and the maximum carboxylation rate of Rubisco (Vcmax). Comparing plants with higher photosynthetic rates to those with lower rates, the table shows that the former have higher Vcmax and Jmax.

Quantifying a plant's photosynthetic characteristics is essential for understanding the biological mechanisms underlying its assimilation rate. This, in turn, aids in predicting how environmental factors influence the plant's capacity for CO_2 absorption. As clarified by [15], elevated CO_2 concentrations have a significant impact on both photochemical and biochemical capacities, including ribulose-1,5-bisphosphate (RuBP) activity. The

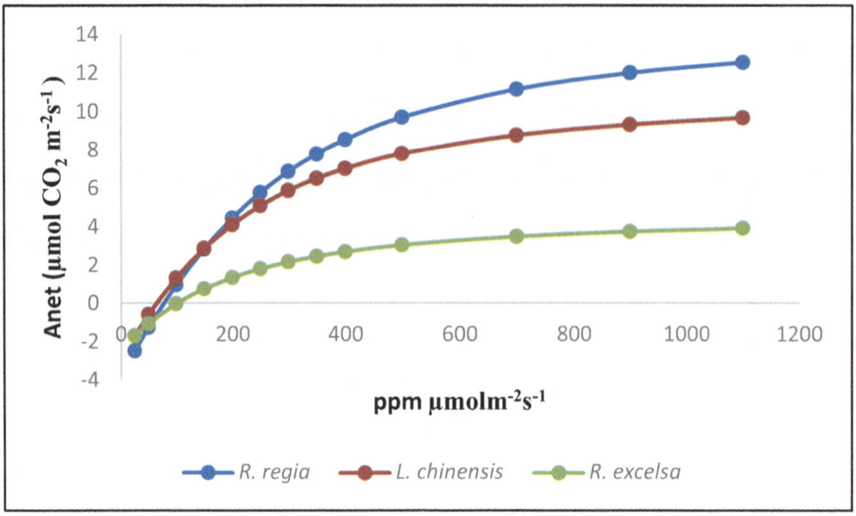

Fig. 4. The *A\Ci* curve of a few of the palms under study

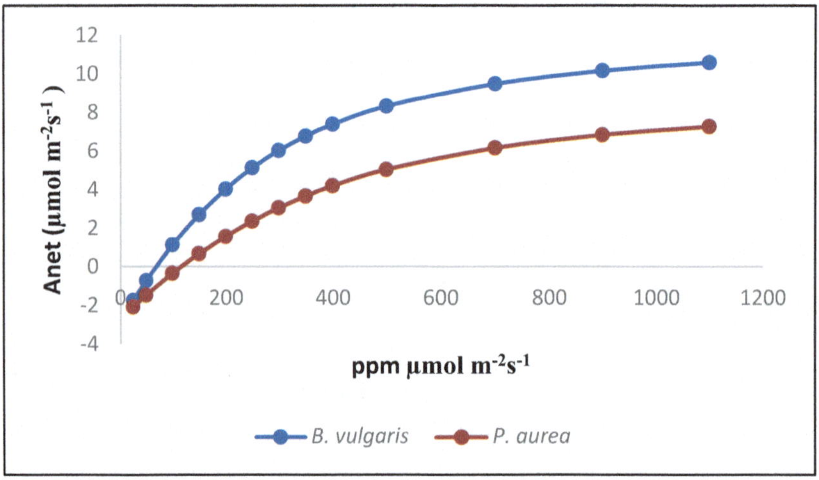

Fig. 5. The *A\Ci* curve of some of the studied Bamboo plants

authors reported that increasing CO_2 levels enhance photosynthesis, as CO_2 serves as a substrate for the Rubisco enzyme, thereby promoting the carboxylation reaction.

In a similar vein, [16] emphasised that higher CO_2 concentrations promote greater electron transport capacity (Jmax), higher maximum net photosynthesis (Amax), and increased carboxylation capacity (Vcmax). At the initial phase of the A/Ci curve, photosynthetic activity is limited by CO_2 availability, emphasizing the role of CO_2 concentration in determining photosynthetic efficiency. Estimating the A/Ci curve of a plant allows for the identification of critical parameters such as Vcmax and Jmax. Ideally, as

Table 1. The highest electron transport rate (Jmax) and maximum carboxylation rate of Rubisco (Vcmax) of a few of the plants under study

	Species	Vcmax $\mu molCO_2\ m^{-2}s^{-1}$	Jmax $\mu mmolCO_2\ m^{-2}s^{-1}$
Trees	*S. campanulata*	52.14400 ± 0.0055076	104.4606 ± 0.011269
	P. pinnata	41.14059 ± 0.0005508	80.4933 ± 0.002082
	C. equisetifolia	22.34555 ± 0.0005774	38.7166 ± 0.001732
Herbs	*Musa sp*	36.93675 ± 0.0015275	78.7224 ± 0.001155
	C. lutea	27.32752 ± 0.0055411	55.2782 ± 0.000153
	C. spicatus	21.13062 ± 0.0061076	42.9767 ± 0.002309
Shrubs	*S. speciosa*	29.79628 ± 0.0020817	60.5465 ± 0.002082
	E. oleina	20.15871 ± 0.0004887	39.1471 ± 0.002309
	T. mantaly Tricolar	14.33337 ± 0.0032146	22.4895 ± 0.000643
Palms	*R. regia*	30.11264 ± 0.0005508	61.9460 ± 0.001677
	L. chinensis	22.16639 ± 0.001182	46.2185 ± 0.011269
	R. excelsa	13.44597 ± 0.0004933	20.4461 ± 0.005774
Bamboo	*B. vulgaris*	24.07629 ± 0.0033779	51.1663 ± 0.000577
	P. aurea	20.45568 ± 0.0017349	10.4606 ± 0.011269

CO_2 levels increase, CO_2 assimilation also rises, aligning with findings by [18], who reported a strong correlation between net photosynthetic rates and relative growth rates across different species over time.

However, non-optimal temperature and nutrient availability can significantly affect assimilation rates [19]. In support of this claim, [21] pointed out that morphological stressors, including dryness, gradually lower CO_2 assimilation because they cause stomatal conductance to decrease. Drought stress also leads to a reduction in leaf size, stem elongation, and root proliferation, disrupts plant water relations, and decreases water-use efficiency. Additionally, such conditions negatively affect photosynthetic pigments and gas exchange processes, ultimately impairing plant growth and productivity.

3.3 Photosynthetic A\Ci Response Curve

Among the selected trees, *S. campunalata* had higher ***A\Ci*** curve compared to *P. pinata* and *C. equisetifolia.* This is because *S. campunalata* was having higher net photosynthesis while *P. pinata* and *C. equisetifolia* were having medium and lowest net photosynthesis among the studied trees (Fig. 6).

The ***A\Ci*** curve of some of the studied herbaceous plants is shown in Fig. 7. The herbaceous having the maximum net photosynthesis (*Musa* sp.) was having a higher curve compared to the herbaceous with medium net photosynthesis (*C. lutea*) and the herbaceous with the least photosynthesis (*C. spicatus*).

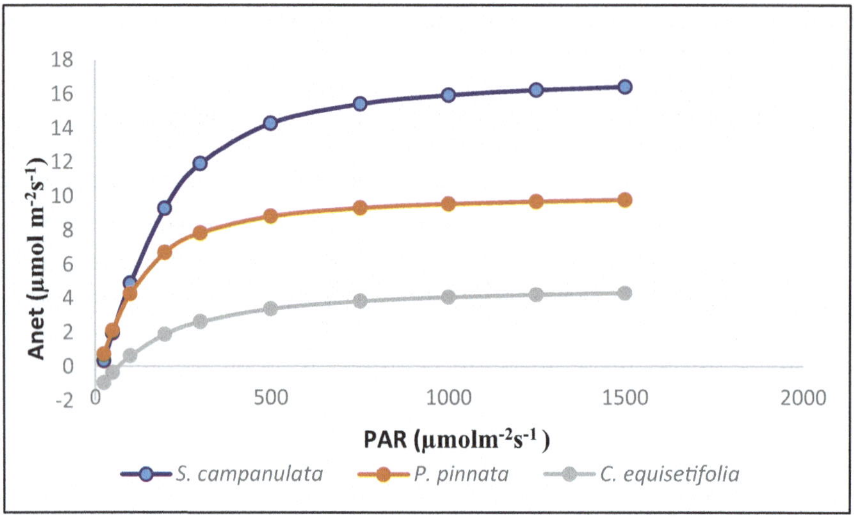

Fig. 6. The ***A\Ci*** reaction curve for a few of the trees under study

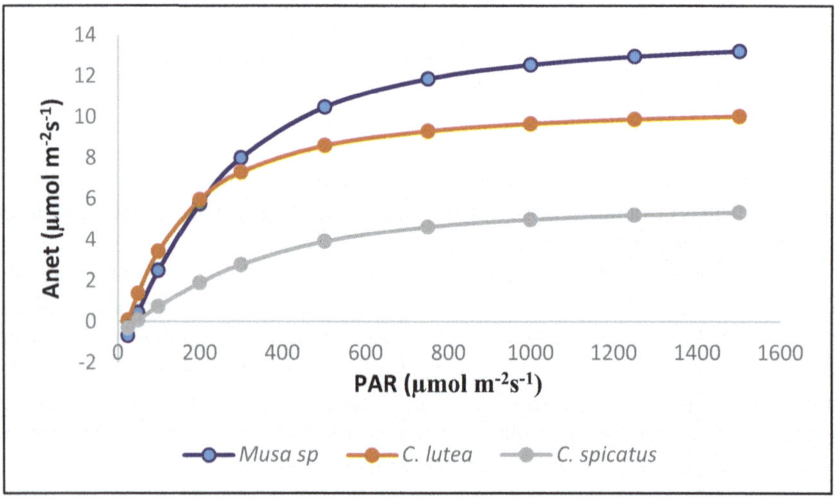

Fig. 7. The ***A\Ci*** sensitivity curve for a few of the herbaceous species under study.

For Shrubs (Fig. 8) and palms (Fig. 9), *S. speciosa* and *R. regia* were having a higher curve. *E. oleina* and *L. chinensis* were having medium curve while *T. buccida* and *R. excelsa* were having the lowest curve.

Among the studied bamboo species, *B. vulgaris* had the highest net photosynthesis, thereby having the highest ***A\Ci*** curve while *P. aurea* was having the medium net photosynthesis (Fig. 10).

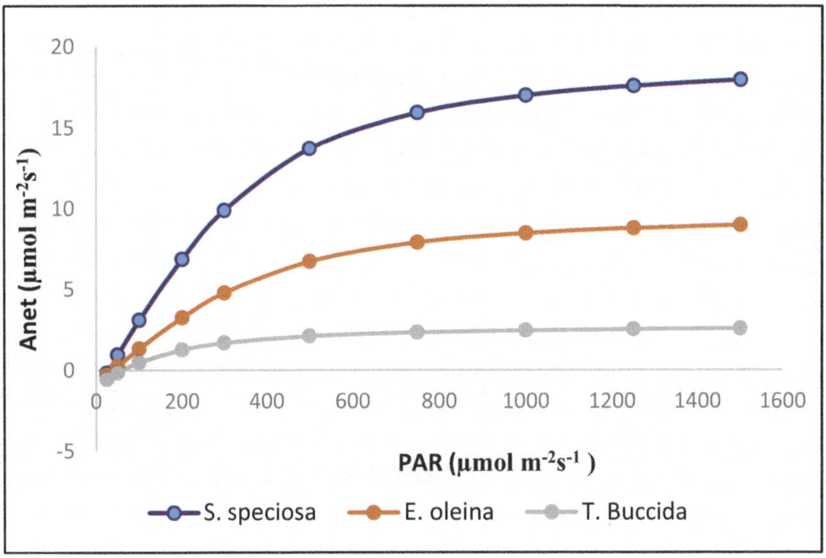

Fig. 8. The ***A\Ci*** responses curve for a few palms under investigation

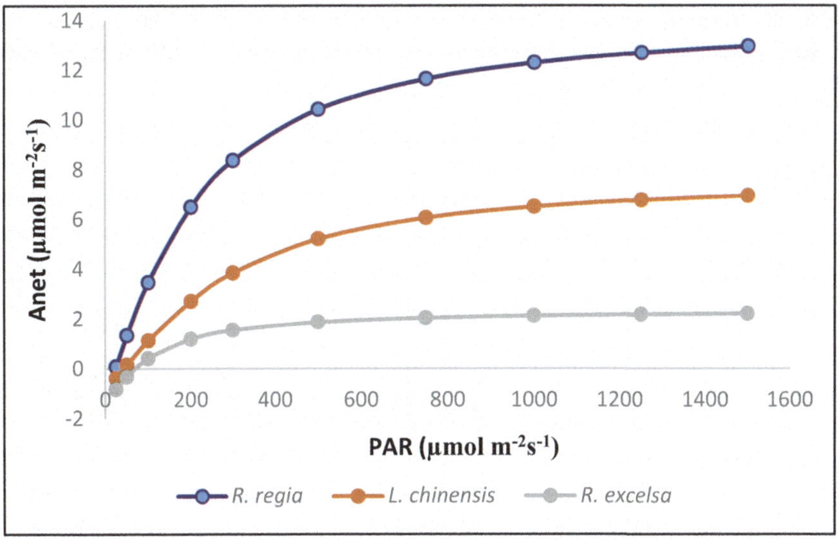

Fig. 9. The ***A\Ci*** sensitivity curve for a few of the palms under investigation

Gas exchange metrics (Table 2) acquired from the developed light response curve include the light compensation point (LCP), light saturation point (LSP), maximal photosynthesis (Amax), and apparent quantum yield (Aqy).

The light compensation point (LCP) serves as a key indicator of a plant's ability to utilize low light efficiently. By reducing their LCP, plants can enhance their capacity for light

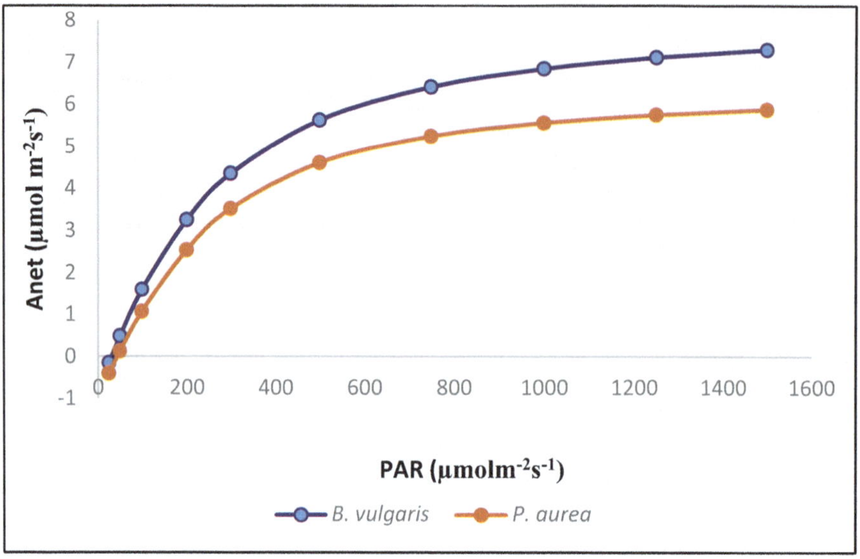

Fig. 10. The ***A\Ci*** reaction curve of a few bamboo species under study

Table 2. Shows the parameters of the light response curve with Aqy (apparent quantum yield), LCP (light compensation point), Amax (maximum photosynthesis), and LSP (light saturation point).

Plant	Species	Aqy (mol CO_2 mol^{-1} photons)	LCP (μmol photons m^{-2} s^{-1})	Amax (μmol. CO_2 m^{-2} s^{-1})	LSP (μmol photons m^{-2} s^{-1})
Trees	*S. campanulata*	0.0720 ± 0.00	19.6033 ± 0.05	18.6733 ± 0.05	801.8699 ± 0.05
	P. pinnata	0.0686 ± 0.00	12.6740 ± 0.00	11.0860 ± 0.01	682.4351 ± 0.05
	C. equisetifolia	0.0295 ± 0.03	66.6506 ± 0.02	6.4656 ± 0.00	1508.882 ± 0.11
Herbs	*Musa sp*	0.0518 ± 0.00	39.7957 ± 0.08	14.3168 ± 0.01	1278.145 ± 0.03
	C. lutea	0.0702 ± 0.01	23.4902 ± 0.05	12.2191 ± 0.05	1018.146 ± 0.1
	C. spicatus	0.0182 ± 0.00	45.0946 ± 0.05	6.6541 ± 0.06	1576.692 ± 0.17
Shrubs	*S. speciosa*	0.0538 ± 0.01	28.3033 ± 0.00	20.9070 ± 0.05	1256.269 ± 0.05
	E. oleina	0.0241 ± 0.00	37.6328 ± 0.04	10.7699 ± 0.05	1326.239 ± 0.06
	T. Buccida	0.0226 ± 0.00	61.2555 ± 0.01	3.8132 ± 0.07	1281.538 ± 0.05
Palms	*R. regia*	0.0583 ± 0.03	23.2734 ± 0.06	14.6215 ± 0.07	1435.896 ± 0.07
	L. chinensis	0.0240 ± 0.02	42.3112 ± 0.05	8.7891 ± 0.01	1557.377 ± 0.01
	R. excelsa	0.0258 ± 0.03	69.6053 ± 0.06	3.8513 ± 0.05	1089.147 ± 0.07
Bamboo	*B. vulgaris*	0.0320 ± 0.05	30.2770 ± 0.10	8.1147 ± 0.05	1508.734 ± 0.10
	P. aurea	0.0261 ± 0.05	43.1343 ± 0.06	7.4494 ± 0.01	1392.087 ± 0.05

absorption, thereby optimizing photosynthesis in shaded environments [23]. As observed in the response curve, an increase in light intensity corresponds to a gradual rise in the photosynthetic rate. When the rate of photosynthesis surpasses the rate of respiration, organic matter accumulates, promoting plant growth and biomass production.

This phenomenon is further supported by [24], who demonstrated that plants allocate biomass between leaves, stems, and roots based on light availability. Their findings suggest that increased light exposure leads to greater root investment, enhancing nutrient uptake and overall plant vigor. Conversely, shading has been shown to reduce both the light compensation and light saturation points in certain plant species. For instance, in *Lithocarpus litseifolius*, reduced light availability was associated with an increased ability to utilize weak light for photosynthesis [28].

The apparent quantum yield (AQY) reflects the efficiency of photosynthesis under low-light conditions. By adjusting the size of their light-harvesting antennae, plants can optimize light capture, thereby improving AQY and overall photosynthetic performance, particularly in shaded environments [29]. These adaptive responses highlight the importance of light availability in regulating plant growth and photosynthetic efficiency.

4 Conclusion

This study confirms that as light intensity increases, the ***A\Ci*** curve also rises until reaching a saturation point, beyond which no further increase in photosynthetic rate occurs. Similarly, the A/Ci curve demonstrates that increasing CO_2 concentration enhances photosynthesis. These findings indicate that plants require high light intensity to achieve their maximum assimilation rate.

A higher availability of atmospheric CO_2 contributes to an increased assimilation rate, leading to greater biomass accumulation, height, and diameter expansion. Consequently, this enhances the overall carbon absorption capacity of plants. Given the sunny climate at the university and the presence of elevated CO_2 concentrations—up to a certain threshold—these environmental conditions can further optimize plant CO_2 absorption. To maximize the effectiveness of carbon sequestration, species with higher assimilation rates should be prioritized for planting, ensuring a more effective mitigation strategy against atmospheric CO_2 accumulation.

Acknowledgement. We express our gratitude to Universiti Tun Hussein Onn Malaysia for providing a conducive research environment that facilitated this study. Additionally, we extend our sincere appreciation to the research supervisor for invaluable guidance and support throughout the research process.

References

1. Mawphlang, O.I., Kharshiing, E.V.: Photoreceptor mediated plant growth responses: Implications for photoreceptor engineering toward improved performance in crops. Front. Plant Sci. **8**, 1181 (2017). https://doi.org/10.3389/fpls.2017.01181
2. Saik, I., Huq, E.: In plant photoreceptors: Multi-functional sensory proteins and their signaling networks. Semin. Cell Dev. Biol. **92**, 114–121 (2019). https://doi.org/10.1016/j.semcdb.2019.03.007

3. Gao, S., DeLuca, T.H.: Use of microdialysis to assess short-term soil soluble N dynamics with biochar additions. Soil Biol. Biochem. **136**, 107512 (2019). https://doi.org/10.1016/j.soilbio.2019.06.008
4. Yang, F., Liao, D., Wu, X., Gao, R., Fan, Y., Ali Raza, M.: Effect of aboveground and below-ground interactions on the intercrop yields in maize-soybean relay intercropping systems. Field Crop Res **203**, 16–23 (2017). https://doi.org/10.1016/j.fcr.2016.12.007
5. Li, C.X., Chang, S.X., Khalil-Ur-Rehman, M., Xu, Z.G., Tao, J.M.: Effect of irradiating the leaf abaxial surface with supplemental light-emitting diode lights on grape photosynthesis. Aust. J. Grape Wine Res. **23**(1), 58–65 (2017)
6. Vialet-Chabrand, S., Matthews, J.S., Simkin, A.J., Raines, C.A., Lawson, T.: Importance of fluctuations in light on plant photosynthetic acclimation. Plant Physiol. **173**, 2163–2179 (2017)
7. Chen, L.A.W.F., Artur, N.G., Jessica, T., Jennie, T.: One Thousand and One Garden Plants in Singapore, p. 824. 3rd Edition. National Parks Board (2003)
8. Victor, J.E., Koekemoer, M., Fish, L., Smithies, S.J., Moessmer, M.: Herbarium essentials: The Southern African Herbarium User Manual (2004)
9. Carter, R., Bryson, C.T., Darbyshire, S.J.: Preparation and use of voucher specimens for documenting research in weed science. Weed Technol. **21**(4), 1101–1108 (2007)
10. Maden, K.: Plant collection and herbarium techniques. Our Nat. **2**(1), 53–57 (2004)
11. Yunusa, A.: Evaluation of carbon dioxide absorption capacity and biomass accumulation of sub-urban plants in Universiti Tun Hussein Onn Malaysia (UTHM). Universiti Tun Hussein Onn Malaysia, Department of Biodiversity (2018)
12. Ethier, G.J., Livingston, N.J.: On the need to incorporate sensitivity to CO2 transfer conductance into the Farquhar-von Caemmerer-Berry leaf photosynthesis model. Plant Cell Environ. **27**(2), 137–153 (2004). https://doi.org/10.1111/j.1365-3040.2004.01140.x
13. Lambreva, M., Stoyanova-Koleva, D., Baldjiev, G., Tsonev, T.: Early acclimation changes in the photosynthetic apparatus of bean plants during short-term exposure to elevated CO2 concentration under high temperature and light intensity. Agr Ecosyst Environ **106**(2–3), 219–232 (2015)
14. Zhang, Z., Yuan, Y., Liu, Q., Yin, H.: Plant nitrogen acquisition from inorganic and organic sources via root and mycelia pathways in ectomycorrhizal alpine forests. Soil Biol. Biochem. **136**, 107517 (2019). https://doi.org/10.1016/j.soilbio.2019.06.013
15. Dusenge, M.E., Duarte, A.G., Way, D.A.: Plant carbon metabolism and climate change: elevated CO2 and temperature impacts on photosynthesis, photorespiration and respiration. New Phytol. (2018). https://doi.org/10.1111/nph.15283
16. Li, A., Li, S., Wu, X., Zhang, J., He, A., Zhao, G., Yang, X.: Effect of light intensity on leaf photosynthetic characteristics and accumulation of flavonoids in Lithocarpus litseifolius (Hance) Chun. (Fagaceae). Open J. For. **6**(5), 445–459 (2016). https://doi.org/10.4236/ojf.2016.65034
17. Marañón, E., Lorenzo, M.P., Cermeño, P., Mouriño-Carballido, B.: Nutrient limitation suppresses the temperature dependence of phytoplankton metabolic rates. ISME J. **12**(7), 1836–1845 (2018). https://doi.org/10.1038/s41396-018-0105-1
18. Warren, J.M., Jensen, A.M., Medlyn, B.E., Norby, R.J., Tissue, D.T.: Carbon dioxide stimulation of photosynthesis in Liquidambar styraciflua is not sustained during a 12-year field experiment. AoB PLANTS **7**(1), 1–13 (2014). https://doi.org/10.1093/aobpla/plu074
19. Anjum, S.A., Xie, X.Y., Wang, L.C., Saleem, M.F., Man, C., Lei, W.: Morphological, physiological and biochemical responses of plants to drought stress. Afr. J. Agric. Res. **6**(9), 2026–2032 (2011)
20. Sharkey, T.D., Bernacchi, C.J., Farquhar, G.D., Singsaas, E.L.: Fitting photosynthetic carbon dioxide response curves for C3 leaves. Plant Cell Environ. **30**(9), 1035–1040 (2007). https://doi.org/10.1111/j.1365-3040.2007.01710.x

21. Zhang, H., Zhu, J., Gong, Z., Zhu, J.K.: Abiotic stress responses in plants. Nat. Rev. Genet. **23**, 104–119 (2022). https://doi.org/10.1038/s41576-021-00413-0
22. Umaña, M.N., Cao, M., Lin, L., Swenson, N.G., Zhang, C.: Trade-offs in above and below-ground biomass allocation influencing seedling growth in a tropical forest. J. Ecol. (2020). https://doi.org/10.1111/1365-2745.13543
23. Evans, J.R., Poorter, H.: Photosynthetic acclimation of plants to growth irradiance: the relative importance of specific leaf area and nitrogen partitioning in maximizing carbon gain. Plant Cell Environ. **24**(8), 755–767 (2001). https://doi.org/10.1046/j.1365-3040.2001.00724.x
24. Lewis, V.R., Farrell, A.D., Umaharan, P., Lennon, A.M.: Genetic variation in high light responses of Theobroma cacao L. accessions. Heliyon **7**(6), e07404 (2021). https://doi.org/10.1016/j.heliyon.2021.e07404
25. Negi, S., et al.: Light regulation of light-harvesting antenna size substantially enhances photosynthetic efficiency and biomass yield in green algae. Plant J. **103**(2), 584–603 (2020). https://doi.org/10.1111/tpj.14751

Engineering Sciences

Agri-Food Supply Chain Collaboration with the Integration of Blockchain Technology: An Application Case Study in Rice Supply Chain

Zhongchao Hao[1,2] and Mohd Yamani bin Yahya[3(✉)]

[1] Universiti Tun Hussein Onn Malaysia, 86400 Parit Raja, Johor, Malaysia
[2] Changzhou Vocational Institute of Industry Technology, Changzhou 213164, Jiangsu, China
[3] Universiti Tun Hussein Onn Malaysia, 86400 Parit Raja, Johor, Malaysia
yamani@uthm.edu.my

Abstract. With the increasing complexity of Agri-food supply chains and the growing demand for food safety and transparency, blockchain technology presents a promising solution. However, effectively integrating blockchain into Agri-food supply chain collaboration and determining the most suitable collaborative model remain pressing challenges. This study examined the practical implementation and collaboration model of a blockchain-integrated Agri-food supply chain in China, using DN Rice Smart Farm as a case study. Adopting a qualitative research approach, data were collected through semi-structured interviews with seven respondents, including farm managers and department directors. Thematic analysis was employed to interpret the findings. The results indicated that implementing a blockchain-based collaboration model transformed the traditional Agri-food supply chain, addressing collaboration challenges in an environment of low trust among participants. This case study provides a practical and referable model for applying blockchain technology to enhance collaboration in Agri-food supply chains, particularly within smart farms.

Keywords: Agri-food Supply Chain Collaboration · Blockchain Technology · Integration · Supply Chain

1 Introduction

The agricultural sector is undergoing a transformative shift with the integration of advanced technologies [1], among which Blockchain Technology (BCT) stands out for its potential to enhance Agri-food Supply Chain (SC) governance. This study takes DN Rice Smart Farm, a pioneering case study that illustrates the integration of BCT within the Agri-food Supply Chain Collaboration (SCC). By analyzing DN Rice Smart Farm, this research aims to explore the practical implementation and collaborative model of BCT in improving the transparency, efficiency, and traceability of Agri-food SC.

BCT offers a decentralized and immutable ledger system that can record transactions and track assets in a SC network. This technology is particularly valuable in the context of agriculture, where issues such as food safety, traceability, and efficiency are

P. Chang et al. (Eds.): SCIEMATHIC 2024, SPPHY 427, pp. 253–263, 2025.
https://doi.org/10.1007/978-981-96-7679-8_20

paramount [2]. Smart contract is an important part of BCT, which refers to the contract defined by digital code, and the parties involved form an agreement, and can automatically identify, judge and execute conditions, touch the contract threshold, and the contract will be automatically verified and executed, without the need for institutions or individuals to adjudicate or execute [3]. Smart contract is like contract mechanism [4]. It is implemented through code programming [5], has a wider range of applications and lower investment costs [6]. Smart contract can automatically identify the execution environment and control processes without interference from human factors [7]. Its execution process is more transparent [8] and the risk of default is small, which can reduce or even eliminate dependence on credit intermediaries [9] and reduce transaction costs.

BCT has been integrated into the Agri-food SC process. (a) In the production of Agri-food, BCT is combined with the IoT technology, and sensors automatically record key information of Agri-food and record it in the BCT [10]. The key information includes [11] environment, the growth process of plants, records of the use of fertilizers and pesticides, relevant information of the warehouse. When an abnormal situation occurs, the smart contract will be automatically triggered [12]. (b) In Agri-food processing, farmers or Agri-food intermediary organizations directly transfer ownership of Agri-food to processing enterprises through BCT [13]. Processing enterprises record new data into the BCT. The data includes [14] environment, relevant information about warehouses, basic information about processing enterprises and related personnel, financial transactions between producer and processing enterprises. (c) In the process of Agri-food storage and transportation, BCT can provide a shared standard for each stakeholder in the SC to record and exchange data at each stage of storage and transportation [15]. Real-time environmental data is recorded into the BCT through sensors [16], and an alarm will be issued immediately when the data exceeds safety standards [17]. By using GPS, the distribution center can locate each transport truck and develop the best transportation route to shorten delivery time [18]. The transportation route, method, and time will be recorded [19]. (d) In Agri-food sales, sales enterprises store detailed information on the quantity and quality of received Agri-food in the BCT [20]. Sensors regularly automatically store information about the sales environment status in the BCT [16]. Some Agri-foods are prone to decay and have a short lifespan [21]. Sales enterprises can use traceability systems to monitor the freshness of the products and replace the products that are close to their shelf life [22]. (e) In Agri-food consumption, enterprises respectively store detailed information about sold Agri-food in the BCT. Consumers can transparently verify the entire history of the Agri-food before purchasing [23]. If a quality and safety accident occur, defective the Agri-food can be immediately found [24]. Consumers can use their phones to connect to Internet or APPs to scan QR code related to the Agri-food and view all relevant information [25].

In China, the application of BCT in the Agri-food SCC has gained significant traction, driven by government initiatives, and increasing consumer demand for food safety and transparency [26]. In the application research of BCT, the integration of BCT and SC is an important research direction [27], forming part of the research results. On the one hand, many scholars have designed and built the operation architecture of the SC through the BCT architecture from a technical angle. For example, Liao [28] proposed the logical architecture of the Agri-food SC based on the BCT; Wang [29], Ning [30], and

Li et al. [31] designed the technical functional architecture and certification traceability process of BCT embedded in the quality and safety management of Agri-food SC. Cao et al. [32] combined decentralization, transparency, fairness and openness of BCT with the concept of energy Internet, and analyzed the feasibility of BCT application in food SC from different angles such as function, subject and performance. On the other hand, some scholars analyzed and deduced the application logic of BCT in specific scenarios from the angle of BCT application. For example, Wu [33] deeply analyzed the market potential and feasibility of BCT in the development of SC finance; Hao & Feng [34] discussed the intelligent management mechanism formed by the BCT coupling SC from three different angles: efficiency, fairness and sustainable development. Jiang [35] summarized several specific scenarios where BCT embedding in Agri-food circulation field can solve the application pain points, including: cross-border Agri-food trade or logistics alliance chain, Agri-food traceability alliance chain, agricultural financing BCT that can efficiently transfer value, etc. The Agri-food SC includes multiple links [36]. Each link involves related sub-links and different organization carriers, with a long and loose chain and a complex organization structure [37]. Due to the complex transaction structure of Agri-food SC [38], the evolution of its organization structure and the changes in the cost and flexibility of its collaboration mechanism will also cause many practical problems. Trace the logical starting point of all problems [39], namely information asymmetry within and outside the SC, and the resulting trust crisis [40]. Therefore, to establish a trust relationship in the Agri-food SC, needs to solve the problem of information asymmetry and improve transparency of Agri-food information [41]. The introduction of BCT into the Agri-food SC can solve information asymmetry and trust crises [42]. BCT can provide immutable and accurate statistical data related to Agri-food production, processing, storage and transportation, sales, consumption, credit scoring, and tracking [43]. It can play a role in all aspects of Agri-food SC.

The adoption of BCT in China's agricultural sector aims to address several challenges. Firstly, it enhances traceability, allowing consumers to access detailed information about the origin and journey of Agri-food, thus ensuring food safety and quality [44]. Secondly, BCT facilitates SCC by providing a shared, transparent platform for all stakeholders, including farmers, processors, distributors, and retailers [45]. Lastly, it improves efficiency by reducing fraud, errors, and administrative costs through automated and transparent transaction records [46].

By using DN Rice Smart Farm as a case study, this research will examine the practical implementation of BCT on Agri-food SCC. The results of the analysis will help to enrich the knowledge on BCT applications in agriculture and provide valuable lessons for other farms and agribusinesses considering similar technological integrations.

2 Methods

This study adopts qualitative research, case study to explore Agri-food SCC with the integration of BCT. DN Rice Smart Farm is located in Jiangsu, China. It is a representative and leading agricultural enterprise in the region. It adopts advanced technologies such as the BCT, Internet of Things, big data, and artificial intelligence. It has a good technical foundation and data accumulation, and is suitable as a test scenario for BCT integration.

DN Rice Smart Farm has rich practical experience in Agri-food SCC and is a successful case of effectively promoting the application of BCT in Agri-food SCC. So, case study is particularly well-suited for this research as it allows for an in-depth, contextual analysis of complex issues within their real-life settings. This approach facilitates a comprehensive understanding of the factors and dynamics at play [47], which might not be captured through other research methods.

The researcher used a purposive sampling technique in selecting the participants that meet the research criteria. Purposive sampling selects the research participants that can provide the most information for the research questions according to the research purpose [48]. The sample criteria of this research are as follows: (a) manager or department director. (b) more than 5 years' work experience in BCT area is required. They have a comprehensive understanding of the entire SC process and the integration of blockchain technology. They can propose solutions to the integration of BCT and Agri-food SCC from a technical perspective, or put forward strategic plans and suggestions for the integration from a strategic level. This study recruited participants from smart farm via email, phone calls, and snowballing. Data for the case study were collected through semi-structured interviews, which allows for the exploration of specific topics while also enables respondents to share insights freely [49]. Semi-structured interviews [50] are particularly effective in capturing rich, detailed data and understanding the perspectives of participants in depth.

The primary method of data analysis employed is thematic analysis. Thematic analysis is a qualitative analytic method for identifying, analyzing, and reporting patterns (themes) within data [51]. Thematic analysis was conducted following Clarke and Braun's six-phase framework [52]. The research aims to uncover underlying themes that shed light on Agri-food SCC with the integration of BCT, providing deeper insights and contributing to the existing body of knowledge.

This methodological combination of case study and thematic analysis ensures a thorough examination of the research subject, capturing both the broad patterns and the nuanced details essential for a holistic understanding.

3 Case Analysis: DN Rice SCC Model with the Integration of BCT

3.1 Demographics of Respondents

Table 1 is the demographic of 7 respondents. Their job positions, years of experience in blockchain and other information are shown.

As shown in Table 1, the 7 interviewees included participants from smart farm, manager, or directors. Manager or directors have more than 5 years of work experience.

3.2 Implementation of DN-Harvest-Cloud

DN Rice smart farm has established Agricultural Machinery Service Professional Cooperative Commune and Organic Rice Planting Cooperative Commune in the past 10 years. It relies on organic rice planting industry as the main body. Local farmers are responsible for planting organic rice and taking care of it. The smart farm is responsible for collecting

Table 1. Demographic of respondents.

No	Position	Code	Years of Total Experience	Years of Experience in Blockchain
1	Smart Farm Manager	R1	16	8
2	Director of Planting Department	R2	15	6
3	Director of Picking and Processing Department	R3	13	7
4	Director of Warehousing Department	R4	8	6
5	Director of Logistics Department	R5	17	6
6	Director of Sales Department	R6	12	8
7	Director of Information Department	R7	14	8

and selling uniformly. The series of soft rice produced by the Smart Farm has created DN brand and has successively passed the certifications of Organic Products and Green Food. DN Rice Smart Farm has become a pioneer in the application of BCT in Agri-food SC in China. This study takes DN Rice Smart Farm as an example to analyze Agri-food SCC with the integration of BCT. DN Rice adopts a BCT & Agriculture cloud service platform, coupled with various information technologies such as the IoT, visualization, big data, AI. It innovatively creates a BCT-based rice SCC system. As shown in Fig. 1, DN Rice SCC model with the Integration of BCT, DN-Harvest-Cloud.

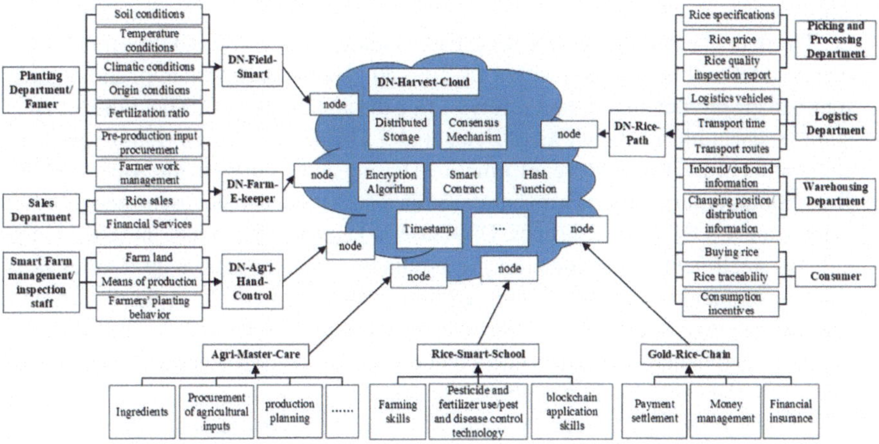

Fig. 1. DN-Harvest-Cloud (Source: DN Rice Smart Farm).

Figure 1 is provided by the Technology Department of DN Rice Smart Farm. It shows how DN Rice Smart Farm builds a rice SCC model based on BCT to achieve modern and standardized production management. Various smart devices are used to collect data on the soil environment, fertilizer use, seedling information, rice processing, storage, transportation, sales and consumption. These data are uploaded to the BCT through process screening and algorithm screening to ensure the objective authenticity of the data. DN Rice can achieve a comprehensive digital mapping of rice SC participants, production and circulation, agricultural assets, and transaction systems. It breaks data islands and forms a consistent BCT-based database for the rice SC. In other words, DN Rice develops BCT-based APPs and digital equipment for digital governance of the rice SC.

Through the thematic analysis of the interview data, it is concluded that the following themes are involved in the implementation of DN-Harvest-Cloud: IoT device, data uplink and tracing, rice management APP, online transactions, farm management tools.

(a) **Rice Production.** The smart farm matches farmers with a BCT-based field planting management IoT device, DN-Field-Smart.

R2 stated, farmers can automatically obtain basic data in the rice planting process through the intelligent sensing equipment of DN-Field-Smart, with data geo-stamps and timestamps.

They can also manually judge the changes in rice seedlings based on experience and supplement the data input. This is similar with the description of Wen et al. [10] about BCT and IoT application in Agri-food SC.

(b) **Rice Processing and Circulation.** Smart Farm developed the DN-Rice-Path. DN-Rice-Path will collect data from logistics vehicles, drying yards, and warehouses embedded with intelligent sensing devices in the SC, and directly uplink these data.

R3 descripted some hands-on process of data uplink. The smart hardware device of the harvester......Smart sensor equipment embedded in the factory and IoT devices held by workers......R5 discussed, the smart hardware of the logistics vehicle......R4 said, IoT devices can also collect rice storage data and upload it to the DN-Rice-Path database.

The data in each link on DN-Rice-Path is mutually verified and proofread. Each rice product will be stamped with a BCT-based geo-stamp, timestamp, and quality-stamp to ensure the quality and safety of rice throughout the SC. The effect of BCT at this stage confirms the views of Feng et al.'s [14]. These descriptions explain the application of BCT to achieve information sharing and security in the Agri-food SC.

(c) **Rice Management.** Smart Farm has developed an APP that matches the needs of rice management, DN-Farm-E-keeper. DN-Farm-E-keeper can conduct integrated, digital, and intelligent management of pre-production input procurement, farmer work management and other content required for rice management. The sales department can also manage rice sales and financial services by DN-Farm-E-keeper. And the operation and management data carried out on DN-Farm-E-keeper will be uploaded to the DN-Harvest-Cloud through the backend.

R1 took the rice sales as an example, illustrated, consumers can order rice grown by farmers online. Farmers/salespeople can track orders. Once a sale is made, the smart contract will automatically distribute the proceeds to each party.

(d) **Rice Transactions.** On the one hand, the smart farm uses BCT to organically combine the circulation of digital assets on farm land with bulk raw grain transactions. Smart farm links the digital assets of agricultural land to the tradable and mortgageable digital planting orders produced by the land. Agricultural land has been digitized through its association with orders and real grains. The research of Jiang [35] described the similar application of BCT in Agri-food SCs. On the other hand, DN Rice promotes consumer-oriented e-commerce sales.

R6 said, consumers can order rice produced on a particular part of farm land online. Farmers can contract orders through DN-Field-Smart.

DN-Harvest-Cloud has established a closed-loop information and traceability certification process for rice. Whether it is rice trading based on BCT or online rice pre-orders, DN Rice uses smart contracts. DN-Harvest-Cloud digitizes and programs rice orders through script codes. Once the smart contract terms are triggered, they will be automatically executed. This is consistent with previous research [6][8][9] in the field of smart contracts and Agri-food SCs. Each party has no right to unilaterally cancel or prevent the execution of the contract, which effectively solves the credit problem in the traditional SC. Through smart contracts, DN Harvest-Cloud presents the rights, responsibilities, and interests in the Agri-food SC in a transparent, open, and fair manner. The originally fragmented SC entities have achieved efficient collaboration and positive incentives through the terms and conditions of smart contracts [54].

(e) **Agri-food production management.** Smart Farm developed DN-Agri-Hand-Control as a handheld IoT terminal for Agri-food production management at farm nodes. DN-Agri-Hand-Control is the main tool for farm supervision and management of farmers' production and business activities. It can conduct comprehensive management of farm land, production materials, and farmers' planting behaviors. All data uploaded and collected by DN-Field-Smart and DN-Farm-E-keeper will be synchronized to DN-Agri-Hand-Control. At the same time, DN-Agri-Hand-Control equipment will be regularly used to collect and verify data in the planting fields.

R7 expressed, if the data uploaded by farmers and the data collected meet the standards……digital signature confirmation. Else……If there are serious violations of standardized planting or serious problems with rice quality, farmers may also lose the land contract qualification.

In addition, Agri-Master-Care was developed to provide integrated services to participants in the SC and unify seed breeding, agricultural input procurement, production planning, warehousing acquisition, logistics distribution, etc. Rice-Smart-School was developed to disseminate farming skills and knowledge, and provide comprehensive technical services such as the use of fertilizers and pesticides, and pest and disease control technologies to farmers. Gold-Rice-Chain was created to form a consistent database based on BCT with financial institutions. Based on big data analysis of rice SC, financial services such as payment settlement, fund management, lending, and insurance are carried out. Provide effective protection for the financial needs of farmers and small or medium-sized enterprises in the rice SC.

DN-Harvest-Cloud has changed the hierarchical mode of traditional Agri-food SC. It solves the problem of unsystematic and unstandardized planting methods in traditional agriculture. It eliminates the abuse of agricultural assets and resources such as processing plants and agricultural input products. It avoids the confusion and lack of Agri-food quality certification caused by the difficulty of quantifying the operation process and opaque planting and processing. The smart farm has completed the technological upgrade of the current Agri-food production and processing process. It has realized the tracking and recording of the entire process of Land-Seed-Farmer-Planting, and guaranteed the production safety and quality of food from the source [55]. Its production, logistics, processing, warehousing, social services and other links have formed an autonomous and trusted SC ecology [56] with decentralization, distribution according to work, and resource sharing in the BCT-based value network system to solve the problem of participation. It solves the collaboration issues of participants in a distrustful environment.

This study has drawn some important conclusions through the case study. (a) The application of BCT in the Agri-food SC has significantly improved information transparency and traceability. By recording information at every link through BCT, every step from production, processing to transportation and sales can be verified and tracked. This not only improves consumers' trust in Agri-food, but also enhances the collaborative efficiency of SC. (b) The decentralized nature of BCT reduces the participation of intermediaries and reduces transaction costs. The automatic execution of transactions through smart contracts greatly improves efficiency and reduces the possibility of human errors and fraud. (c) The application of BCT in Agri-food SCC also promotes data security and privacy protection. The data encryption and distributed storage methods of BCT ensure the immutability and security of data. This provides higher protection for sensitive data involved in the Agri-food SC.

4 Conclusion

A case study on the Agri-food SCC of Rice Smart Farm shows that the innovative implementation of the Agri-food SCC model has significantly changed the traditional mode. The key difference between DN Rice Smart Farm's BCT cloud platform application and previous related research is that it innovatively integrates the core characteristics of BCT with the collaborative mechanism of the Agri-food SC. Through the two-way empowerment of technology and mechanism, it realizes the systematic combination of technical tools and institutional design, promotes the full chain of agricultural resources and asset digitization. It realizes the traceability of the entire chain of Agri-food. It breaks through the limitations of traditional research on the separation of technology application and mechanism optimization, and provides an integrated solution for the Agri-food SCC. This case study provides a valuable practical guidance model and reference basis for the application of BCT in smart farm Agri-food SCC.

This study has revealed many advantages of BCT in the model of Agri-food SCC. However, there are still many aspects that deserve further exploration. More empirical research is needed to verify the application effect of BCT in different Agri-food SC. The promotion of BCT in Agri-food SC also requires the support and supervision of policies and regulations. The establishment of a legal framework and standard specifications need to be explored to adapt to the development of BCT. In short, the application prospects of BCT in Agri-food SCC are broad, but to achieve its comprehensive promotion and efficient operation, more in-depth and systematic research is needed at the theoretical and practical levels.

References

1. Fuentes-Peñailillo, F., Gutter, K., Vega, R., Silva, G.C.: Transformative technologies in digital agriculture: leveraging internet of things, remote sensing, and artificial intelligence for smart crop management. J. Sens. Actuator Netw. **13**(4), 39 (2024)
2. Xu, X., Chen, H., Lu, Q.: Blockchain technology and its implications for food safety and traceability in the supply chain. Food Control **132**, 108351 (2022)
3. Mohanta, B.K., Panda, S.S. Jena, D.: An overview of smart contract and use cases in blockchain technology. In: 2018 9th International Conference on Computing, Communication and Networking Technologies (ICCCNT), pp. 1–4 (2018)
4. Tang, Z.: Advancing blockchain ecosystems: smart contract security and scalability examined. In: 2024 2nd International Conference on Image, Algorithms and Artificial Intelligence (ICIAAI 2024), pp. 366–379. Atlantis Press (2024)
5. Karanjai, R., Xu, L., Shi, W.: Teaching machines to code: smart contract translation with LLMs. arXiv preprint arXiv:2403.09740 (2024)
6. Elias, O., et al.: The evolution of green fintech: leveraging AI and IoT for sustainable financial services and smart contract implementation. World J. Adv. Res. Rev. **23**(1), 2710–2723 (2024)
7. Wu, C., Xiong, J., Xiong, H., Zhao, Y., Yi, W.: A review on recent progress of smart contract in blockchain. IEEE Access **10**, 50839–50863 (2022)
8. Keskin, N.B., Li, C., Song, J.S.J.: The blockchain newsvendor: value of freshness transparency and smart contracts. Manage. Sci. (2024)
9. Wu, H., et al.: Blockchain for finance: a survey. IET Blockchain (2024)
10. Wen, Q., Gao, Y., Chen, Z., Wu, D.: A blockchain-based data sharing scheme in the supply chain by IoT. In: 2019 IEEE International Conference on Industrial Cyber Physical Systems (ICPS), pp. 695–700. IEEE (2019)
11. Kaur, A., Singh, G., Kukreja, V., Sharma, S., Singh, S., Yoon, B.: Adaptation of IoT with blockchain in food supply chain management: an analysis-based review in development, benefits and potential applications. Sensors **22**(21), 8174 (2022)
12. Jiao, T., Xu, Z., Qi, M., Wen, S., Xiang, Y., Nan, G.: A survey of ethereum smart contract security: attacks and detection. Distrib. Ledger Technol. Res. Pract. **3**(3), 1–28 (2024)
13. Marchese, A., Tomarchio, O.: A blockchain-based system for agri-food supply chain traceability management. SN Comput. Sci. **3**(4), 279 (2022)
14. Feng, H., Wang, X., Duan, Y., Zhang, J., Zhang, X.: Applying blockchain technology to improve agri-food traceability: a review of development methods, benefits and challenges. J. Clean. Prod. **260**, 121031 (2020)
15. Chandan, A., John, M., Potdar, V.: Achieving UN SDGs in food supply chain using blockchain technology. Sustainability **15**(3), 2109 (2023)
16. Sathiya, V., Nagalakshmi, K., Raju, K., Lavanya, R.: Tracking perishable foods in the supply chain using chain of things technology. Sci. Rep. **14**(1), 21621 (2024)
17. Su, L., et al.: Chinese experts' consensus on the application of intensive care big data. Front. Med. **10**, 1174429 (2024)
18. Laranjeiro, P.F., et al.: Using GPS data to explore speed patterns and temporal fluctuations in urban logistics: the case of São Paulo, Brazil. J. Transp. Geogr. **76**, 114–129 (2019)

19. Shahmohammad, M., Salamattalab, M.M., Sohn, W., Kouhizadeh, M., Aghamohmmadi, N.: Opportunities and obstacles of blockchain use in pursuit of sustainable development goal 11: a systematic scoping review. Sustain. Cities Soc. 105620 (2024)
20. Adamashvili, N., Zhizhilashvili, N., Tricase, C.: The integration of the internet of things, artificial intelligence, and blockchain technology for advancing the wine supply chain. Computers **13**(3), 72 (2024)
21. Moussavi, S.E., Sahin, E., Riane, F.: A discrete event simulation model assessing the impact of using new packaging in an agri-food supply chain. Int. J. Syst. Sci. Oper. Logistics **11**(1), 2305816 (2024)
22. Tagarakis, A.C., Benos, L., Kateris, D., Tsotsolas, N., Bochtis, D.: Bridging the gaps in traceability systems for fresh produce supply chains: overview and development of an integrated iot-based system. Appl. Sci. **11**(16), 7596 (2021)
23. Hasan, I., Habib, M.M., Mohamed, Z., Tewari, V.: Integrated agri-food supply chain model: an application of IoT and blockchain. Am. J. Ind. Bus. Manag. **13**(2), 29–45 (2023)
24. Zhao, G., et al.: Risk analysis of the agri-food supply chain: a multi-method approach. Int. J. Prod. Res. **58**(16), 4851–4876 (2020)
25. Bhatia, S., Albarrak, A.S.: A blockchain-driven food supply chain management using QR code and XAI-faster RCNN architecture. Sustainability **15**(3), 2579 (2023)
26. Zhou, G., Hu, J.: The role of blockchain in enhancing transparency and efficiency in China's agricultural supply chains. J. Clean. Prod. **350**, 131456 (2022)
27. Younis, H., Bwaliez, O.M., Al-Okaily, M., Tanveer, M.I.: Revolutionizing supply chain management: a critical meta-analysis of empowerment and constraint factors in blockchain technology adoption. Bus. Process Manage. J. (2024)
28. Liao, D.A.: Research and implementation of agricultural product traceability trusted platform based on blockchain technology. Hubei Agric. Sci. **07**, 140–144 (2022)
29. Wang, Y.: Analysis of agricultural product quality and safety traceability system based on blockchain. Qinghai Agric. For. Technol. (03), 59–61+80 (2020)
30. Ning, D.D.: Research on the construction of quality management system of fresh agricultural products supply chain based on blockchain. Logistics Eng. Manage. **12**, 70–73 (2022)
31. Li, P., Zhang, L.T., Li, Y.F.: Research on credit mechanism construction strategy of fresh agricultural products supply chain based on blockchain technology. Food Ind. **01**, 90–92 (2022)
32. Cao, Y., Tao, L., Wu, K., Wan, G.: Coordinating joint greening efforts in an agri-food supply chain with environmentally sensitive demand. J. Clean. Prod. **277**, 123883 (2020)
33. Wu, T.: The Application of Blockchain in the Field of Money and Finance: Theory and Demonstration (Central University of Finance and Economics) (2020)
34. Hao, L., Feng, L.: Risk control mechanism of fresh food supply chain driven by internet + smart contract. Logistics Technol. **03**, 90–98 (2022)
35. Jiang, X.: Research on the application mode of blockchain technology in the circulation of agricultural products. China Agric. Account. **04**, 89–92 (2022)
36. Raimbekov, Z., et al.: The impact of agri-food supply channels on the efficiency and links in supply chains. Economies **11**(8), 206 (2023)
37. Alshurideh, M., Kurdi, B., Alzoubi, H., Obeidat, B., Hamadneh, S., Ahmad, A.: The influence of supply chain partners' integrations on organizational performance: the moderating role of trust. Uncertain Supply Chain Manage. **10**(4), 1191–1202 (2022)
38. Hernandez, J.E., Mortimer, M., Panetto, H.: Operations management and collaboration in agri-food supply chains. Prod. Plan. Control **32**(14), 1163–1164 (2021)
39. Rocha, T., Costa, P., Sousa, V., Coelho, P., Sousa, F., Cardoso, N.: SmartAgriChain: a blockchain based solution for agri-food certification and supply chain management. Int. J. Environ. Agric. Biotechnol **6**, 235–245 (2021)

40. DuHadway, S., Mena, C., Ellram, L.M.: Let the buyer beware: how network structure can enable (and prevent) supply chain fraud. Int. J. Oper. Prod. Manag. **42**(2), 125–150 (2022)
41. Mardenli, A., Sackmann, D., Fiedler, A., Rhein, S., Alghababsheh, M.: Determinants of information asymmetry in agri-food supply chains. Int. J. Logistics Manage. (2024)
42. Bosona, T., Gebresenbet, G.: The role of blockchain technology in promoting traceability systems in agri-food production and supply chains. Sensors **23**(11), 5342 (2023)
43. Baralla, G., Pinna, A., Tonelli, R., Marchesi, M., Ibba, S.: Ensuring transparency and traceability of food local products: a blockchain application to a smart tourism region. Concurrency Comput. Pract. Experience **33**(1), e5857 (2021)
44. Li, X., Zhang, Y., Liu, H.: Enhancing food traceability with blockchain technology: a case study of China's agricultural sector. Int. J. Inf. Manage. **64**, 102456 (2023)
45. Wang, Y., Han, J.H., Zhang, P.: Blockchain technology for agricultural supply chains: benefits, challenges, and future research directions. Supply Chain Manage. Int. J. **27**(4), 512–526 (2022)
46. Chen, S., Wang, J.: Blockchain in agri-food supply chains: a review of the benefits and challenges. J. Agric. Food Chem. **71**(3), 234–245 (2023)
47. Alam, M.K.: A systematic qualitative case study: questions, data collection, NVivo analysis and saturation. Qual. Res. Org. Manage. Int. J. **16**(1), 1–31 (2021)
48. Li, W.Y., Tu, W.Y., Xu, X.Q.: Research on the influence of sample size on the interpretation of qualitative research topics and phenomena. J. Nurs. **03**, 62–65 (2020)
49. Chen, S., Lou, V.W., Leung, R., Doris, S.F.: Meaning-making of dementia caregiving: a systematic review of qualitative studies. Int. J. Nurs. Stud. 104848 (2024)
50. Karatsareas, P.: Semi-structured interviews. Res. Methods Lang. Attitudes, 99–113 (2022)
51. Clarke, V., Braun, V.: Thematic analysis. J. Posit. Psychol. **12**(3), 297–298 (2017)
52. Braun, V., Clarke, V., Hayfield, N.: 'A starting point for your journey, not a map': Nikki Hayfield in conversation with Virginia Braun and Victoria Clarke about thematic analysis. Qual. Res. Psychol. **19**(2), 424–445 (2022)
53. Liu, Y., et al.: An overview of blockchain smart contract execution mechanism. J. Ind. Inf. Integr. 100674 (2024)
54. Liu, S., Liu, Z., Chen, B., Pan, X.: Construction and application of online learning resource incentive mechanism driven by smart contract. IEEE Access (2024)
55. Granillo-Macías, R., Gonzalez Hernandez, I.J., Olivares-Benítez, E.: Logistics 4.0 in the agri-food supply chain with blockchain: a case study. Int. J. Logistics Res. Appl. **27**(10), 1766–1786 (2024)
56. Zrelli, I., Rejeb, A.: A bibliometric analysis of IoT applications in logistics and supply chain management. Heliyon, **10**(16) (2024)

Omics Data Analysis for Cancer Diagnosis and Prognosis: A Comparative Study of Deep Learning Techniques

Logenthiran Machap[1](✉), Weng Howe Chan[2], Kauthar Mohd Daud[3], Noriza Satam[4], and Wan Nor Munirah Ariffin[1]

[1] Computational Intelligence and Data Science Focus Group (CIDaS), Faculty of Applied Sciences and Technology, Universiti Tun Hussein Onn Malaysia (Pagoh Campus), 84600 Muar, Johor, Malaysia
{logen,normunirah}@uthm.edu.my

[2] Faculty of Computing, Universiti Teknologi Malaysia, 81310 Johor Bahru, Johor, Malaysia
cwenghowe@utm.my

[3] Center for Artificial Intelligence Technology, Fakulti Teknologi Dan Sains Maklumat, Universiti Kebangsaan Malaysia, 43600 Bangi, Selangor, Malaysia
kauthar.md@ukm.edu.my

[4] Department of Computing and Information Technology, Tunku Abdul Rahman University of Management and Technology, Johor Branch, 85000 Segamat, Johor, Malaysia
norizas@tarc.edu.my

Abstract. Recent advancements in omics technologies have generated vast amounts of complex biological data, providing a deeper molecular understanding of cancer. Leveraging this data for cancer classification and prediction is crucial for early detection, effective treatment, and improved patient outcomes. This study compares various deep learning models, including Convolutional Neural Networks (CNNs), Recurrent Neural Networks (RNNs), Deep Belief Networks (DBNs), and Autoencoders, using diverse types of omics data—genomics, transcriptomics, proteomics, and metabolomics. Omics data is valuable for uncovering new biomarkers and insights into cancer, but it is challenging to analyze due to its high dimensionality and noise. Our comparative analysis involves preprocessing steps such as normalization, imputation, and feature selection, and evaluates model performance using metrics like accuracy, precision, recall, F1-score, and AUC-ROC. Results indicate that deep learning models generally outperform traditional methods, with each model demonstrating unique strengths: CNNs excel at spatial pattern recognition, RNNs are effective with time-series data, DBNs integrate multi-omics data well, and Autoencoders are adept at dimensionality reduction. Case studies highlight these models' potential to improve cancer diagnosis, predict patient outcomes, and identify novel therapeutic targets. Despite their promise, challenges remain, including data quality, model interpretability, generalizability, and computational demands. Future research should focus on enhancing model transparency, expanding datasets, and optimizing computational efficiency. Overall, this study underscores the transformative potential of deep learning in cancer research and the need for interdisciplinary collaboration to advance precision oncology.

P. Chang et al. (Eds.): SCIEMATHIC 2024, SPPHY 427, pp. 264–277, 2025.
https://doi.org/10.1007/978-981-96-7679-8_21

Keywords: Deep Learning · Omics Data · Biomarker · Cancer Oncology · Precision Oncology

1 Introduction

Cancer remains a major global health problem, with millions of new cases diagnosed each year [1]. Accurate diagnosis and prognosis are crucial for guiding treatment and improving outcomes. Traditional methods like examining tissues and medical imaging provide helpful information, but they often have limitations in understanding the complex molecular aspects of cancer [2]. Recently, omics technologies have revolutionized cancer research by offering a complete picture of the molecular changes involved in cancer development. This data provides a detailed, multi-layered view of the biological processes, helping researchers discover new biomarkers, understand disease mechanisms, and develop targeted treatments [3, 4]. However, the massive amount of data generated by omics technologies presents significant challenges, including dealing with its complexity, noise, and heterogeneity [4].

Traditional statistical and machine learning methods often struggle to handle these challenges effectively. Deep learning, a type of machine learning, offers a promising alternative. Unlike traditional methods, deep learning models can automatically learn patterns from raw data, eliminating the need for manual feature engineering. These models can capture intricate relationships within complex datasets, making them well-suited for analyzing omics data. Additionally, deep learning models can handle large datasets, further enhancing their applicability in cancer research [5].

This study aims to compare various deep-learning approaches for analyzing omics data in the context of cancer diagnosis and prognosis. By evaluating the performance of different models across multiple types of omics data, we seek to identify the most effective strategies for improving diagnostic accuracy and prognostic predictions. The findings of this study will provide valuable insights into the strengths and limitations of each approach, contributing to the development of more robust and interpretable models for cancer research.

2 Omics Data and Cancer

Omics data offers a complete picture of biological information at different molecular levels, providing a deep understanding of how cells function and how diseases develop. There are four main types of omics data: genomics, transcriptomics, proteomics, and metabolomics, each offering unique insights into cancer research [6]. Genomics studies an organism's entire set of Deoxyribonucleic Acid (DNA), including its genes, focusing on identifying changes that might contribute to cancer development. Transcriptomics analyzes the RNA transcripts produced by the genome, revealing which genes are active and when helping us understand the regulatory mechanisms that might be altered in cancer. Proteomics studies proteins, the molecules that carry out most biological activities in cells. By examining changes in protein expression and interactions, proteomics provides critical insights into the functional consequences of genetic mutations in cancer.

Metabolomics studies the complete set of small molecules, such as sugars, amino acids, and lipids, within a biological sample. This helps identify metabolic changes associated with cancer cell growth and survival, which are essential for understanding cancer biology and developing targeted treatments [7, 8].

Omics data is crucial in cancer research because it allows us to examine the molecular details of cancer in great depth. Genomic alterations, such as mutations and structural variations, are often the main causes of cancer. By studying these alterations, researchers can identify genes that promote cancer development (oncogenes) and genes that prevent it (tumour suppressor genes). These genes can be potential targets for treatment or used as biomarkers for early detection and predicting the course of the disease [9]. Gene expression profiles derived from transcriptomic data provide insights into the changes that occur in cancer cells compared to normal cells. This helps us identify pathways that are disrupted in cancer and might be targeted for treatment. Proteomics extends this understanding by revealing the functional impact of genetic changes, showing how proteins involved in cancer-related pathways might be potential biomarkers for detection or targets for therapy [10]. Furthermore, cancer cells often undergo metabolic reprogramming, changing their metabolic processes to support rapid growth and survival. Metabolomic studies can identify these shifts, offering insights that can be exploited for developing new treatment strategies [11].

Despite the great potential of omics data, analyzing it comes with significant challenges. Omics datasets are often very large and complex, with many features compared to the number of samples. This can lead to overfitting in statistical models and make it difficult to distinguish real biological signals from noise. Additionally, omics data can be noisy due to technical errors in measurement and biological differences between samples, making it harder to draw meaningful conclusions [12, 13]. Another challenge is data heterogeneity. Different types of omics data measure different aspects of cellular biology and combining them is complex. Each type of omics data has its own noise characteristics, measurement scales, and missing data patterns, making data integration a difficult task [14]. Furthermore, cancer is a complex disease with significant variability not only between different types of cancer but also among patients with the same type of cancer. This makes it difficult to identify universal biomarkers or therapeutic targets, requiring sophisticated analytical methods that can account for this variability [13].

To overcome these challenges, researchers need to use advanced computational techniques that can handle large and complex data, combine different types of data, and separate important information from noise. Deep learning methods, which can model complex, non-linear relationships and automatically extract relevant features from raw data, have become powerful tools for analyzing omics data in cancer research. By using these techniques, researchers can better utilize the full potential of omics data, improving our understanding of cancer biology and clinical outcomes [15].

3 Deep Learning Techniques for Omics Data Analysis

Deep learning has transformed the way we analyze omics data in cancer research by offering powerful tools to understand complex and large datasets. These techniques can automatically learn patterns and features within the data, which is especially helpful for extracting valuable insights from the massive amounts of information generated in omics studies. The following sections will discuss the main deep learning architectures used in omics data analysis, their applications in cancer research, and their strengths and weaknesses. The focus on CNNs, RNNs, DBNs, and Autoencoders in omics data analysis is justified by their ability to address specific challenges in cancer research. Each model brings unique strengths: CNNs are effective at recognizing spatial patterns, RNNs excel with sequential data, DBNs can integrate multimodal omics datasets, and Autoencoders reduce data complexity and noise. These models are also more interpretable and efficient for handling high-dimensional, noisy omics data compared to general-purpose DNNs. Additionally, their established success in previous omics studies and their ability to reduce overfitting make them better suited for these complex tasks.

3.1 Convolutional Neural Networks (CNNs)

Convolutional Neural Networks (CNNs) are commonly used in omics data analysis because they can identify patterns and relationships within the data [16]. Originally designed for image recognition, CNNs use convolutional layers to detect features like motifs in genomic sequences or groups of co-expressed genes. In cancer research, CNNs have been successfully used to predict DNA-binding proteins, classify cancer subtypes based on gene expression data, and analyze images of tissue samples to detect tumour regions [17]. Their strength is in their ability to recognize local patterns, making them well-suited for applications that require the detection of specific motifs or local structures within omics data. However, CNNs may be less effective when the data doesn't have spatial hierarchies or when temporal relationships are more important, such as in time-series analysis [18].

3.2 Recurrent Neural Networks (RNNs)

Recurrent Neural Networks (RNNs) are especially good for analyzing data that follows a sequence, which is common in studies that track gene expression or the progression of mutations over time. RNNs have a unique structure that allows them to remember previous inputs, enabling them to model temporal dependencies and dynamic changes in biological processes. This makes RNNs ideal for applications such as analyzing time-series gene expression data, modelling patient trajectories, and predicting cancer progression. However, traditional RNNs can have problems learning long-term dependencies. Variants like Long Short-Term Memory (LSTM) networks and Gated Recurrent Units (GRUs) have been developed to address these issues, making them more robust for modelling long sequences of data [19]. Despite these advancements, RNNs and their variants require careful tuning and can be computationally intensive, especially when dealing with large amounts of omics data [20].

3.3 Deep Belief Networks (DBNs)

Deep Belief Networks (DBNs) are a type of generative model that consists of multiple layers of Restricted Boltzmann Machines (RBMs). They are particularly useful for learning features and reducing dimensionality, which are crucial for handling large and complex omics datasets. DBNs can learn complex patterns within data, which can help identify patterns that might not be obvious using simpler methods [21]. In cancer research, DBNs have been used to combine different types of omics data, such as gene expression, methylation, and copy number variation, to identify biomarkers and predict clinical outcomes [22]. The main advantage of DBNs is their ability to learn complex, high-level features from large-scale data. However, training DBNs can be time-consuming and computationally expensive, and they often require a significant amount of labelled data for supervised fine-tuning, which can be a limitation in many biomedical applications [23].

3.4 Autoencoders

Autoencoders are unsupervised neural networks designed to learn efficient representations of data by compressing and then reconstructing input data. This process allows autoencoders to reduce dimensionality and extract features, which are valuable for denoising omics data and identifying underlying biological structures. In cancer research, autoencoders have been used to classify cancer subtypes, predict patient outcomes, and discover biomarkers. Variants such as denoising autoencoders and variational autoencoders improve the robustness and interpretability of these models, making them more effective for tasks like cancer subtype classification and biomarker discovery [24]. However, autoencoders can struggle to learn useful representations if the network architecture or the latent space dimensionality is not well-chosen, and they are generally less interpretable compared to other models. This can be a drawback when trying to understand the biological relevance of the features learned by the model [25].

3.5 Summary

In conclusion, deep learning offers a variety of tools for analyzing omics data in cancer research, each with its advantages and disadvantages. By carefully selecting the right model based on the specific characteristics of the data and the research goals, researchers can use these advanced computational methods to gain new insights into cancer biology and develop better diagnostic and prognostic models. Figure 1 shows the general process of omics data analysis with computation methods.

4 Comparative Study of Deep Learning Techniques for Omics Data Analysis

A comparative study of deep learning techniques in omics data analysis involves evaluating different models on common tasks to determine their effectiveness in cancer diagnosis and prognosis. This process requires a well-defined experimental setup, including selecting appropriate datasets, preprocessing methods, model architecture, hyperparameters, and evaluation metrics. By comparing different models, researchers can identify the most suitable approaches for specific types of omics data and gain insights into the strengths and weaknesses of each method.

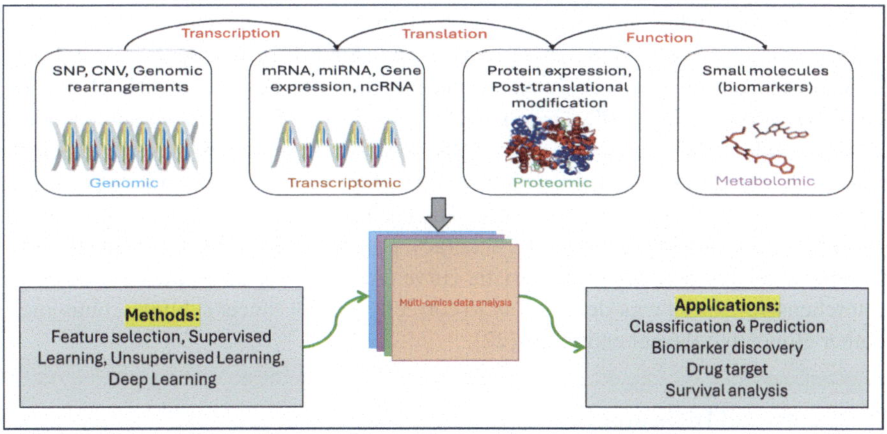

Fig. 1. General process flow of the omics integration for precision oncology.

4.1 Datasets

The datasets used in comparative studies are typically either publicly available or proprietary. Publicly available datasets, such as those from The Cancer Genome Atlas (TCGA), the Gene Expression Omnibus (GEO), and the International Cancer Genome Consortium (ICGC), provide a vast resource of omics data, including genomics, transcriptomics, proteomics, and epigenomics, from various cancer types [26]. These datasets often include data from thousands of patients, allowing for a comprehensive evaluation of deep-learning models across different types of cancers and various stages of disease progression. Proprietary datasets, on the other hand, may offer more specific or high-quality data tailored to research needs but are typically limited in availability and scope.

4.2 Method

Preparing omics data for deep learning analysis is a crucial step. Omics data is often high-dimensional and complex, so it needs careful preparation to ensure its quality and relevance. Common steps include normalization, which adjusts the data to a common scale without affecting the differences in value ranges; imputation, which fills in missing data points; and feature selection, which reduces the dimensionality of the data by selecting the most important features. For example, in gene expression data, normalization might involve transforming read counts into transcripts per million (TPM) or fragments per kilobase of transcript per million mapped reads (FPKM) to account for differences in sequencing depth and gene length.

The experimental setup for comparative study typically involves defining model architectures, selecting hyperparameters, and establishing evaluation metrics. The choice of model architectures—such as Convolutional Neural Networks (CNNs) [21], Recurrent Neural Networks (RNNs) [18], Deep Belief Networks (DBNs) [22], and autoencoders [25]—depends on the nature of the data and the specific analysis goals. Hyperparameters, including learning rate, batch size, number of layers, and number of neurons per layer, are tuned to optimize model performance. The evaluation metrics used to assess model performance can vary but often include accuracy, precision, recall, F1-score, and the area under the receiver operating characteristic curve (AUC-ROC). These metrics provide a comprehensive view of a model's ability to correctly predict cancer subtypes, outcomes, or other clinically relevant endpoints [27].

4.3 Results and Discussion

A comparative analysis can reveal the strengths and weaknesses of each deep learning technique in handling different types of omics data. For example, CNNs might be better at tasks involving spatial data patterns, like analyzing genomic sequences, while RNNs could excel at modelling changes in gene expression data over time. DBNs and autoencoders might be particularly useful for combining multiple types of omics data to identify complex patterns and biomarkers. Performance metrics like AUC-ROC and F1-score can help quantify these differences, providing a basis for selecting the most appropriate model for a given type of data. For example, a study might find that CNNs achieve high accuracy and AUC-ROC scores in classifying cancer subtypes based on gene expression data, while RNNs show superior performance in predicting patient outcomes based on time-series proteomics data [28].

The findings from a comparative study can provide valuable insights into the suitability of different deep learning techniques for specific types of omics data and cancer research applications. For example, if CNNs consistently outperform other models in analyzing genomic data, it suggests that these models are particularly well-suited for capturing spatial patterns and local dependencies within the data. On the other hand, if RNNs demonstrate superior performance in longitudinal studies of cancer progression, it highlights their strength in modelling temporal dependencies. These insights can guide future research efforts and inform the development of more effective diagnostic and prognostic tools in cancer research [29, 30].

Comparative studies of deep learning techniques in omics data analysis are crucial for advancing our understanding of cancer biology and improving patient outcomes. By rigorously evaluating different models on a common set of tasks, researchers can identify the most effective approaches for specific applications and drive innovation in cancer diagnosis and prognosis (Table 1).

Table 1. Comparative study results of deep learning techniques used in omics data analysis.

Deep Learning Technique	Convolutional Neural Networks (CNNs)	Recurrent Neural Networks (RNNs)	Deep Belief Networks (DBNs)	Autoencoders
Data Type	Genomics, Histopathology	Time-series (Transcriptomics, Proteomics)	Multi-omics (Genomics, Proteomics, Epigenomics)	Multi-omics, Noisy data
Key Applications	DNA-binding prediction, Cancer subtype classification, Tumor region detection	Gene expression analysis over time, Patient trajectory modelling, Cancer progression prediction	Biomarker discovery, Integration of multi-omics data	Dimensionality reduction, Cancer subtype classification, Biomarker discovery
Performance Metrics	High accuracy, High AUC-ROC in spatial data tasks	Superior recall, Precision in temporal analysis	High F1-score in multi-omics integration tasks	Good accuracy, Moderate AUC-ROC in denoising tasks
Strengths	Effective for spatial pattern recognition and motif detection	Captures temporal dependencies, suitable for sequential data	Learns hierarchical features, effective for complex data patterns	Effective for noise reduction and feature extraction
Limitations	Less effective for non-spatial data and temporal dependencies	Vanishing/exploding gradient issues require complex tuning	Requires large datasets, computationally intensive	Can struggle with learning useful representations, less interpretable than other models

5 Case Studies: Application of Deep Learning in Specific Cancer Subtypes

Case studies offer valuable insights into the practical application of deep learning techniques for analyzing omics data in specific cancer subtypes. In the context of cancer research, "subtypes" refer to distinct groups or categories of a particular type of cancer that exhibit unique biological, genetic, or molecular characteristics. These subtypes often differ in terms of their underlying genetic mutations, patterns of gene expression, protein markers, and responses to treatments. Identifying cancer subtypes is crucial for understanding the diversity within a single cancer type, as it helps in developing more targeted and personalized treatment strategies. These studies typically focus on types of cancer, such as breast cancer, lung cancer, or glioblastoma, to explore how deep learning models can uncover critical biomarkers, predict patient outcomes, and enhance our understanding of the disease's molecular mechanisms. By leveraging different omics data types—such as genomics, transcriptomics, proteomics, and metabolomics—researchers can develop robust models that aid in cancer diagnosis, prognosis, and personalized treatment strategies.

One prominent case study involves the use of Convolutional Neural Networks (CNNs) for the classification of breast cancer subtypes based on histopathological images and gene expression profiles. For instance, a study by Coudray, Ocampo [31] utilized CNNs to analyze whole-slide images of lung cancer biopsies, achieving high accuracy in distinguishing between adenocarcinoma and squamous cell carcinoma. This study demonstrated that CNNs could not only match but often exceed the performance of pathologists in diagnosing cancer from histopathological images, highlighting the potential of deep learning to augment human expertise in clinical settings [31]. Similarly, in breast cancer, CNNs have been employed to classify different molecular subtypes using gene expression data, providing insights into the heterogeneity of the disease and aiding in the development of targeted therapies [32].

Recurrent Neural Networks (RNNs) have shown promise in modelling temporal dynamics in cancer progression, particularly in cancers where treatment response and disease progression are highly variable, such as in glioblastoma. A study by Esteva, Kuprel [33] applied RNNs to longitudinal gene expression data to predict patient survival times and identify key temporal patterns associated with treatment response. The ability of RNNs to capture the temporal evolution of gene expression allowed for more accurate predictions of patient outcomes, demonstrating the potential of deep learning to provide dynamic, real-time insights into cancer biology [33]. These findings suggest that RNNs could be particularly valuable in personalized medicine, where understanding the temporal dynamics of tumour evolution is crucial for optimizing treatment strategies.

Deep Belief Networks (DBNs) have been effectively used to integrate multi-omics data to predict clinical outcomes in various cancer types. In one notable study on ovarian cancer, DBNs were employed to integrate genomics, transcriptomics, and proteomics data to predict patient survival and identify potential therapeutic targets [34]. By learning hierarchical, non-linear relationships among different omics layers, DBNs were able to uncover complex interactions that are not easily detectable with traditional statistical methods. This approach not only improved the accuracy of survival predictions but also highlighted novel biomarkers that could be targeted in personalized therapies, showcasing the potential of DBNs to advance precision oncology [35].

Autoencoders have been widely used for the dimensionality reduction of high-dimensional omics data, aiding in the identification of distinct cancer subtypes and potential biomarkers. For example, a study on prostate cancer utilized variational autoencoders to reduce the dimensionality of multi-omics data, including genomics, proteomics, and metabolomics, to uncover novel cancer subtypes associated with distinct clinical outcomes [36]. The reduced dimensionality provided by the autoencoders allowed researchers to visualize the data in a lower-dimensional space, making it easier to identify clusters of patients with similar molecular profiles. This method helped in identifying previously unrecognized cancer subtypes, which could inform more targeted therapeutic approaches and improve patient stratification in clinical trials [25, 36].

In summary, these case studies illustrate the diverse applications of deep learning techniques in analyzing omics data for specific cancer subtypes. Each deep learning model brings unique strengths to the table: CNNs excel in spatial pattern recognition in histopathology and gene expression, RNNs are adept at modelling temporal dynamics in gene expression and patient outcomes, DBNs effectively integrate multi-omics data for predictive modelling, and autoencoders are powerful tools for dimensionality reduction and visualization of high-dimensional data. By applying these models to specific cancer types, researchers can develop more accurate diagnostic tools, identify novel therapeutic targets, and ultimately improve patient outcomes in cancer care.

6 Challenges and Future Directions in Deep Learning for Omics Data Analysis

Despite its promising results, deep learning for cancer research still faces significant hurdles. Key obstacles include data quality, interpretability, generalizability, and computational requirements. Overcoming these challenges will necessitate creative solutions and strategic breakthroughs in both computing methods and biological knowledge.

One of the biggest hurdles in using deep learning to analyze omics data is the quality and availability of that data. Omics datasets are often massive, noisy, and incomplete, making it difficult to train effective deep-learning models. For instance, gene expression data might be missing due to technical issues with sequencing, while proteomics data can be affected by variations in sample preparation and measurement. Moreover, most publicly available datasets are too small to train deep learning models properly, which need large amounts of data to avoid overfitting and generalize well to new data. To overcome these challenges, we need to develop robust methods for preprocessing and normalizing data, as well as techniques for handling missing values, such as imputation algorithms and data augmentation strategies [23, 37].

The next biggest obstacle in using deep learning for omics data analysis is understanding how the models make their predictions. While deep learning algorithms are incredibly powerful at recognizing patterns in data, they can be like a "black box" - it's hard to know how they're making decisions. This lack of transparency can be a major problem in clinical settings, where doctors and researchers need to be able to understand why a model is making a particular prediction. Without clear explanations, it's difficult to trust the results or identify the key factors that are driving them. To overcome this challenge, researchers are working on developing more interpretable models and incorporating techniques like attention mechanisms, feature importance scores, and layer-wise relevance propagation to provide insights into how the models work [38].

Deep learning models are amazing, but they can be tricky to use in real-life situations. The problem is that they're often trained for specific groups of people, and they might not work as well when applied to new groups or different types of data. This is especially true when it comes to cancer, which is a super complex and varied disease. To make these models more reliable, researchers are experimenting with new techniques that allow them to learn from multiple sources of data. By doing this, they're hoping to create models that can handle different types of data and make better predictions for patients.

On top of that, deep learning models have the potential to revolutionize omics data analysis, but they can be a challenge to use in real-life research due to their demanding computational requirements. The problem is that they require a lot of computing power and storage space to run, which can be a problem for smaller research teams or institutions that don't have access to super-fast computers. This can be frustrating for scientists who want to use these models to analyze large amounts of data but don't have the resources to do so. To make things more manageable, researchers are working on making these models more efficient and easier to use. They're developing new algorithms that can handle larger datasets without breaking the bank and creating cloud-based platforms that allow anyone to access powerful computing resources without having to buy their own. By making these technologies more accessible, scientists can focus on using them to make discoveries, rather than struggling to get them to work [39, 40].

To overcome these obstacles, we need a team effort that combines the expertise of machine learning, computational biology, and clinical research. In the future, deep learning for omics data analysis should focus on creating models that are easier to understand and provide practical information about cancer. We also need to build larger and more diverse datasets that reflect the global population. Furthermore, models that can combine data from different omics sources will provide a complete picture of cancer. Finally, working closely together, computational scientists and clinicians can ensure that deep learning models are not only technically sound but also useful in real-world medical practice. Addressing these challenges will allow deep learning to transform omics data analysis, leading to better cancer diagnosis, prognosis, and personalized treatment plans (Table 2).

Table 2. The key finding of deep learning techniques is on omics data.

Cancer Subtype	Deep Learning Technique	Omics Data Type	Key Findings	Clinical Implications
Breast Cancer	Convolutional Neural Networks (CNNs)	Histopathological images, Gene expression	CNNs were able to classify molecular subtypes of breast cancer with high accuracy, often surpassing the performance of pathologists (Coudray et al., 2018)	Enhanced diagnostic accuracy and identification of targeted therapies for specific subtypes
Lung Cancer	CNNs	Histopathological images	CNNs successfully differentiated between adenocarcinoma and squamous cell carcinoma with high accuracy, providing a robust diagnostic tool (Coudray et al., 2018)	Supports precise diagnosis and guides personalized treatment strategies
Glioblastoma	Recurrent Neural Networks (RNNs)	Longitudinal gene expression data	RNNs captured temporal patterns in gene expression, predicting patient survival and treatment responses more accurately (Esteva et al., 2019)	Offers potential for personalized treatment planning based on predicted disease progression

(continued)

Table 2. (*continued*)

Cancer Subtype	Deep Learning Technique	Omics Data Type	Key Findings	Clinical Implications
Ovarian Cancer	Deep Belief Networks (DBNs)	Genomics, Transcriptomics, Proteomics	DBNs effectively integrated multi-omics data to predict patient survival and identify novel therapeutic targets, uncovering complex interactions (Zhou et al., 2014)	Facilitates precision oncology by identifying key biomarkers and informing targeted therapies
Prostate Cancer	Autoencoders	Multi-omics data (genomics, proteomics, metabolomics)	Autoencoders reduced data dimensionality, enabling visualization of patient clusters and identification of novel subtypes associated with distinct clinical outcomes (Tan et al., 2015)	Enhances the stratification of patients for more personalized treatments and aids in identifying new cancer subtypes

7 Conclusion

The study compares different deep-learning techniques for analyzing cancer data, specifically focusing on genomics, transcriptomics, proteomics, and metabolomics data. It finds that these advanced computational methods can identify intricate patterns and relationships in cancer data that are often missed by traditional statistical methods. Each deep learning architecture has its strengths: CNNs are well-suited for analyzing spatial data like histopathological images, RNNs can model temporal dynamics to understand disease progression, and DBNs and Autoencoders can integrate multi-omics data and reduce dimensionality.

While deep learning models generally outperform traditional methods in terms of accuracy, precision, recall, and other performance metrics, they are not without limitations. Issues such as data quality, interpretability, generalizability, and computational demands need to be addressed to fully realize their potential in clinical settings. Future research should focus on enhancing model interpretability and robustness, developing efficient algorithms to handle high-dimensional data, and creating larger, more diverse datasets to improve model generalizability across different populations and clinical contexts.

The integration of deep learning into clinical practice for cancer diagnosis and prognosis is not just a technical challenge but also a translational one. Clinicians, data scientists, and researchers must collaborate closely to ensure that deep learning models are not only accurate and efficient but also interpretable and actionable in real-world clinical settings. By addressing these challenges and continuing to innovate, deep learning has the potential to significantly impact cancer care, leading to earlier detection, more accurate prognoses, and personalized treatment strategies that improve patient outcomes. This study lays the groundwork for future research and development in this rapidly evolving field, emphasizing the importance of multidisciplinary approaches to tackle the complexities of cancer and advance precision oncology.

References

1. Ferlay, J., et al.: Cancer statistics for the year 2020: an overview. Int. J. Cancer **149**(4), 778–789 (2021)
2. Saikia, B., Gupta, K., Saikia, U.N.: The modern histopathologist: in the changing face of time. Diagn. Pathol. **3**(1), 25 (2008)
3. Dai, X., Shen, L.: Advances and trends in omics technology development. Front. Med. **9**, 911861 (2022)
4. Zhao, Y., Chang, C., Long, Q.: Knowledge-guided statistical learning methods for analysis of high-dimensional -omics data in precision oncology. JCO Precis. Oncol. **3**, 1–9 (2019)
5. Correa-Aguila, R., Alonso-Pupo, N., Hernández-Rodríguez, E.W.: Multi-omics data integration approaches for precision oncology. Mol. Omics **18**(6), 469–479 (2022)
6. Kang, M., Ko, E., Mersha, T.B.: A roadmap for multi-omics data integration using deep learning. Briefings Bioinform. **23**(1), bbab454 (2022)
7. Ogunjobi, T.T., et al.: Bioinformatics applications in chronic diseases: a comprehensive review of genomic, transcriptomics, proteomic, metabolomics, and machine learning approaches. Medinformatics (2024)
8. Aebersold, R., Mann, M.: Mass-spectrometric exploration of proteome structure and function. Nature **537**(7620), 347–355 (2016)
9. Stratton, M.R., Campbell, P.J., Futreal, P.A.: The cancer genome. Nature **458**(7239), 719–724 (2009)
10. Petricoin, E.F., et al.: Use of proteomic patterns in serum to identify ovarian cancer. Lancet **359**(9306), 572–577 (2002)
11. Vander Heiden, M.G., Cantley, L.C., Thompson, C.B.: Understanding the Warburg effect: the metabolic requirements of cell proliferation. Science, **324**(5930), 1029–1033 (2009)
12. Leek, J.T., et al.: Tackling the widespread and critical impact of batch effects in high-throughput data. Nat. Rev. Genet. **11**(10), 733–739 (2010)
13. Hanahan, D.: Hallmarks of cancer: new dimensions. Cancer Discov. **12**(1), 31–46 (2022)
14. Cui, M., Cheng, C., Zhang, L.: High-throughput proteomics: a methodological mini-review. Lab. Invest. **102**(11), 1170–1181 (2022)
15. Min, S., Lee, B., Yoon, S.: Deep learning in bioinformatics. Brief. Bioinform. **18**(5), 851–869 (2016)
16. Chuang, Y.-H., et al.: Convolutional neural network for human cancer types prediction by integrating protein interaction networks and omics data. Sci. Rep. **11**(1), 20691 (2021)
17. Zompola, A., et al.: Omics-CNN: a comprehensive pipeline for predictive analytics in quantitative omics using one-dimensional convolutional neural networks. Heliyon, **9**(11) 2023
18. Tsimenidis, S., Vrochidou, E., Papakostas, G.A.: Omics data and data representations for deep learning-based predictive modeling. Int. J. Mol. Sci. **23**(20), 12272 (2022)
19. Lipton, Z.C., et al.: Learning to Diagnose with LSTM Recurrent Neural Networks. CoRR, abs/1511.03677 (2015)
20. Majid, H.A., et al.: Classifying ovarian cancer using deep learning method based on multi-omics data (2023)
21. Martorell-Marugán, J., et al.: Deep learning in omics data analysis and precision medicine. Exon Publications, pp. 37–53 (2019)
22. Hassanzadeh, H.R., Wang, M.D.: An integrated deep network for cancer survival prediction using omics data. Front. Big Data **4**, 568352 (2021)

23. Kaur, P., Singh, A., Chana, I.: Computational techniques and tools for omics data analysis: state-of-the-art, challenges, and future directions. Arch. Comput. Methods Eng. **28**(7), 4595–4631 (2021)
24. Franco, E.F., et al.: Performance comparison of deep learning autoencoders for cancer subtype detection using multi-omics data. Cancers **13**(9), 2013 (2021)
25. Ma, T., Zhang, A.: Integrate multi-omics data with biological interaction networks using multi-view factorization autoencoder (MAE). BMC Genomics **20**(11), 944 (2019)
26. Weinstein, J.N., et al.: The cancer genome atlas pan-cancer analysis project. Nat. Genet. **45**(10), 1113–1120 (2013)
27. Chicco, D., Jurman, G.: The advantages of the Matthews correlation coefficient (MCC) over F1 score and accuracy in binary classification evaluation. BMC Genomics **21**(1), 6 (2020)
28. Zeng, H., et al.: Convolutional neural network architectures for predicting DNA–protein binding. Bioinformatics **32**(12), i121–i127 (2016)
29. LeCun, Y., Bengio, Y., Hinton, G.: Deep learning. Nature **521**(7553), 436–444 (2015)
30. Lipton, Z.: Learning to diagnose with LSTM recurrent neural networks. arXiv preprint arXiv: 1511.03677 (2015)
31. Coudray, N., et al.: Classification and mutation prediction from non–small cell lung cancer histopathology images using deep learning. Nat. Med. **24**(10), 1559–1567 (2018)
32. Ektefaie, Y., et al.: Integrative multiomics-histopathology analysis for breast cancer classification. NPJ Breast Cancer **7**(1), 147 (2021)
33. Esteva, A., et al.: Dermatologist-level classification of skin cancer with deep neural networks. Nature **542**(7639), 115–118 (2017)
34. Nguyen, T.M., et al.: Deep learning for human disease detection, subtype classification, and treatment response prediction using epigenomic data. Biomedicines **9**(11), 1733 (2021)
35. Wang, C.-W., Khalil, M.-A., Firdi, N.P.: A survey on deep learning for precision oncology. Diagnostics **12**(6), 1489 (2022)
36. Ma, Y., et al.: Predicting cancer drug response by proteomic profiling. Clin. Cancer Res. **12**(15), 4583–4589 (2006)
37. Arjmand, B., et al.: Machine learning: a new prospect in multi-omics data analysis of cancer. Front. Genet. **13**, 824451 (2022)
38. Menyhárt, O., Győrffy, B.: Multi-omics approaches in cancer research with applications in tumor subtyping, prognosis, and diagnosis. Comput. Struct. Biotechnol. J. **19**, 949–960 (2021)
39. Han, S., et al.: Learning both weights and connections for efficient neural network. In: Advances in Neural Information Processing Systems, vol. 28 (2015)
40. Alzubaidi, A., Tepper, J., Lotfi, A.: A novel deep mining model for effective knowledge discovery from omics data. Artif. Intell. Med.Intell. Med. **104**, 101821 (2020)

Reliability Analysis of a 1 MW Grid-Connected Solar PV System Using Fault Tree Approach

A. Hassan(✉), V. Dhineswari, and A. S. Hazirah

Department of Physics and Chemistry, Faculty of Applied Sciences and Technology, Universiti Tun Hussein Onn Malaysia (UTHM), Pagoh, Johor, Malaysia
hassan@uthm.edu.my

Abstract. Grid connected solar photovoltaic (PV) plants are currently play the vital role in the generation of electrical power in many countries, such as Malaysia, which is transitioning toward the use of renewable energy. This paper presents a method and analysis for assessing the reliability of a 1 MW grid-connected solar PV system. Theoretically, all components in the system can fail randomly and their effect to the system can be elucidated by using Fault Tree Analysis (FTA) method. In this study, FTA is used to comprehend the interdependencies of all the components that make up the system. Without any component repair, the system's failure in year is approximately 17% whereas that with repairs is shown to decrease to 1.4%. Importance Measure (IM) analysis is performed to gain a deeper understanding of the system's critical components. It is shown that PV module is the key component to the reliability of the system, as it is the main critical component in the system.

Keywords: Solar PV · Reliability · Fault Tree Analysis · Importance Measure Analysis · Failure rate · Repair rate

1 Introduction

Solar photovoltaic (PV) systems have gained its importance to the public and industries, as one of the sustainable and renewable energy sources, which are alternatives to the fossil fuels for electricity generation. Malaysia is fortunate to be one of the countries that have potentials to generate large electricity output by wide spreading the development of grid-connected solar PV systems. This region is subjected to daily high level of solar irradiance and high number of sunny days.

A grid connected solar PV system consists of several components, including PV modules, inverters, step-up transformers and the balance of system components. PV modules generate DC power, which is converted into AC power by inverters. The AC power produced is fed to the grid through the interface between the AC output circuits of the PV system and the electric utility network. For a given PV system, the percent of power supplied to the grid depends on the instantaneous power output from the PV system as well as the reliability of the PV system. A few studies have been reported to evaluate the performance of PV systems within this region, but not the reliability [1–4].

P. Chang et al. (Eds.): SCIEMATHIC 2024, SPPHY 427, pp. 278–289, 2025.
https://doi.org/10.1007/978-981-96-7679-8_22

Conversely, various methods have been utilized elsewhere to access their PV system reliability [5–8]. Reliability is an essential measure for estimating system availability because it aims to identify critical components that have significant impact. A system reliability can be evaluated by using Fault Tree (FT) approach. Fault tree analysis (FTA) analyses the failure events at component levels and predicts how likely a system will fail. It uses logic gates and the possible component failure events to deduce the system failure (top-down approach) [9–11]. Type of event in FT are namely top events, intermediate events and basic events.

The works in this study is an attempt to perform the FTA for a 1 MW grid-connected solar PV system. To this end, a generic design of a solar PV system is proposed. Based on the design, failure mode and effect analysis (FMEA) is conducted, which subsequently will provide inputs (top event, intermediate events, basic events and logic gates) for the FT development. The reliability of the system is determined from the calculation of the FT top event probability, which is accomplished by using a computer software. Critical components in the system are determined via Fussell-Vesely (FV), Risk Achievement Worth (RAW) and Risk Reduction Worth (RRW) Importance Measure (IM). The next sections of this manuscript addresses the methodologies involved and the outcomes.

2 Methodology

2.1 1 MW Grid-Connected Solar PV System

The number of components in a 1 MW grid connected solar PV system could vary depending on the specifications of the components installed, especially the types of PV modules and the architecture of the PV system [12–14]. This study considers the string structure architecture for the proposed PV system. For a 1 MW system, the number of PV modules could range from 2320 to 4764, while the maximum number of inverters is 42, and the maximum number of transformers is 4, reportedly installed [1, 15–25].

2.2 Failure Mode and Effect Analysis (FMEA)

In a grid-connected solar PV system, component failure can occur either at the direct current (DC) side or at the alternating current (AC) side. A component may have multiple modes of failure, and the mode may or may not causing the system failure. In FMEA, the consequences of each components' failures are identified. Even though there are many causes for a component failure, this study only considers the random failure events. The failure of these events is modelled as basic events or intermediate events in the FT.

2.3 Modelling and Quantification of FT

With the aid of TopEvent FTA software (2016), the graphical modelling of the FT is performed. The calculation of the FT top event i.e. the failure probability of the grid-connected solar PV system is also by means of the software. The top event for this FT is defined as: AC power generation is less than 100% of the nominal output (<1 MW) or no power at all (i.e. total power loss). This strictly means that when the system does not produce 100% output, it is considered as fail. The failure probability is obtained from the combination basic events via Boolean operations that produce minimal cut sets (MCSs), depending on the logic gates modelled. At the basic event level, a component failure probability at time t, $Q(t)$ is [26]:

$$Q(t) = 1 - e^{-\lambda t} \tag{1}$$

$$Q(t) = \frac{\lambda}{\lambda + \mu}\left(1 - e^{-(\lambda+\mu)t}\right) \tag{2}$$

with λ is the component failure rate and μ is the component repair rate, with assumption that the component failure and repair process follow the exponential probability distribution function. A component reliability at time t is therefore:

$$R(t) = 1 - Q(t) \tag{3}$$

As this study is not performed for a specific site's solar PV system, generic failure rates λ and repair rates μ of the components are obtained from available literatures and used in the calculations.

2.4 Importance Measure Analysis

Fussell-Vesely (FV) Importance or Critical Importance Factor (CIF) measure for a component x is calculated according to Eq. (4) with $Q(base)$ is the top event probability and $Q(x = 0)$ is the top event probability when component x is completely available [27]. This value shows the fractional contribution of the component x to the overall system failure.

$$FV = \frac{Q(base) - Q(x = 0)}{Q(base)} \tag{4}$$

The Risk Achievement Worth (RAW) evaluates the relative increase in system failure if the component x is known to be in failed state:

$$RAW = \frac{Q(x = 1)}{Q(base)} \tag{5}$$

$Q(x = 1)$ is the top event probability when component x is completely unavailable. The Risk Reduction Worth (RRW) estimates the relative reduction in system failure when the component is made completely dependable:

$$RRW = \frac{Q(base)}{Q(x = 0)} \tag{6}$$

3 Results and Discussion

3.1 1 MW Solar PV System

The hypothetical solar PV system for the reliability analysis in this study is as in Fig. 1(a). It consists of a step-up transformer which receives input from the AC main switchboard. The AC main-switchboard receives input from 3 main AC junction boards (ACJB NO.1 – 3). Each ACJB receives input from sub-AC junction boards (SUB ACJB). For instance, MAIN ACJB NO.1 receives input from SUB ACJB NO.1 – 3. As shown in the figure, the main components in the AC main-switchboard or junction boards are switch and circuit breakers (MCCB or MCB).

3 inverters provide the generated AC power to a SUB ACJB. For example, INVERTER 1, 2, and 3 provide the power to SUB ACJB NO.1. An inverter is connected to a PV array. As in Fig. 1(b), PV ARRAY NO.1 is linked to INVERTER 1. In the figure, the total number of PV strings in an array is 12, and a string may consist of 11 or 12 PV modules. Thus, an array consists of 134 modules. In total, there are 134 modules per array, an inverter per array, and 3 inverters per SUB ACJB. With 9 SUB ACJBs, 26 inverters are needed, and thus the total number of PV modules are $134 \times 27 = 3618$. The power rating assumed for each PV module is 280 W , therefore 280 W/module $\times$ 3618 modules $\approx$ 1 MW is the total power output from the system. For an inverter, the needed power rating is presumably $=$ 280 W/module $\times$ 134 modules $=$ 37 kW $\approx$ 40kW. Since components voltage rating is omitted in the current design considerations, the current rating for fuses, breakers and switch remains undetermined.

3.2 FMEA

Table 1 lists the components, their failure modes, and the undesirable outcomes (effects) to the PV system. Since the defined FT's top event corresponds to the system's lower power output or total loss of power output, only the failure modes that lead to these outcomes are listed here and this will assist the development of a coherent FT.

As in Table 1, only the first component type in a subsystem is indicated, and for the rest component of the same type, they are considered with the similar manner. From the FMEA, it is understood that any single component or subsystem failure will cause the PV system to produce power output less than the nominal value or no power output at all.

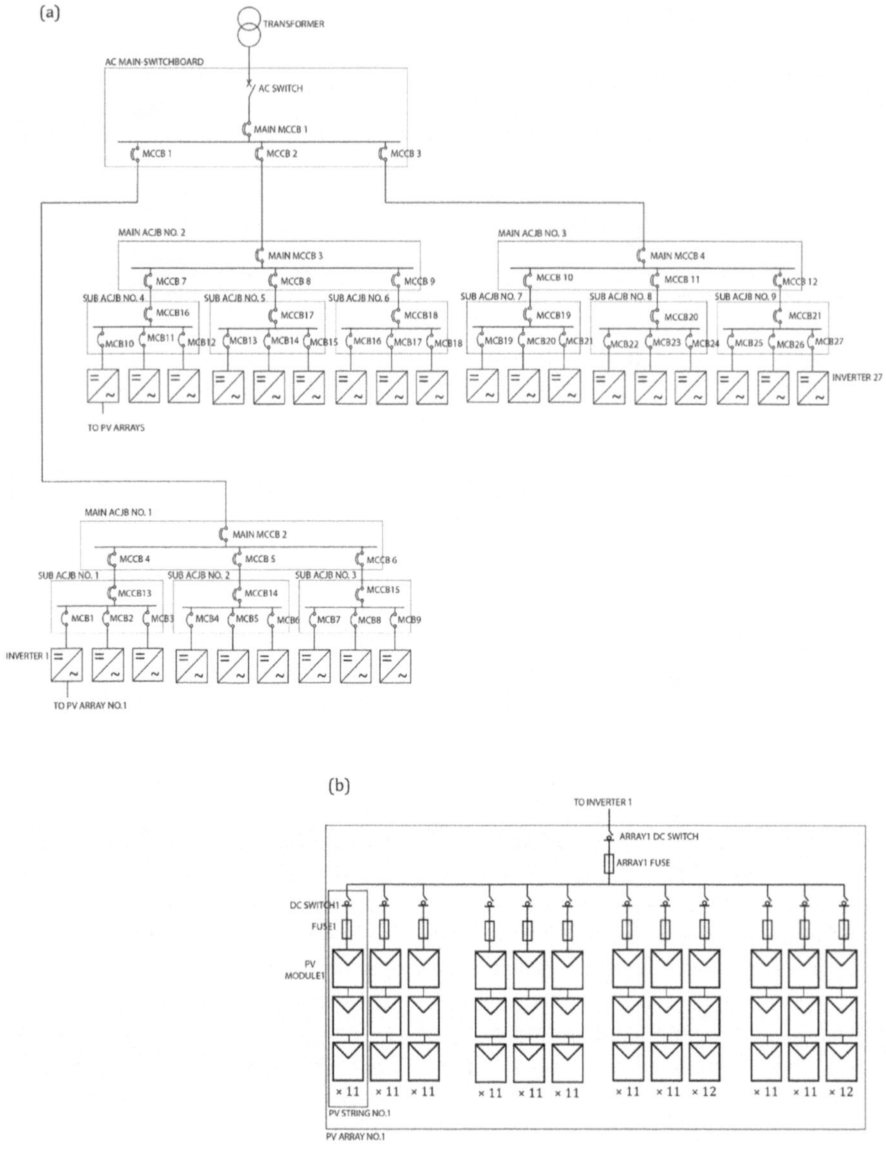

Fig. 1. Schematic diagram of 1 MW grid-connected solar PV system

Table 1. The grid connected solar PV system FMEA

No	Component	Failure Mode	Effect
1	PV module(s) (in PV STRING NO.1)	Fail to function	Lower or total loss of power from PV STRING NO.1
2	PV module(s) (in other PV STRING, of PV ARRAY NO. 1)	Fail to function	Lower or total loss of power from that PV STRING of PV ARRAY NO.1
3	PV module(s) (in other PV STRING, of other PV ARRAY)	Fail to function	Lower or total loss of power from the PV STRING, of that PV ARRAY
4	Fuse in PV STRING	Spurious open	Total loss of power from the PV STRING
5	DC SWITCH in PV STRING	Spurious open	Total loss of power from the PV STRING
6	ARRAY FUSE	Spurious open	Total loss of power from the ARRAY to INVERTER
7	ARRAY DC SWITCH	Spurious open	Total loss of power from the ARRAY to INVERTER
8	INVERTER 1	Fail to function	No power to MCB1 in SUB ACJB NO.1
9	Other inverter(s)	Fail to function	No power to their corresponding MCBs in SUB ACJB
	SUB ACJB NO.1:		
10	MCB(s)	Spurious open	Lower power output from SUB ACJB NO.1 to MAIN ACJB NO. 1
11	MCCB	Spurious open	No power output from SUB ACJB NO.1 to MAIN ACJB NO. 1
	SUB ACJB NO.2	Same as in SUB ACJB NO.1	Same as in SUB ACJB NO.1
	SUB ACJB NO.3	Same as in SUB ACJB NO.1	Same as in SUB ACJB NO.1
	MAIN ACJB NO. 1:		
12	MCCB(s)	Spurious open	Lower or zero power output to MAIN MCCB 2
13	MAIN MCCB 2	Spurious open	No power output from MAIN ACJB NO.1 to AC MAIN-SWITCHBOARD
	MAIN ACJB NO. 2	Same as in MAIN ACJB NO.1	Same as in MAIN ACJB NO.1
	MAIN ACJB NO. 3	Same as in MAIN ACJB NO.1	Same as in MAIN ACJB NO.1
	AC MAIN-SWITCH BOARD:		
14	MCCB(s)	Spurious open	Lower or zero power output to MAIN MCCB 1
15	MAIN MCCB 1	Spurious open	No power output to TRANSFORMER
16	AC SWITCH	Spurious open	No power output to TRANSFORMER
17	TRANSFORMER	Fail to function	Lower or zero power output grid

3.3 Modelling and Quantification of FT

The arrangement from the top event to the basic events in the FT of the PV system (Fig. 2) is according to the PV system design in Fig. 1. At the top, there is TRANSFORMER (as basic event), followed by AC MAIN-SWITCHBOARD (as intermediate event) and the branches continued to the lower-level subsystems and components. A drawback of the current TopEvent FTA software is that the total events in a FT cannot be more than 30 events. To overcome this restriction and complete the study's main objective, the main FT is connected to other FTs of lower events, that are modelled and managed in separate files. These FTs and as in Fig. 3, 4, 5 and 6. This approach allows the calculation of the PV system top event probability but losing the connections with other basic events from the other FTs.

The failure and repair rate of components for the *Q(t)* calculations are obtain from the values reported by other studies, as in Table 2. The time, *t* for the calculation is 1 year.

Table 2. Component reliability data for the FT calculation

Component	Failure rate, $\times 10^{-6}$	Repair rate	Reference
PV module	2.3 (yr^{-1}) (calculated median)	0.0057 (yr^{-1}) (calculated median)	[7]
DC switch	0.2(yr^{-1}) (calculated median)	0.0208 (yr^{-1}) (calculated median)	[7]
AC switch	0.034 (yr^{-1}) (calculated median)	0.0208 (yr^{-1}) (calculated median)	[7]
Circuit breaker (AC)	5.7 (yr^{-1}) (calculated median)	0.0208 (yr^{-1}) (calculated median)	[7]
Inverter	25 (yr^{-1}) (calculated median)	0.0041 (yr^{-1}) (calculated median)	[7]
Transformer	0.4963 (hr^{-1})	-	[28]
Fuse	0.049 (hr^{-1}) (highest value)	- (using DC switch's repair rate)	[5]

Figure 2 is the main FT of the system, with top event probability 0.1707, without considering any component repair. This means that at one year time, chance of having the PV system in lower of zero output is approximately 17%. The gates in the FT are only OR gates. Thus, any basic events are the MCSs of the FT. In the subsequent FTs, there are also only OR gate events. In Fig. 2, the main contributor to the top event probability are the basic events described as "Less or no power at MCCB4 - 12", with *Q(t)* = 0.0184, which are not actually basic events. Their values are calculated separately, as in FT for MCCB4 (Fig. 3). In rank, the contributors to the top event probability are "Less or no power at MCCB4", "TRANSFORMER", "MCCB" and "AC SWITCH".

Figure 3 shows the FT for events at MCCB4. This also represents MCCB5 – MCCB12. At 1 year time, the probability of failure is approximately 1.8×10^{-2}. In rank, the contributors are the ARR1FUSE (ARR2FUSE, ARR3FUSE too) followed by Inverters, MCB/MCCBs and DC SWITCHs. ARR1 (or 2, 3)FUSE have higher probability which is due to the contribution of its lower subsystems' component failures, which is quantified in FT: "FT for ARRAY1 FUSE" as in Fig. 4.

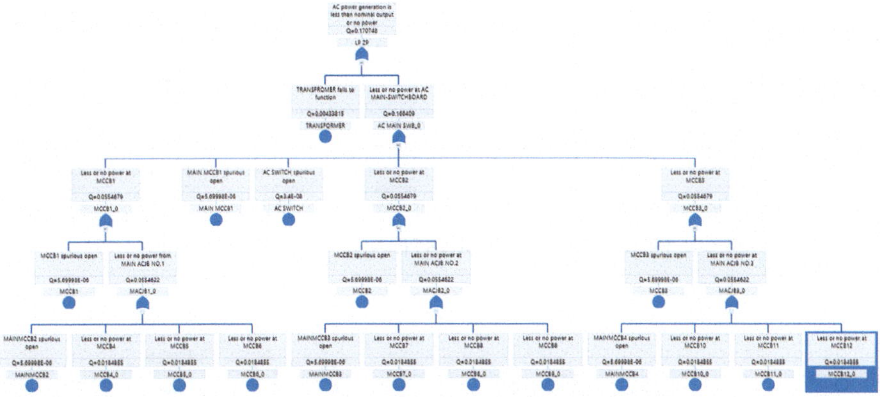

Fig. 2. FT for 1 MW grid-connected solar PV system

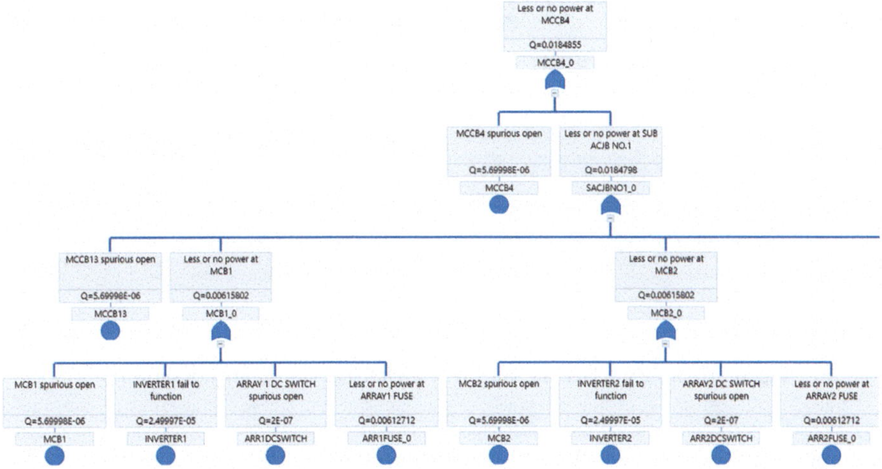

Fig. 3. FT for MCCB4 (MCB3 is not shown)

Figure 4 shows the FT for events at ARRAY1 FUSE. At 1 year time, the probability of failure is approximately 6.1×10^{-3}. In rank, the contributors are the String with 12 PV modules (i.e. STRING 9 and 12, but are not shown in Fig. 4), followed by String with 11 PV modules and fuse. String with 12 modules has higher *Q(t)* than that of 11 modules. The probability of basic event "Lower of zero power at STRING1" and the rest are obtained from FT for PV STRING NO.1 and NO.9, as in Fig. 5 and 6, respectively.

Figure 5 shows the FT for events at PV STRING NO.1. At 1 year time, the probability of failure is approximately 4.7×10^{-4}. In rank, the contributors are FUSE1, DC SWITCH1 and PV MODULES. The similar trend is observed for STRING NO. 9, with 12 strings (Fig. 6).

The previous paragraphs explain the system's top event probability, without considering component repairs. To be realistic, with the similar approach and considering

Fig. 4. FT for ARRAY1 FUSE (STRING8 – 12 are not shown)

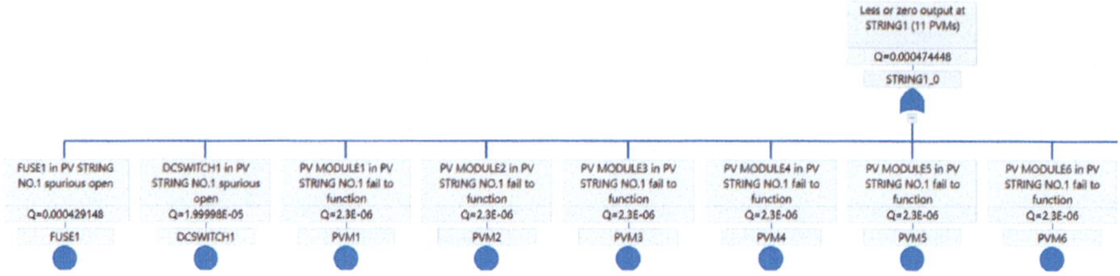

Fig. 5. FT for PV STRING NO.1 (PV MODULES7 – 11 are not shown)

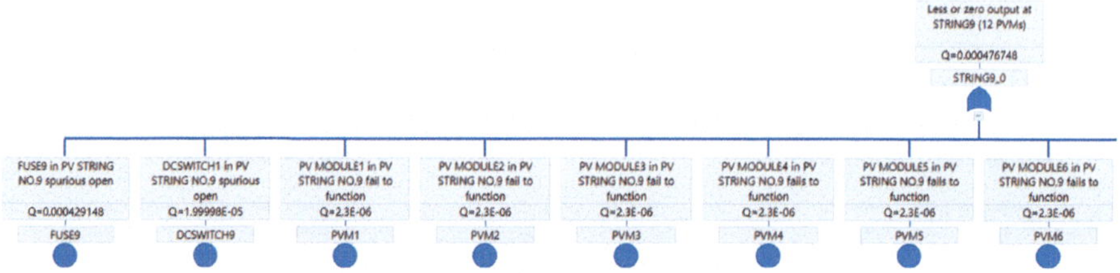

Fig. 6. FT for PV STRING NO.9 (PV MODULES7 – 12 are not shown)

the components' repair rate in Table 3, the top event probabilities are as in Table 3. The chance of having the PV system in lower of zero output in one year time is approximately 1.4%.

Table 3. FT top events' probability with repair rate considered

FT	*Q(t)*	Remarks
FT for 1 MW grid-connected solar PV system (main top event)	1.4E-02	No repair considered for TRANSFORMER (no repair rate data)
FT for MCCB4	1.03E-03	
FT for ARRAY1 FUSE	3.11E-04	-
FT for PV STRING NO.1	2.55E-05	-
FT for PV STRING NO.9	2.78E-05	-

3.4 Importance Measure (IM) Analysis

A more holistic and integrated quantitative insights can be obtained in the IM analysis if the overall system is modelled in a single FT which is not the case in the current study. Perhaps, some insights can be obtained by considering the basic components at the system's bottom hierarchy i.e. the FT for PV STRING NO.9. The results of IM analysis are as in Table 4. PV modules (PVM1 ~ 12) significantly contribute to the top event probability, followed by DC SWITCH and lastly, FUSE, which are in line with FV and RRW. There is no different in the RAW values among the events, but these values are in the order of 10^{-4}. Overall, the trends observed from the table can be inferred to the component's failure rate, in Table 2.

Table 4. Importance measure analysis for components in FT for PV STRING NO.9

No	MCS	Probability	% contribution to top event probability	*Q(x = 0)*	*Q(x = 1)*	FV	*RAW*	*RRW*
-	Top event, *Q(base)*	2.78E–05	-	-	-	-	-	-
1	PVM1	2.29E–006	8.24E + 00	2.55E–05	1.00E + 00	8.24E-02	3.60E + 04	1.09E + 00
2	PVM2	2.29E-06	8.24E + 00	2.55E-05	1.00E + 00	8.24E-02	3.60E + 04	1.09E + 00
3	PVM3	2.29E-06	8.24E + 00	2.55E-05	1.00E + 00	8.24E-02	3.60E + 04	1.09E + 00
4	PVM4	2.29E-06	8.24E + 00	2.55E-05	1.00E + 00	8.24E-02	3.60E + 04	1.09E + 00
5	PVM5	2.29E-06	8.24E + 00	2.55E-05	1.00E + 00	8.24E-02	3.60E + 04	1.09E + 00
6	PVM6	2.29E-06	8.24E + 00	2.55E-05	1.00E + 00	8.24E-02	3.60E + 04	1.09E + 00
7	PVM7	2.29E-06	8.24E + 00	2.55E-05	1.00E + 00	8.24E-02	3.60E + 04	1.09E + 00
8	PVM8	2.29E-06	8.24E + 00	2.55E-05	1.00E + 00	8.24E-02	3.60E + 04	1.09E + 00
9	PVM9	2.29E-06	8.24E + 00	2.55E-05	1.00E + 00	8.24E-02	3.60E + 04	1.09E + 00
10	PVM10	2.29E-06	8.24E + 00	2.55E-05	1.00E + 00	8.24E-02	3.60E + 04	1.09E + 00
11	PVM11	2.29E-06	8.24E + 00	2.55E-05	1.00E + 00	8.24E-02	3.60E + 04	1.09E + 00
12	PVM12	2.29E-06	8.24E + 00	2.55E-05	1.00E + 00	8.24E-02	3.60E + 04	1.09E + 00
13	DCSWITCH9	1.98E-07	7.12E-01	2.76E-05	1.00E + 00	7.12E-03	3.60E + 04	1.01E + 00
14	FUSE9	4.85E-08	1.74E-01	2.78E-05	1.00E + 00	1.74E-03	3.60E + 04	1.00E + 00

4 Conclusions

This study evaluates the reliability of a 1 MW grid-connected solar PV system by using FTA. A hypothetical system design has been developed and used in FMEA and for the FT development. To complete the FT quantification, failure rates and repair rates reported in other literatures are used.

The defined top event implies that the system has zero margin for failure, which is far from real situations. Practically, a PV system is designed with more output capacity than its stated nominal output, to allow some margin for component failures. In this study, the FT(s) comprise of only OR gates, which can aid us to a quick conclusion: large failure rate components are the critical components in the system mission. More precisely, this will correspond to components with large failure rate and large repair rate. Therefore, from Table 2, if we neglect the repair rates for a quick comparison, inverter, circuit breakers and PV modules are the critical components. This also emphasizes that having a good and reliable reliability data are very important in a FTA.

From the developed FTs, it is realized that the top event probability can be determined by using Reliability Block Diagram (RBD) too. With the consideration of Rare Event Approximations, the PV system's top event is simply the summation the failure probability of all components involved.

Acknowledgments. This research was supported by Universiti Tun Hussein Onn Malaysia (UTHM) through Grant Q410.

References

1. Zaini, N.H., et al.: Lightning surge analysis on a large scale grid-connected solar photovoltaic system. Energies **10**, 2149 (2017)
2. Khan, R., Go, Y.I.: Assessment of Malaysia's large-scale solar projects: power system analysis for solar PV grid integration glob. Challenges **4**, 1–17 (2020)
3. Zainuddin, H., Salikin, H.R., Shaari, S., Hussin, M.Z., Manja, A.: Revisiting solar photovoltaic roadmap of tropical Malaysia: past, present and future Pertanika. J. Sci. Technol. **29**, 1567–1578 (2021)
4. Ahmed, T., Mekhilef, S., Shah, R., Mithulananthan, N.: An assessment of the solar photovoltaic generation yield in Malaysia using satellite derived datasets. Int. Energy J. **19**, 61–76 (2019)
5. Sonawane, P.R., Bhandari, S., Patil, R.B., Al-Dahidi, S.: Reliability and criticality analysis of a large-scale solar photovoltaic system using fault tree analysis approach. Sustainability **15**(5), 4609 (2023)
6. Sarita, K., Saket, R.K., Khan, B.: Reliability, availability, and condition monitoring of inverters of grid-connected solar photovoltaic systems. IET Renew. Power Gener. **17**, 1635–1653 (2023)
7. Sayed, A., El-Shimy, M., El-Metwally, M., Elshahed, M.: Reliability, availability and maintainability analysis for grid-connected solar photovoltaic systems. Energies **12**(7), 1213 (2019)
8. Ahadi, A., Ghadimi, N., Mirabbasi, D.: Reliability assessment for components of large scale photovoltaic systems. J. Power. Sources **264**, 211–219 (2014)
9. Volkanovski, A., Čepin, M., Mavko, B.: Application of the fault tree analysis for assessment of power system reliability. Reliab. Eng. Syst. Saf. **94**, 1116–1127 (2009)
10. Vesely, W.E., Goldberg, F.F., Roberts, N.H., Haasl, F.D.: Fault tree handbook (U.S Nuclear Regulatory Commission) (1981)
11. Jianing, W., Shaoze, Y.: Reliability analysis of the solar array based on fault tree analysis J. Phys. Conf. Ser. **305** (2011)
12. Kolantla, D., Mikkili, S., Pendem, S.R., Desai, A.A.: Critical review on various inverter topologies for PV system architectures IET Renew. Power Gener. **14**, 3418–3438 (2020)
13. Liu, W., Niazi, K.A.K., Kerekes, T., Yang, Y.: A review on transformerless step-up single-phase inverters with different DC-link voltage for photovoltaic applications. Energies **12**, 1–17 (2019)

14. Cabrera-Tobar, A., Bullich-Massagué, E., Aragüés-Peñalba, M., Gomis-Bellmunt, O.: Topologies for large scale photovoltaic power plants. Renew. Sustain. Energy Rev. **59**, 309–319 (2016)
15. Yalawar, A., Nayak, S.: Evaluation of performance parameters and economic analysis of 1 MW Grid connected solar PV plant. In: 2018 4th International Conference for Convergence in Technology I2CT 2018, pp. 1–5 (2018)
16. Almarshoud, A.F.: Technical and economic performance of 1 MW Grid-connected PV system in Saudi Arabia. Int. J. Eng. Res. Appl. **07**, 09–17 (2017)
17. Zini, G., Mangeant, C., Merten, J.: Reliability of large-scale grid-connected photovoltaic systems. Renew. Energy **36**, 2334–2340 (2011)
18. Kazem, H.A., Albadi, M.H., Al-Waeli, A.H.A., Al-Busaidi, A.H., Chaichan, M.T.: Techno-economic feasibility analysis of 1 MW photovoltaic grid connected system in Oman. Case Stud. Therm. Eng. **10**, 131–141 (2017)
19. Sahay, K.B., Anand, P., Yadav, V.: Modeling and simulation of grid connected 1 MW solar PV power plant. In: Proceedings of Conference Industrial Commerce Use Energy, ICUE, pp. 1–5 (2018)
20. Abbood, A.A., Salih, M.A., Mohammed, A.Y.: Modeling and simulation of 1 MW grid connected photovoltaic system in Karbala city. Int. J. Energy Environ. **9**, 153–168 (2018)
21. Macit, E., Vural, A.M.: Modelling and simulation of 1 MW grid-connected PV system regulated by sliding mode control, model predictive control and PI control. Gazi Univ. J. Sci. **35**, 1433–1452 (2022)
22. Bano, T., Rao, K.: Performance analysis of 1 MW grid connected photovoltaic power plant in Jaipur, India. In: 2016 International Conference on Energy Efficient Technologies for Sustainability ICEETS 2016, pp. 165–70 (2016)
23. Louy, M., Tareq, Q., Al-Jufout, S., Alsafasfeh, Q., Wang, C.: Effect of dust on the 1-MW photovoltaic power plant at Tafila Technical University. In: 2017 8th International Renewable Energy Congress IREC, pp. 1–4 (2017)
24. Alblooshi, A., Masoud, M.I.: Design of a 1 MW grid-tied photovoltaic system. In: 7th International Conference on Engineering and Emerging Technologies (ICEET), pp. 1–6 (2021)
25. Umar, N., Bora, B., Banerjee, C., Panwar, B.S.: Comparison of different PV power simulation softwares: case study on performance analysis of 1 MW grid-connected PV solar power plant. Int. J. Eng. Sci. Invent. **7**, 2319–6734 (2018)
26. Rausand, M., Høyland, A.: System Reliability Theory: Models, Statistical Methods, and Applications. John Wiley and Sons (2004)
27. Van der Borst, M., Schoonakker, H.: An overview of PSA importance measures. Reliab. Eng. Syst. Saf. **72**, 241–245 (2001)
28. Falade, A.J., Fabiyi, S.D., Abdulkarim, A.: Reliability assessment of power transformers for Oshogbo area control center in Nigeria USEP J. Res. Inf. Civ. Eng. **14** (2016)

Artificial Intelligence

Driving Ninth SDG with Document Interaction: Integrating Natural Language Processing for Enhanced Question-Answering on PDF Documents

Xu Xiang Ng[1(✉)], Chee Kent Lim[1], Kean Lok Chee[1], Chi Wee Tan[1(✉)], Gloria Jennis Tan[2], Khai Yin Lim[3], and Chai Lim Chin[1]

[1] Tunku Abdul Rahman University of Management and Technology (TAR UMT), Jalan Genting Kelang, Setapak, 53300 Kuala Lumpur, Malaysia
ngxx-wm21@student.tarc.edu.my, chiwee@tarc.edu.my

[2] Universiti Teknologi MARA (UiTM), Jalan Ilmu 1/1, 40450 Shah Alam, Selangor, Malaysia

[3] Department of Computing and Information Technology, Tunku Abdul Rahman University of Management and Technology (TAR UMT), Penang, Malaysia

Abstract. In the digital age, managing and interacting with document content has become increasingly complex due to information overload, particularly from PDFs and other formats. This project aims to enhance user experience and efficiency in document interaction by implementing advanced search algorithms, improving navigation structures, and integrating intelligent question-answering capabilities. We assess six question-answering models: BERT Large (uncased) fine-tuned on SQuAD, BERT Base (uncased), DistilBERT Base (uncased), ALBERT Base v2, Google ELECTRA Base Discriminator, and RoBERTa Base fine-tuned on SQuAD2. Performance is measured using confidence scores to determine accuracy and reliability. Additionally, the project involves developing a chatbot interface that leverages natural language processing (NLP) techniques to facilitate seamless and intuitive user interactions with document content. This includes extracting and structuring text from PDFs and other documents to improve accessibility and collaboration, particularly for remote teams. By addressing document format complexities and accessibility issues, this project aims to streamline workflows and reduce information fatigue, ultimately leading to higher productivity and satisfaction in document management tasks. The outcomes of the comparative study will inform the selection of the most effective QA model, thereby enhancing the integration of document content across diverse workflows and tools.

1 Introduction

In the digital age, getting and discussing document content has become increasingly difficult and critical. People are continually assaulted with information, resulting in information overload, to which PDFs and other documents contribute considerably. Extracting essential information from lengthy texts is difficult, resulting in information fatigue and reduced productivity. The intricacy of multiple document formats, such as

P. Chang et al. (Eds.): SCIEMATHIC 2024, SPPHY 427, pp. 293–315, 2025.
https://doi.org/10.1007/978-981-96-7679-8_23

PDFs, Word documents, spreadsheets, and presentations, makes it difficult to create universal procedures for accessing and debating their contents. PDFs, while widespread, are notoriously difficult to navigate due to their rigid layout and poor text accessibility. Traditional document engagement methods, such as reading entire documents, are time-consuming and inefficient, making it difficult to locate specific information or engage in meaningful discussions. Collaboration workflows are also impacted, as teams working on projects or researchers debating findings require good communication of document information. Managing debates and ensuring everyone has access to important parts of a document is extremely difficult for large teams or remote workers. Another key issue is document content accessibility since badly formatted documents or ones without alternate text explanations create considerable impediments for users with visual impairments or other disabilities. Additionally, connection with existing tools is critical, as many employees rely on email, messaging apps, and project management software. To properly incorporate document material into these workflows, systems and formats must be seamlessly interoperable.

The primary objective is to enhance the user experience and efficiency when interacting with documentation, particularly by minimizing the time required for users to locate relevant information. This involves optimising the interface and functionality of documentation systems to ensure users can quickly and intuitively find the answers they need. Key strategies include implementing advanced search algorithms, improving navigation structures, and integrating intelligent question-answering capabilities. By reducing the time and effort required to access information, users can work more efficiently and experience less frustration, leading to higher overall satisfaction with the documentation.

The second objective focuses on evaluating the performance of various question-answering models to identify the most effective one for documentation purposes. Six models will be assessed: the BERT large model (uncased) fine-tuned on SQuAD, the BERT Base model (uncased), the DistilBERT Base model (uncased) distilled for SQuAD, the ALBERT Base v2 model, the Google ELECTRA Base Discriminator model, and the RoBERTa Base model fine-tuned on SQuAD2.

The performance of these models will be assessed using a confidence score, which shows the correctness and dependability of their replies. The confidence score is calculated by examining the model's probability distribution of probable solutions. Each model calculates a confidence level depending on how probable it feels a particular response is right, usually using softmax probabilities on the output logits. Higher confidence scores suggest that the model is confident in its response, whereas lower values indicate that the answer is unclear or ambiguous. These scores are frequently validated with benchmark datasets like SQuAD, in which models are compared to known right responses.

The purpose of comparing these confidence ratings is to discover which model provides the best balance of precision and efficiency, thereby supporting the first objective of increasing user experience and efficiency in documentation interaction (Fig. 1).

Fig. 1. Adobe PDF logo

2 Contemporary Study

PDF is referring to Portable Document Format. Adobe's file format protects document structure and formatting regardless of program, hardware or operating system. PDFs ensure that reports, manuals, forms, and ebooks look the same on every platform, making them popular for sharing. PDFs allow for universal viewing: independent of the operating system or application that created them, they may be opened on almost any device that has a PDF reader. As a result, they provide a safe method of document distribution while guaranteeing that the formatting and style are seen by everybody. Furthermore, PDFs avoid the messy formatting changes that may happen when transferring files between various apps by maintaining the original layout, fonts, images, and graphics just as the developer intended. Three types of PDFs can be distinguished based on compatibility: native, semi-native, and non-native. Because a native PDF is produced straight from an electronic source like a word processor, it has the best quality and formatting retention. A digital file is used to create a semi-native PDF, however, it may go through several conversion steps that could slightly change its original formatting. A non-native PDF is produced by scanning a physical document; because the original digital formatting is missing, this frequently leads to a lower quality and less editable file.

Natural Language Processing (NLP)

Natural Language Processing (NLP) is a field of artificial intelligence that enables computers to understand, interpret, and generate human language, forming the foundation of applications like chatbots, virtual assistants, and text analysis tools. Machine learning techniques such as Recurrent Neural Networks (RNNs) and Transformers, including models like BERT and GPT, enhance NLP capabilities by processing and analyzing text efficiently. These techniques power applications such as Named Entity Recognition (NER) for identifying key terms, Document Categorization for sorting text, Text Summarization for condensing information, and Sentiment Analysis for understanding emotions in text. By leveraging these models, NLP systems can accurately process user queries, generate meaningful responses, and improve user experience through more natural and intelligent interactions (Bajaj, 2023).

Document Processing

Document processing techniques focus on extracting and organizing text from PDF files to improve accessibility and usability. This involves Optical Character Recognition (OCR) for scanned documents, Natural Language Processing (NLP) for understanding content, and Named Entity Recognition (NER) for identifying key information such as names, dates, and organizations. Additionally, indexing techniques enhance searchability, while information retrieval methods enable users to quickly locate relevant content. By leveraging these techniques, document processing systems can efficiently extract, structure, and present text, making it easier for users to analyze and interact with large volumes of data (Rakshit, Mehta, & Dasgupta, 2023).

Chatbot Development

Chatbot development for document interaction involves creating conversational agents that enable users to retrieve, summarize, and explore document content through natural language queries. This process includes Natural Language Understanding (NLU) to interpret user input, Natural Language Generation (NLG) to generate coherent responses, and dialogue management to maintain context throughout interactions. The chatbot is integrated with a backend document processing system that extracts and organizes text from PDFs or other sources, ensuring accurate responses. By designing effective conversation flows and leveraging AI-driven language models, chatbots provide seamless, efficient, and user-friendly access to document content (Hiremath, Hajare, Bhosale, Nanaware, & Wagh, 2018).

Interactive Chat Application

This covers any app (such as customer support, information retrieval, entertainment, and productivity) that allows texting in real-time. Chatbots for websites and mobile apps, as well as messenger apps like Facebook Messenger, Slack, and WhatsApp, are a few examples. Applications in software development can be categorized as non-native, semi-native, or native. Better performance and integration are achieved when native apps like those made for iOS and Android with Swift or Objective-C for iOS and Java or Kotlin for Android are developed for a particular platform or device using platform-specific languages and technologies. With the help of cross-platform frameworks like React Native or Flutter, developers can create semi-native apps that combine native components for better performance and allow code to be written once and delivered across platforms. Although they can operate on a variety of devices and operating systems with little modification, non-native apps such as browser-based web apps or hybrid mobile apps that use HTML, CSS, and JavaScript in a native shell might not function as well or interact with device features as smoothly (CircleCI, 2024).

2.1 Existing Approaches

Bidirectional Encoder Representations from Transformers (BERT)

Bidirectional Encoder Representations from Transformers (BERT) is a sophisticated question-answering tool capable of producing cutting-edge findings on a variety of datasets. Here are some important reasons why the BERT model is the favored choice for question-answering models:

1. **Bidirectional Understanding:**
   BERT is designed to interpret language bidirectionally, reading in both directions at the same time, as opposed to typical models that read sequentially. This bidirectionality improves its ability to grasp long-term connections and contextual relationships in text, making it extremely useful for comprehending complicated issues and settings.
2. **Pretraining and Fine-Tuning:**
   BERT is pre-trained on a big dataset, such as Wikipedia text and then fine-tuned on question-and-answer datasets like SQuAD (Stanford Question Answering Dataset). This pretraining and fine-tuning procedure enables BERT to adapt to individual question-answering tasks and datasets, hence increasing its performance and accuracy (Devlin, Chang, Lee, Google, & Language, 2019) (Fig. 2).

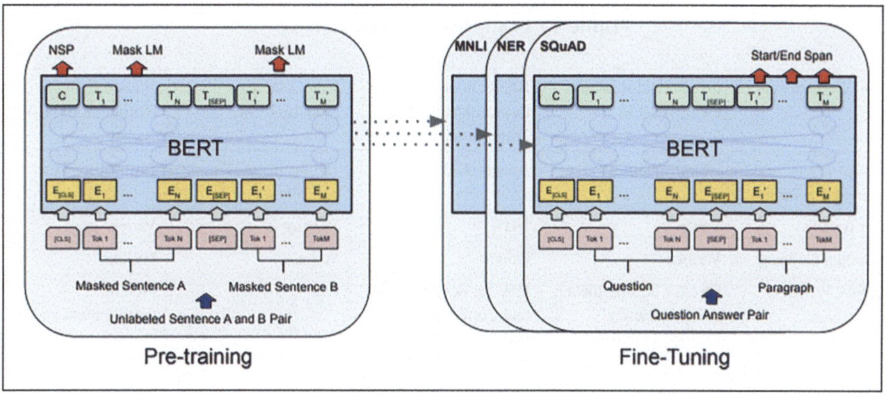

Fig. 2. Diagram of BERT Model's Pre-Training and Fine-Turning

3. **Transformer Architecture:**
   BERT is based on transformer models, allowing it to process data in any order, as opposed to traditional models, which require predefined sequences. This data processing versatility enables BERT to efficiently handle a wide range of queries and circumstances, which contributes to its performance in question-answering tasks.
4. **State-of-the-Art Performance:**
   BERT has produced cutting-edge results in a variety of natural language understanding challenges, giving it a reliable candidate for question answering. Its capacity to grasp context, understand language nuances and offer accurate responses has established it as a prominent model in the field of NLP.

In conclusion, the BERT model is preferred for question-answering models due to its bidirectional understanding, pretraining and fine-tuning capabilities, transformer design and superior performance in natural language processing tasks.

2.2 Analysis of Existing Technology

2.2.1 Natural Language Generation

In Table 1 the authors conduct an original comparative overview of different Natural Language Generation (NLG) tools, focusing on their licensing, supported features, common applications, advantages, and limitations. NLG is a crucial aspect of chatbot development and automated content creation, enabling systems to generate human-readable text from structured data. The tools compared include Conditional Random Fields (CRFs), Bidirectional Encoder Representations from Transformers (BERT) (Malingan, 2023), Large Language Models (LLM) (Kancharla, 2023), and Named Entity Recognition (NER). Understanding the strengths and weaknesses of each tool is essential for selecting the most appropriate one for a given task.

Table 1. Comparison of NLG Tools

Tools comparison	Conditional Random Fields (CRFs)	Bidirectional Encoder Representations from Transformers (BERT)	Large Language Models (LLM)	Named Entity Recognition (NER)
Type of license and open-source license	various licenses	Apache License 2.0	various licenses	various licenses
Year founded	Varies	2018	Various	Varies
Founding company	Various	Google	Various	Various
License Pricing	Free for development and Commercial licenses available	Free for development and Commercial licenses available (Google Cloud AI Platform)	Free	Free for development and Commercial licenses available
Supported features	Sequence labeling (e.g., part-of-speech tagging) and Sentence segmentation, information extraction	Bidirectional knowledge of word relationships from text	Language understanding	Entity identification and classification (e.g., names, locations)
Common applications	Text segmentation, pattern recognition	Various NLP tasks, including text extraction from PDFs	NLP tasks, document processing	Invoices, receipts, contact information
Advantages	Exploits the sequential nature of language	Understands complex word relationships, even in shuffled order	Understanding language	Excellent at identifying specific entities, functional for structured data
Limitations	Requires annotated training data, may not generalize well to new domains	Computationally intensive, may require significant resources for training and inference	Computationally intensive	May struggle with nuanced contexts, may require fine-tuning for specific domains

2.2.2 Library of Data Preprocessing

In Table 2 the authors compare several popular data preprocessing libraries commonly used in NLP projects. Data preprocessing is a critical step in preparing text data for machine learning models. The table highlights the licensing, features, applications, advantages, and limitations of NLTK, spaCy, and Gensim. Choosing the right library depends on the specific needs of the project, considering factors such as speed, ease of use, and the availability of pre-trained models.

Table 2. Comparison of Data Preprocessing Tools

Tools comparison	NLTK	spaCy	Gensim
Type of License	Apache License 2.0	MIT License	Apache License 2.0
Year Founded	2001	2015	2008
Founding Company	University of Pennsylvania	Explosion AI	Radim Řehůřek
License Pricing	Free	Free/Commercial	Free
Supported Features	Tokenization, Lemmatization, POS Tagging	Tokenization, Lemmatization, POS Tagging	Tokenization, Lemmatization
Common Applications	NLP, Text Analysis	NLP, Text Analysis	NLP, Topic Modeling
Advantages	Established library, Wide community	Fast, Efficient, Pre-trained models	Scalability, Efficient
Limitations	Slower compared to spaCy	Less language support, Smaller community	Limited functionality for NLP tasks

2.2.3 Word Embedding

In Table 3 study conducted by the authors finds that Word Embeddings are essential for representing words as numerical vectors, capturing semantic relationships between words. This table compares various word embedding techniques, including Word2Vec, GloVe, FastText, BERT (Karanjit, 2021), ELMo, and Named Entity Recognition (NER). The comparison covers licensing, features, applications, advantages, and limitations. Selecting the appropriate word embedding technique depends on factors such as the size of the dataset, the desired level of contextual understanding, and computational resources.

Table 3. Comparison of Word Embedding Tools

Tools comparison	Word2vec	Glove	FastText	BERT	ELMO	Named Entity Recognition (NER)
Type of license and open source license	Free (Apache License)	Free (MIT License)	Free (MIT License	Free (Apache License)	Free (Apache License)	various licenses
Year founded	2013	2014	2016	2018	2018	Varies

(continued)

Table 3. (*continued*)

Tools comparison	Word2vec	Glove	FastText	BERT	ELMO	Named Entity Recognition (NER)
Founding company	Google	Stanford	Facebook	Google	Allen Institute	Various
License Pricing	Free	Free	Free	Free	Free	Free for development and Commercial licenses available
Supported features	Efficient training Supports pre-trained models	Vector arithmetic Pre-trained word vectors	Subword embeddings Supports multiple languages	Bidirectional context Supports fine-tuning	Deep contextualized word representations Supports character embeddings	Entity identification and classification (e.g., names, locations)
Common applications	NLP tasks, Information retrieval, Recommendation systems	NLP tasks, Word analogy, Semantic similarity	NLP tasks, Text classification, Language modeling	NLP tasks, Question answering, Sentiment analysis	NLP tasks, Named entity recognition, Language modeling	Invoices, receipts, contact information
Advantages	Simple and efficient Works well with large datasets	Captures global word co-occurrence statistics Fast training time	Handles out-of-vocabulary words Efficient for rare words	Contextual understanding of words State-of-the-art performance	Captures complex word meanings Effective in transfer learning	Excellent at identifying specific entities, functional for structured data
Limitations	Requires a significant amount of training data May not capture semantic relationships well in some cases	Dependency on co-occurrence matrix size Memory intensive for large datasets	Slower inference time compared to Word2Vec Larger model size	High computational requirements Large model size can be challenging to deploy	Slower training compared to simpler models Requires large amounts of annotated data	May struggle with nuanced contexts, may require fine-tuning for specific domains

3 Proposed Methodology

The dataset comprises resumes from various channels such as recruitment websites, job fairs, or company HR departments. The data dictionary includes the name, details like living location, contact number and email, place of birth, driving license, skills, education, employment history, hobbies, and languages as shown in Table 4. Each resume is stored as a separate PDF file. The pdf file is retrieved from ResumeViking (Resume Viking, 2021). The same document is also available over CourseHero (CourseHero, 2024).

Table 4. Content of the Resume

Data	Description	Data Type	Example
Name	Applicant's full name	Text	Mandy Campbell
Living Location	Applicant's current place of residence	Text	1515 Pacific Ave, Los Angeles, CA 90291, United States
Contact Number	Applicant's contact number for communication	Integer	3868683442
Email	Applicant's email address for communication	Text	email@email.com
Place of Birth	Applicant's place of birth	Text	San Antonio
Driving License	Whether the applicant possesses a valid driving license	Text	Full
Skills	Technical or professional skills possessed by the applicant	Text	Cardio Training - Expert
Education	Details of educational qualifications, including degree, institution and graduation year	Text	Bachelor's Degree in Exercise Science, Concordia University
Employment History	Previous work experience, including job titles, company names, dates of employment and job responsibilities	Text	Jan 2019 - Jul 2021 Certified Bootcamp Instructor, Dominist Fitness
Hobbies	Interests or hobbies of the applicant	Text	Hockey
Languages	Languages spoken or proficiency levels of languages known by the applicant	Text	English

3.1 Proposed System Flowchart

The system flowchart in Fig. 3 illustrates the workflow of a document-processing chatbot that extracts and answers user queries based on document content. The process begins with the uploading of a PDF document, followed by preprocessing techniques such as chunking, tokenization, lemmatization, and stopword removal to refine the text.

Once preprocessing is complete, the system reads the user's query and applies different NLP models to extract relevant answers. The model identifies the most accurate response based on confidence scores and displays the result. The flowchart also incorporates decision points to handle multiple queries on the same document or process a new document file if needed. If no further input is provided, the system terminates the process.

This structured approach ensures efficient document data extraction and question answering, enhancing the chatbot's reliability in assisting users.

3.2 Proposed Test Plan/hypothesis

Test Plan: Assessing Consistency of Different Models in Answering Resume Questions.

Objective: To evaluate the consistency of different NLP models in answering resume-related questions by comparing their responses using similarity scores.

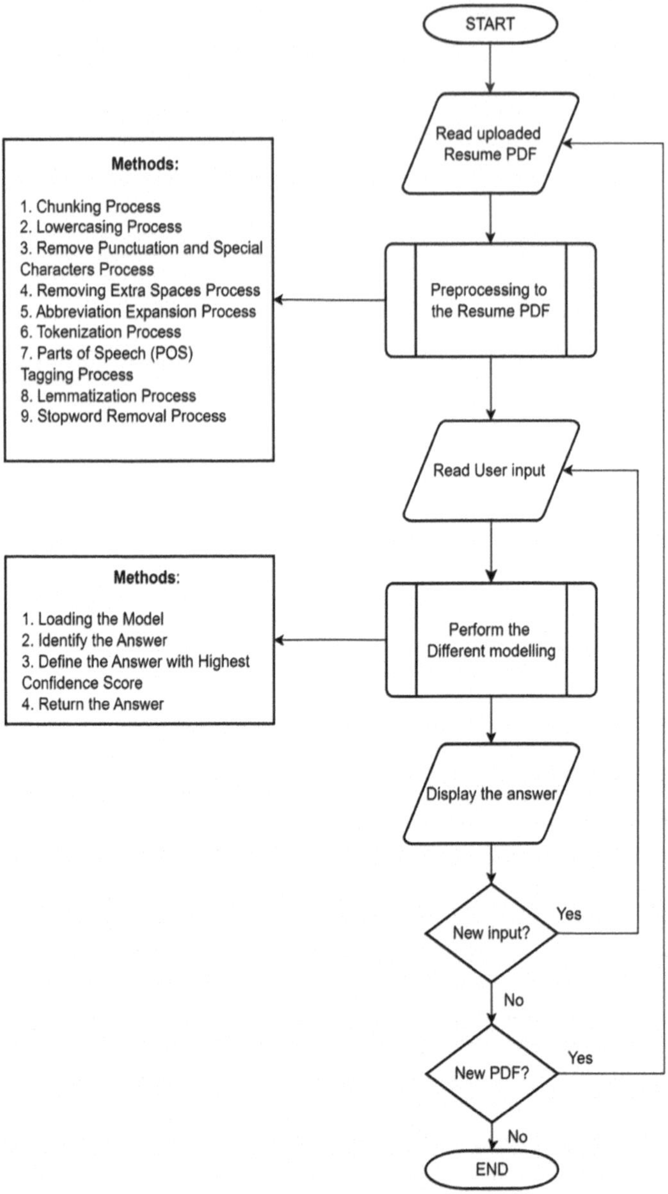

Fig. 3. System Flowchart

Hypothesis: The hypothesis is that NLP models trained on specific tasks Bidirectional Encoder Representations from Transformers (BERT) with different BERT libraries and Google model libraries will exhibit varying degrees of consistency in providing answers to resume questions.

Test Cases:

Table 5. The Sample Question for Test Plan

No	Type of Question	Question		Expected Answer Outline	Ground Truth
1	Personal Information	001	"What is your full name?"	Applicants provide their complete name	Mandy Campbell
		002	"Where do you currently reside?"	Applicant states their current place of residence	1515 Pacific Ave, Los Angeles, CA 90291, United States
		003	"What is your contact number?"	Applicants provide their primary contact number	3868683442
		004	"What is your email?"	Applicants share their email address	email@email.com
		005	"Where were you born?"	Applicants specify their place of birth	San Antonio
		006	"How is your driving license?"	Applicant states whether they have a driver's license or not	Full
2	Professional Details	007	"What skills do you possess relevant to this position?"	Applicants list their relevant skills and abilities	Cardio Training Expert Fitness Routines Expert HIIT Expert Client Assessments Expert Health & Safety Expert
		008	"What is your education?"	Applicants provide details of their academic qualifications	Bachelor's Degree in Exercise Science, Concordia University Advanced First Aid Diploma, National CPR Foundation High School Diploma, Barlow High School
		009	"How is your working experience?"	Applicants describe their previous job roles, responsibilities and achievements	Passionate, entry-level Bootcamp Instructor with experience in freelancing assisting clients to reach tough fitness goals Assisted one client to decrease chronic A1C blood sugar levels from 230 mg/dl to 110 mg/dl. Assisted another client in designing a routine to decrease body mass into healthy ranges over eight months
		010	"What is your sector?"	Applicants provide the sector or industry where they are coming from	Bootcamp Instructor

(continued)

Table 5. (*continued*)

No	Type of Question	Question		Expected Answer Outline	Ground Truth
3	Personal Interests	011	"What are your hobbies?"	Applicants share their personal interests and hobbies	Skiing, Hockey, Knitting
4	Language Proficiency	012	"What languages are you proficient in?"	Applicants list the languages they can speak and their proficiency level in each	English Spanish

Table 5 presents a structured set of sample questions used in the test plan to assess the consistency of NLP models in generating resume-related responses. It categorizes questions into four key sections: Personal Information, Professional Details, Personal Interests, and Language Proficiency. Each question is accompanied by an expected answer outline, which defines the type of response anticipated from applicants, and a ground truth answer, serving as the benchmark for evaluation.

This table helps ensure a standardized approach in measuring the accuracy of different NLP models by comparing their responses against predefined correct answers. Through this structured evaluation, the test plan aims to determine how reliably the models can generate precise and contextually appropriate responses in the resume-building process.

4 Result and Discussion

4.1 Result of Models

Based on the results in Table 6 highlight the performance differences among various transformer-based models in generating answers. The BERT Large model (fine-tuned on SQuAD) performed the best, achieving high confidence scores and the highest acceptance in human language processing, but at the cost of longer processing times. On the Other hand, the BERT Base model struggled with accuracy, often producing incomplete or irrelevant answers, though it was faster than BERT Large. DistilBERT, known for its efficiency, was the fastest model but failed to provide high-confidence answers in most cases, often returning "No answer with high confidence found." Similarly, ALBERT Base v2 showed low confidence scores and poor acceptance, making it unreliable despite its speed.

The Google ELECTRA Base Discriminator model also performed poorly, generating mostly low-confidence responses with little accuracy. Lastly, RoBERTa Base (fine-tuned on SQuAD2) demonstrated some potential, correctly answering a few questions, but frequently returned irrelevant or missing responses (" < s>"), limiting its overall effectiveness. These findings suggest a trade-off between accuracy and processing time, where larger models like BERT Large provide better responses but require more computation, while smaller, faster models prioritise speed but struggle with precision.

Table 6. Results of Different Models

Model	Question	Answer Generated	Confident Score	Time Taken, second (s)	Human Language Processing	Acceptance
BERT large model (uncased) with fine-tuned on SQuAD	001	mandy campbell	5.96	14.15	✓	✓
	002	los angeles	1.48	6.29	×	×
	003	3868683442	1.95	7.81	✓	✓
	004	emailcom place birth san antonio	6.49	7.71	✓	✓
	005	san antonio	7.26	7.73	✓	✓
	006	full l n k linkedin	5.68	7.93	✓	✓
	007	health safety expert	1.26	7.97	✓	✓
	008	bachelor ' s degree exercise science	4.88	7.35	✓	✓
	009	entry - level bootcamp instructor experience	4.68	7.76	✓	✓
	010	sale associate fast - paced fitness	2.00	7.84	×	×
	011	skiing, hockey, knit	0.70	7.78	✓	✓
	012	english native speakerspanish v ery	1.10	7.40	✓	✓
DistilBERT Base model (uncased) distilled for SQuAD	001	jul 2021 present bachelor ' s degree exercise science, concordia university concordia surpass expectation personal training coursework crossfit level 1 instructor coach prep certify jan 2020 jul 2021 advance first aid diploma, national cpr foundation miami jan 2018 dec 2018 high school diploma, barlow high school ennessee v arsity track athlete leader varsity cycling club k l l cardio training expert fitness routine expert hiit expertclient assessment expert health safety expert h b b e skiing, hockey	0.75	3.69	×	×
	002	[CLS] where do you currently reside ? [SEP] high - intensity training education design instruct stimulate group cycle workout base five heart rate zone e u c n jul 2021 present bachelor '	0.56	3.12	×	×
	003	##rtify jan 2020 jul 2021 advance	0.89	2.91	×	×
	004	client	0.61	2.27	×	×

(*continued*)

Table 6. (*continued*)

Model	Question	Answer Generated	Confident Score	Time Taken, second (s)	Human Language Processing	Acceptance
	005	jul 2021 certify bootcamp instructor, dominist fitness datteln run daily task, staff, marketing group fitness	0.68	2.30	×	×
	006	##5 pacific ave, los angeles, ca 90291, united state, 3868683442, email @ emailcom place birth san antoniodriving license full l n k linkedin, pinterest, resume emplates, build template p r f l e passionate, entry - level boo	0.87	2.31	✓	×
	007	##rtify jan 2020 jul 2021 advance first aid diploma, national cpr foundation miami jan 2018 dec 2018 high school diploma, barlow high school ennessee v arsity track athlete leader varsity cycling club k l l cardio training expert fitness routine expert hiit expertclient assessment expert health safet	0.55	2.35	×	×
	008	[SEP] participation class number 110% manage train 140 client stair climb los angeles event 100% client compete challenge develop, implement, supervise four training program, profile result 70 client ook advantage social medium platform like instagram, recruit five new client	0.57	2.36	×	×
	009	how is your working experience ? [SEP] mandy campbell, bootcamp instructor 1515 pacific ave, los angeles, ca 90291, united state, 3868683442, email @ emailcom place birth san antoniodriving license full l n k linkedin, pinterest, resume emplates, build template p r f l e passionate, entry - level bootcamp instructor experience freelance assist client reach tough fitness goal assist one client decrease chronic a l	0.60	2.41	×	×

(*continued*)

Table 6. (*continued*)

Model	Question	Answer Generated	Confident Score	Time Taken, second (s)	Human Language Processing	Acceptance
	010	expert h b b e skiing, hockey, knit l n g	0.50	2.48	×	×
	011	- level bootcamp instructor experience freelance assist client reach tough fitness goal assist one client decrease chronic a1c blood sugar level	0.53	2.34	×	×
	012	ponta por give new gym member comprehensive outline equipment, go safety advantage wide array exercise rain individual	0.58	3.40	×	×
ALBERT Base v2 model	001	intensity training education design instruct stimulate group cycle workout base five heart rate zone e u c n jul 2021 present bachelor 's degree exercise science, concordia university concordia surpass expectation personal training coursework crossfit level 1 instructor coach prep certify jan 2020 jul 2021 advance first aid diploma, national cpr foundation miami jan 2018 dec 2018 high school diploma, barlow high school ennessee v arsity track athlete leader varsity cycling club k 11 cardio training expert fitness routine expert hiit	1.36	4.64	×	×
	002	currently reside?[SEP] mandy campbell, bootcamp instructor 1515 pacific ave, los angeles, ca 90291, united state, 3868683442, email <unk> emailcom place birth san antoniodriving license full l	1.41	3.06	✓	✓
	003	2, email <unk> emailcom place birth san antoniodriving license full l n k linkedin, pinterest, resume emplates	1.97	3.28	×	×

(*continued*)

Table 6. (*continued*)

Model	Question	Answer Generated	Confident Score	Time Taken, second (s)	Human Language Processing	Acceptance
	004	No answer with high confidence found	0	2.78	×	×
	005	ve, los angeles, ca 90291, united state, 3868683442, email <unk> emailcom place birth san antoniodriving license full l n k linkedin, pinterest, resume emplates, build template p r f l e passionate, entry-level bootcamp instructor experience freelance assist client reach tough fitness goal assist one client decrease chronic a1c	1.71	2.89	×	×
	006	291, united state, 3868683442, email <unk> emailcom place birth san antoniodriving license full l n k linkedin, pinterest, resume emplates, build template p r f l e passionate, entry-level bootcamp instructor experience freelance assist client reach tough fitness goal assist one client decrease chronic a1c blood sugar level 230 mgdl 110 mgdl assist another client design routine decrease body mass healthy range eight month e p l oy e n h ry jan 2019 jul 2021 certify bootcamp instructor, dominist fitness datteln run daily task, staff, marketing group fitness program uphold active rapport member	1.62	3.25	✓	×
	007	dl assist another client design routine decrease body mass healthy range eight month e p l oy e n h ry jan 2019 jul 2021 certify bootcamp instructor, dominist fitness datteln run daily task, staff, marketing group	1.59	2.91	×	×
	008	level 1 instructor coach prep certify jan 2020 jul 2021 advance first aid diploma, national cpr foundation miami jan 2018 dec 2018 high	1.20	2.97	×	×

(*continued*)

Table 6. (*continued*)

Model	Question	Answer Generated	Confident Score	Time Taken, second (s)	Human Language Processing	Acceptance
	009	110% manage train 140 client stair climb los angeles event 100% client compete challenge develop, implement, supervise four training program, profile result 70 client ook advantage social medium platform like instagram, recruit five new client weekly six month develop three advance circuit-	1.65	2.79	×	×
	010	heart rate zone e u c n jul 2021 present bachelor 's degree exercise science, concordia university concordia surpass expectation personal training coursework crossfit level 1 instructor coach prep certify jan 2020 jul 2021 advance first aid diploma, national cpr foundation miami jan 2018 dec 2018 high school diploma, barlow high school ennessee v arsity track athlete leader varsity cycling club k l l cardio training expert fitness routine expert hiit expertclient assessment expert health safety expert h b b e skiing, hockey, knit l n g u	1.05	3.06	×	×
	011	jul 2021 present bachelor 's degree exercise science, concordia university concordia surpass expectation personal training coursework crossfit level 1 instructor coach prep certify jan 2020 jul 2021 advance first aid diploma, national cpr foundation miami jan 2018 dec 2018 high school diploma, barlow high school ennes	2.28	3.71	×	×
	012	level 1 instructor coach prep certify jan 2020 jul 2021 advance first	0.83	3.15	×	×

(*continued*)

Table 6. (*continued*)

Model	Question	Answer Generated	Confident Score	Time Taken, second (s)	Human Language Processing	Acceptance
Google ELECTRA Base Discriminator model	001	state	0.60	3.70	×	×
	002	coursework crossfit level 1 instructor coach prep certify jan 2020 jul 2021 advance first aid diploma, national cpr foundation miami jan 2018 dec 2018	0.64	4.01	×	×
	003	, implement, supervise four training program, profile result 70 client ook advantage social medium platform like instagram, recruit five new client weekly six month develop three advance circuit - style program, integrate variation level fitness [SEP]	0.46	4.01	×	×
	004	is your email ? [SEP] instructor sale associate fast - paced fitness business environment perform personalized fitness assessment client receive letter thanks 50 happy client surpass fitness goal exceed sale goal 29% average per quarter due create incredibly positive interaction client, produce active word -	0.35	2.33	×	×
	005	stimulate group cycle workout base five heart rate zone e u c n jul 2021 present bachelor ' s degree exercise science, concordia university concordia surpass expectation personal training coursework crossfit level 1 instructor coach prep certify jan 2020 jul 2021 advance first aid diploma, national cpr foundation miami jan 2018 dec 2018 high school diploma, barlow high school ennessee v arsity track athlete leader varsity cycling club k 1 1	0.34	2.32	×	×

(*continued*)

Table 6. (*continued*)

Model	Question	Answer Generated	Confident Score	Time Taken, second (s)	Human Language Processing	Acceptance
	006	[SEP] mandy campbell, bootcamp instructor 1515 pacific ave, los angeles, ca 90291, united state, 3868683442, email @ emailcom place birth san antoniodriving license full l n k linkedin, pinterest, resume emplates, build template p r f l e passionate, entry - level bootcamp instructor experience freelance assist client reach tough fitness goal assist one client decrease chronic a1c blood sugar level 230 mgdl 110 mgdl assist another client design routine decrease body mass healthy range eight month e p l oy e n h ry jan 2019 jul 2021 certify bootcamp instructor, dominist fitness datteln run daily task, staff, marketing group fitness program uphold active rapport member improved group -	0.36	3.47	✓	×
	007	No answer with high confidence found	0	2.32	×	×
	008	of - mouth	0.54	2.32	×	×
	009	rain individual client, help achieve fitness goal combination healthy exercise,	0.38	2.48	×	×
	010	, implement,	0.61	2.68	×	×
	011	assist one client decrease chronic a1c blood sugar level 230 mgdl 110 mgdl assist another	0.77	3.81	×	×
	012	No answer with high confidence found	0	2.40	×	×
RoBERTa Base model fine-tuned on SQuAD2	001	<s>	3.21	2.68	×	×
	002	<s>	3.01	3.61	×	×
	003	<s>	3.16	2.47	×	×
	004	<s>	2.61	2.45	×	×
	005	<s>	3.33	2.46	×	×
	006	full l n k linkedin	4.35	2.73	✓	✓
	007	<s>	1.83	2.56	×	×
	008	high-intensity training	4.65	2.50	×	×
	009	freelance	2.78	2.55	×	×
	010	<s>	3.21	2.45	×	×
	011	<s>	3.23	2.59	×	×
	012	<s>	2.58	3.14	×	×

4.2 Discussion/Interpretation

Table 7. Summary between Question and the Question Answering Model

Question	BERT large model (uncased) whole word masking finetuned on SQuAD	BERT Base model (uncased)	DistilBERT Base model (uncased) distilled for SQuAD	ALBERT Base v2 model	Google ELECTRA Base Discriminator model	RoBERTa Base model fine-tuned on SQuAD2
1	✓	×	×	×	×	×
2	×	✓	×	×	×	×
3	✓	×	×	×	×	×
4	✓	×	×	×	×	×
5	✓	×	×	×	×	×
6	✓	✓	✓	✓	✓	✓
7	✓	×	×	×	×	×
8	✓	×	×	×	×	×
9	✓	×	×	×	×	×
10	×	×	×	×	×	×
11	✓	×	×	×	×	×
12	✓	×	×	×	×	×

Based on the evaluation results in Table 7, BERT Large (uncased) with whole word masking fine-tuned on SQuAD outperforms all other models, achieving correct answers in 10 out of 12 cases, making it the most reliable choice. In comparison, BERT Base (uncased) is the only other model to produce at least one correct result, while DistilBERT, ALBERT, ELECTRA, and RoBERTa consistently fail, except in Row 6, where all models succeed. This suggests that Row 6 represents a relatively easier test case that all models could handle.

These findings highlight that BERT Large fine-tuned on SQuAD is the best option for question-answering tasks, particularly when accuracy is a priority. In contrast, lighter models like DistilBERT and ALBERT struggle significantly, making them less suitable for complex reasoning. While no model achieved perfect accuracy, BERT Large remains the top performer, demonstrating its effectiveness in handling this specific evaluation.

Ultimately, this analysis reinforces the idea that while smaller models like DistilBERT and ALBERT offer efficiency benefits, they may not be ideal for tasks requiring high precision and deep understanding. Fine-tuned BERT Large continues to be the superior choice, supporting previous research that larger, well-trained models tend to perform better on NLP tasks, provided sufficient computational resources are available.

Table 8. Compare Question Answering Model with Performance Evaluation

Question Answering Model	Correct Answering	False Answering	Average Confidence Score	Average Time Taken	Acceptance
BERT large model (uncased) whole word masking finetuned on SQuAD	**83% (Highest)**	**16.7% (Lowest)**	**43.44 (Highest)**	**1 min 38 s**	✓
BERT Base model (uncased)	(1/12)	(11/12)	7.69	32 s	×
DistilBERT Base model (uncased) distilled for SQuAD	(0/12)	(12/12)	0.51	18 s	×
ALBERT Base v2 model	(0/12)	(12/12)	16.67	39 s	×
Google ELECTRA Base Discriminator model	(0/12)	(12/12)	5.05	36 s	×
RoBERTa Base model fine-tuned on SQuAD2	(0/12)	(12/12)	37.95	36 s	×

Based on the experiment, the RoBERTa Base model fine-tuned on SQuAD2 achieved the second highest Average Confidence score for its answers and exhibited a lower Average Time taken compared to the BERT Large model (uncased) with whole word masking fine-tuned on SQuAD. However, the answers provided by the RoBERTa Base model were deemed not useful, with some responses including unexplained tokens such as <s> which may confuse users. Consequently, the RoBERTa Base model fine-tuned on SQuAD2 was disqualified from further evaluation of model performance. The BERT Large model (uncased) with whole word masking fine-tuned on SQuAD demonstrated strong performance in question-answering tasks. According to Tables 7 and 8, this model consistently provided answers that were closely related to the text in the PDFs, as confirmed by human evaluators. Table 8 further supports this by showing that the BERT Large model frequently produced accurate answers based on the PDF content. Table 8 reveals that the BERT Large model has the highest Average Confidence score, indicating a high level of certainty or likelihood that its answers are correct. This score reflects the model's trust in its own outputs. Despite having a higher average time taken to produce answers, the BERT Large model successfully answered 10 out of 12 questions accurately, demonstrating its reliability and effectiveness in answering questions based on given texts.

5 Conclusion

This project aimed to evaluate the performance of various question-answering models using a standardized PDF document. The models tested include BERT Large model (uncased) with fine-tuned on SQuAD, BERT Base model (uncased), DistilBERT Base model (uncased) distilled for SQuAD, ALBERT Base v2 model, Google ELECTRA Base Discriminator model, and RoBERTa Base model fine-tuned on SQuAD2. The results indicate that the BERT Large model (uncased) with whole word masking fine-tuned on SQuAD significantly outperforms the other models in terms of accuracy, confidence, and reliability. It correctly answered 10 out of 12 questions, achieving the highest average confidence score of 43.44 and demonstrating robust performance across different types of questions. Despite taking longer on average (1 min 38 s) to generate answers, its higher accuracy and confidence make it a superior choice for question-answering tasks based on PDF content.

Other models, such as BERT Base, DistilBERT, ALBERT, Google ELECTRA, and RoBERTa, did not perform as well. The BERT Base model only answered 1 question correctly, and the others failed to provide accurate answers for any of the questions. The RoBERTa model, despite having the second highest average confidence score, was disqualified due to providing non-useful responses, including unexplained tokens.

In conclusion, the BERT Large model (uncased) with whole word masking, fine-tuned on the SQuAD dataset, emerges as the preeminent model for accurately extracting information from PDF documents. This model's exceptional performance not only highlights the profound impact of meticulous fine-tuning and sophisticated model architecture on the accuracy and reliability of question-answering capabilities but also underscores its pivotal role in advancing natural language processing tasks. The precision and depth of information retrieval achieved by this model set a benchmark for future endeavors in the field.

However, while the BERT Large model's superior accuracy is commendable, there remains an imperative to address the latency associated with its implementation. Future research should, therefore, concentrate on optimizing the model's response time, ensuring that its high accuracy does not come at the expense of efficiency. Additionally, exploring the potential for enhancements in alternative models, particularly in balancing the trade-offs between speed and accuracy, will be crucial. These efforts will contribute to the ongoing evolution of QA systems, ensuring they meet the increasingly complex demands of real-world applications.

Acknowledgement. We would like to extend our sincere appreciation to TAR UMT for their invaluable support throughout the duration of this project. We are grateful for their unwavering commitment to advancing knowledge and fostering innovation.

References

Bajaj, C.: Building a Resume Parser Using NLP. Medium (2023). https://medium.com/@2020.chetaniya.bajaj/building-a-resume-parser-using-nlp-dd36c2afbce9

Devlin, J., Chang, M.-W., Lee, K., Google, K., Language, A.: BERT: pre-training of deep bidirectional transformers for language understanding (2019). https://arxiv.org/pdf/1810.04805

Dyshel, M.: Which flavor of BERT should you use for your QA task? Tangible AI (2020). https://tangibleai.com/benchmarking-bert-based-transformers-for-question-answering-and-reading-comprehension-tests/

Hiremath, G., Hajare, A., Bhosale, P., Nanaware, R., Wagh, K.S.: Chatbot for education system. Int. J. Adv. Res. Ideas Innovations Technol. **4**(3), 37–43 (2018)

Huggingface.co. (n.d.). Question answering - Hugging Face NLP Course. https://huggingface.co/learn/nlp-course/en/chapter7/7

Karanjit, S.: Question and answering using BERT (2021). https://red.mnstate.edu/cgi/viewcontent.cgi?article=1018&context=sac

Lutkevich, B.:. What is BERT (Language Model) and How Does It Work? SearchEnterpriseAI (2020). https://www.techtarget.com/searchenterpriseai/definition/BERT-language-model

Malingan, N.: Question-answering with BERT. Scaler Topics (2023). https://www.scaler.com/topics/nlp/bert-question-answering/

Mahbub, M., et al.: Question-answering system extracts information on injection drug use from clinical notes. Commun. Med. **4**(1), 61 (2024). https://doi.org/10.1038/s43856-024-00470-6

Kancharla, N.: PDF Summarizer using LLM - Naresh Kancharla - Medium. Medium; Medium (2023). https://medium.com/@naresh.kancharla/pdf-summarizer-using-llm-108881921189

Pryor, J.J.: How Do Large Language Models (LLMs) Actually Work? A Beginner's Guide. TLG (2023). https://threwthelookingglass.com/how-do-large-language-models-llms-work/

Salnik, R.: Building LLM — based Application Using Langchain and Open AI. Brocoders Team (2023). https://medium.com/brocoders-team/one-more-example-of-building-llm-based-application-using-langchain-and-open-ai-a7888cadb300

Todeschini, S.: How to Chunk Text Data — A Comparative Analysis. Medium (2023). https://towardsdatascience.com/how-to-chunk-text-data-a-comparative-analysis-3858c4a0997a

Tiwari, V.: Unlocking the Potential of PDFs with PDF-Chatbot. Medium (2024). https://medium.com/@vaibhavtiwari.945/unlocking-the-potential-of-pdfs-with-pdf-chatbot-2c2ab35e86a9

Vu, K.: Question answering tutorial with Hugging Face BERT - DataScienceCentral.com. Data Science Central (2024). https://www.datasciencecentral.com/question-answering-tutorial-with-hugging-face-bert/

Zhu, M., Cole, J.M.: PDFDataExtractor: a tool for reading scientific text and interpreting metadata from the typeset literature in the portable document format. J. Chem. Inf. Model. **62**(7), 1633–1643 (2022). https://doi.org/10.1021/acs.jcim.1c01198

CourseHero. Coursehero.com (2024). https://www.coursehero.com/file/132585681/London-Resume-Template-Professionalpdf/

Rakshit, A., Mehta, S., Dasgupta, A.: A Novel Pipeline for Improving Optical Character Recognition through Post-processing Using Natural Language Processing (2023). arXiv preprint arXiv:2307.04245. https://arxiv.org/abs/2307.04245

CircleCI. Native vs. cross-platform mobile app development. CircleCI Blog (2024). https://circleci.com/blog/native-vs-cross-platform-mobile-dev/

Resume Viking. Free resume templates. Resume Viking (2021). https://www.resumeviking.com/templates/

Interactive Conversational Chat Application for Question and Answering Enabled by a Large Language Model

Qi Zheng Tai[1(✉)], Kai Le Kong[1], Xue Wen Tan[1], Chi Wee Tan[1(✉)], Khai Yin Lim[2], and Gloria Jennis Tan[3]

[1] Tunku Abdul Rahman University of Management and Technology (TAR UMT), Jalan Genting Kelang, Setapak, 53300 Kuala Lumpur, Malaysia
taiqz-wm20@student.tarc.edu.my, chiwee@tarc.edu.my

[2] Department of Computing and Information Technology, Tunku Abdul Rahman University of Management and Technology (TAR UMT), Penang, Malaysia

[3] Universiti Teknologi MARA (UiTM), Jalan Ilmu 1/1, 40450 Shah Alam, Selangor, Malaysia

Abstract. The rapid expansion of digital communication and document management necessitates the development of advanced solutions that seamlessly integrate conversational interfaces with document processing capabilities. This paper introduces an Interactive Chat Application that leverages Large Language Model (LLM) technology to enrich conversations with PDF and text file content. The application provides functionalities such as question answering, streamlining information retrieval and enhancing collaborative discussions within a unified platform. Significant technical challenges, including balancing computational demands with low-latency communication, are addressed through effective management of LLM resources and performance optimization. The application holds potential across various sectors, including enterprise search, customer support, legal research, and education, promising to revolutionise internal document search processes and enhance AI-powered customer support. Additionally, it promotes social benefits such as improved accessibility, multilingual conversation facilitation, and the democratisation of knowledge. The evaluation of the application's question-answering capabilities is conducted using ROUGE metrics and assessments of Human-Level Performance (HLP), demonstrating its effectiveness and accuracy in understanding and generating responses.

Keywords: Question and Answering · Natural Language Processing · Large Language Model

1 Introduction

In an era of digital transformation, efficient handling and analysis of PDF documents have become essential across various industries, including administration, law, academia, and research. The integration of artificial intelligence (AI) with document processing has revolutionized information management, enabling more seamless and intelligent interactions with textual data [11]. Traditional chat applications often struggle to provide

P. Chang et al. (Eds.): SCIEMATHIC 2024, SPPHY 427, pp. 316–325, 2025.
https://doi.org/10.1007/978-981-96-7679-8_24

smooth access to document content, leading to fragmented interactions and limited contextual understanding. Addressing this gap, the proposed Interactive Chat Application leverages Large Language Model (LLM) technology to enhance user engagement with embedded PDF and text file content.

By incorporating sophisticated Natural Language Processing (NLP) techniques, the application enables users to extract critical information, summarize content, identify relevant keywords, and navigate documents effortlessly. This cutting-edge AI-driven approach transforms document analysis, offering a more intuitive and interactive experience. The tool's question-answering functionalities facilitate efficient information retrieval, fostering meaningful discussions within a single platform. Through advanced LLM optimization, the application ensures real-time, low-latency interactions, making it a powerful solution for both professional and academic settings [11].

The primary objectives of this project are to integrate advanced Natural language Processing (NLP) models into the chat application to enable effective question-answering functionalities and to evaluate their performance using ROUGE metrics and Human-Level Performance (HLP) assessments. Additionally, the application's design incorporates features for seamless document upload and interactive processing, allowing users to engage with textual content dynamically. By offering an intuitive and user interface, this tool aims to support a wide range of professional, academic, and research-based use cases, enhancing productivity and knowledge retrieval in various domains.

The potential impact of this application spans multiple sectors, including enterprise search, customer support, legal research, and education. It offers transformative possibilities for internal document search processes, AI-driven customer support, and interactive legal research tools [11]. Beyond its professional applications, the Interactive Chat Application also provides significant social benefits. By improving accessibility for individuals with deafness and hearing loss, facilitating multilingual conversations, and democratising access to knowledge, the application supports inclusivity and global collaboration [6]. The evaluation of the application's question-answering capabilities using ROUGE metrics and Human-Level Performance assessments underscores its effectiveness and accuracy, marking a significant advancement in the integration of conversational AI with document processing.

The paper is structured to provide a comprehensive overview of the Interactive Chat Application, covering its development, key functionalities, technical challenges, potential applications, and evaluation metrics. Each section contributes to a thorough understanding of the project's objectives, methodology, findings, and implications for future research and application development.

2 Literature Review

2.1 Text Extraction from Portable Document Format (PDF)

Extracting text from PDF files is essential for accessing valuable content programmatically, but it poses challenges due to the diverse structures and formats of PDF documents. Libraries such as PyPDF2, pdfplumber, and pdfminer enable developers to parse and manipulate text data within PDF files [9]. PyPDF2 is commonly used for basic text extraction, while pdfplumber and pdfminer offer more advanced capabilities, including the extraction of text, words, and tables [2]. These tools utilise parsing algorithms and OCR technology to extract text accurately, facilitating subsequent NLP tasks such as question answering.

2.2 Text Preprocessing

Text preprocessing is a fundamental step in the NLP workflow, aiming to improve the quality and relevance of text data for analysis [8]. This stage involves operations such as lowercasing, stemming, stopword removal, and correcting spelling errors to optimise data representation and enhance semantic understanding [12]. Effective preprocessing eliminates extraneous content, merges semantically similar words, and enriches text data, thereby preparing it for more accurate and meaningful analysis in tasks like sentiment analysis and information retrieval [10] (Table 1).

2.3 Question Answering

Question answering (QA) systems enable automated responses to user queries by extracting relevant knowledge and formulating answers in real-time [15]. These systems use lexical matching and heuristic extraction to identify relevant information, followed by synthesising evidence to provide accurate answers [13]. By leveraging structured and unstructured data sources, QA systems enhance information retrieval and conversational interactions, finding applications in customer service, education, and knowledge management [1, 3, 4, 7, 14] (Table 2).

2.4 Performance Measurement

According to Chen et al. in the year of 2019 [5], different techniques such as BLEU, ROUGE, METEOR, and F1 are used to evaluate question answering abilities. The tool used in this project to evaluate QA is the ROUGE score. ROUGE-N measures the overlap of N-grams between the generated and reference, providing a robust evaluation of content accuracy and relevance. Specifically, ROUGE-1, ROUGE-2, and ROUGE-L are the metrics applied. ROUGE-1 evaluates the overlap of unigrams (single words), which gives an indication of the basic content similarity. ROUGE-2 evaluates the overlap of bigrams (two-word sequences), capturing the ability of the model to generate pairs of words in the correct order. ROUGE-L measures the longest common subsequence (LCS), which assesses the fluency and coherence of the generated text by considering the longest sequence of words that appears in both the generated and reference texts. ROUGE scores are essential for validating the effectiveness of QA algorithms, ensuring high-quality and meaningful output.

Formulas used in the ROUGE score:-

$$\text{RECALL} = \frac{Overlapping\ number\ of\ n - grams}{Number\ of\ n - grams\ in\ the\ reference} \tag{1}$$

$$\text{PRECISION} = \frac{Overlapping\ number\ of\ n - grams}{Number\ of\ n - grams\ in\ the\ candidate} \tag{2}$$

$$\text{F1} = \frac{2 \times Precision \times Recall}{(Precision + Recall)} \tag{3}$$

2.5 Contemporary Study

Tokenization Algorithms Comparison

Table 1. Comparison of Tokenisation Methods

Tool Comparison	NLTK Tokenisation	Spacy Tokenisation	StanfordNLP Tokeniser	CoreNLP Tokeniser
Library / Tool	NLTK	spaCy	StanfordNLP	Stanford CoreNLP
Language	Python	Python	Java	Java
Advantages	Easy to use and well-documented	Fast and efficient tokenisation	Highly accurate tokenisation	Handles complex tokenisation scenarios well
Disadvantages	May not handle complex tokenisation scenarios effectively	Requires language model downloads, which can be large	Setting up can be complex	Slower compared to other tokenisation tools

Comparison of Question Answering Method

Table 2. Question Answering Algorithms Comparison

Tool Comparison	TinyRoBERTa	Pre-trained RoBERTa (SQuAD 2.0)	BERT	DistilBERT	Electra (SQuAD 2.0 fine-tuned)	CamemBERT (French question answering fine-tuned model)
Features	Distilled version of RoBERTa model fine-tuned for extractive question answering (e.g., SQuAD 2.0)	Handles unanswerable questions	Bidirectional context understanding	Reduced parameters and computational complexity	Provides an intuitive API for loading pre-trained models	Well-suited for tasks requiring comprehension and answering of questions in French
Accuracy	81.91%	82.91%	93.15%	86.90%	90.77%	89.14%
Advantages	Offers faster inference speeds and Lower computational resource requirements	Recognizes unanswerable questions and accurate answer extraction	Versatile and widely adopted	Faster inference times	With a vast repository of pre-trained models and tokenizers contributed by the community	Enhances its performance by exposing it to a diverse range of question types, contexts, and linguistic variations

(continued)

Table 2. (*continued*)

Tool Comparison	TinyRoBERTa	Pre-trained RoBERTa (SQuAD 2.0)	BERT	DistilBERT	Electra (SQuAD 2.0 fine-tuned)	CamemBERT (French question answering fine-tuned model)
Disadvantages	Potentially less robust when dealing with highly complex or ambiguous queries	Complexity in handling unanswerable questions	Longer training and inference times	Slightly reduced performance compared to BERT	Computationally intensive and require substantial computational resources	Heavily relies on the quality and diversity of the training data, potentially limiting its performance

3 Proposed Methodology

3.1 Dataset Description

The testing dataset utilised for this project comprises essays extracted from past-year reading question papers of the Malaysian University English Test (MUET). MUET Malaysia is an English proficiency test administered by the Malaysian Examination Council (MEC). The primary goal of the MUET proficiency exam is to assess pre-university students' English language competency in order to facilitate their admission to nearby public universities (MUET Exam in Malaysia - Papers and Fees, 2023). The reading component of the test typically includes passages or essays that test takers are required to comprehend, analyse, and respond to. The data size of one MUET essay or passage in the reading session usually does not exceed 500 words.

The source of the dataset can be traced back to official MUET question papers, which are made available by the Malaysian Examination Council or other authorised educational bodies responsible for conducting the MUET examinations. These question papers contain a variety of reading passages, including essays on diverse topics ranging from social issues to scientific discoveries, which are intended to assess candidates' reading comprehension skills and critical thinking abilities.

For example, the first past-year reading question paper used as the dataset in this project consists of 362 words. The dataset titled "Apology for not attending lectures" comprises an email exchange between Abdul Raza and Dr. Rashidah. Abdul Raza apologises for missing lectures due to illness and attending his sister's wedding, explaining his inability to inform earlier. He requests an opportunity to retake a missed test and proposes alternatives. Abdul Raza maintains a respectful tone, promises future attendance, and expresses regret for any inconvenience caused.

3.2 Question Answering

The selected algorithms and techniques for question answering in this project enable the LLM to comprehend and respond to user queries based on the input text data. Tiny-RoBERTa is a compact, distilled version of the RoBERTa model specifically optimized for extractive question answering. In essence, it's designed to extract answers from texts quickly and accurately while being much more efficient than its larger counterparts.

The pre-trained RoBERTa models fine-tuned on the SQuAD 2.0 datasets are specifically trained for question answering tasks, utilising the robust RoBERTa architecture to extract answers from passages of text. These models excel at understanding context and semantics, allowing them to accurately identify and extract relevant information to provide precise answers to user questions.

Similarly, BERT (Bidirectional Encoder Representations from Transformers) is another powerful model used for question answering in the project. BERT employs bidirectional transformers to capture the contextual relationships between words in a given text, enabling it to comprehend the nuances of language and accurately infer answers from passages. Its versatility and effectiveness make it a popular choice for various natural language processing tasks, including question answering.

DistilBERT, a distilled version of BERT, offers comparable performance with reduced computational resources, making it more efficient for deployment in resource-constrained environments. Despite its smaller size, DistilBERT retains much of the functionality and accuracy of its larger counterpart, making it a suitable option for question answering tasks where computational efficiency is a priority.

Additionally, the Electra model fine-tuned on the SQuAD 2.0 dataset and the CamemBERT model fine-tuned on multiple French question answering datasets further expand the project's capabilities by providing support for specific languages and datasets. These models leverage transformer-based architectures to process input text and generate accurate responses to user queries, catering to a diverse range of linguistic contexts and datasets.

Overall, the selected algorithms and techniques collectively empower the LLM to perform question answering tasks effectively and accurately, enhancing its utility and versatility for users.

4 Results and Discussion

Figure 1 displays the graphical user interface (GUI) of the proposed solution. This interface has been meticulously developed using Flask, a micro web framework for Python. Flask serves as the bridge connecting the front-end and back-end components of the application. The front-end is constructed using HTML for the structural framework, CSS for styling and visual design, and JavaScript for dynamic interactions. Meanwhile, Python is employed on the back-end to handle data processing and business logic. This cohesive integration ensures a seamless and efficient user experience, highlighting the robustness and versatility of the proposed solution.

4.1 Evaluation Metrics

The effectiveness of the question answering was evaluated using ROUGE evaluation metrics. To test the feasibility of the proposed method, this project also extended to the experiment to test it against Human Level Performance (HLP). The results of each model's performance are compared based on their ability to correctly answer questions, as detailed in Fig. 2 and Table 3.

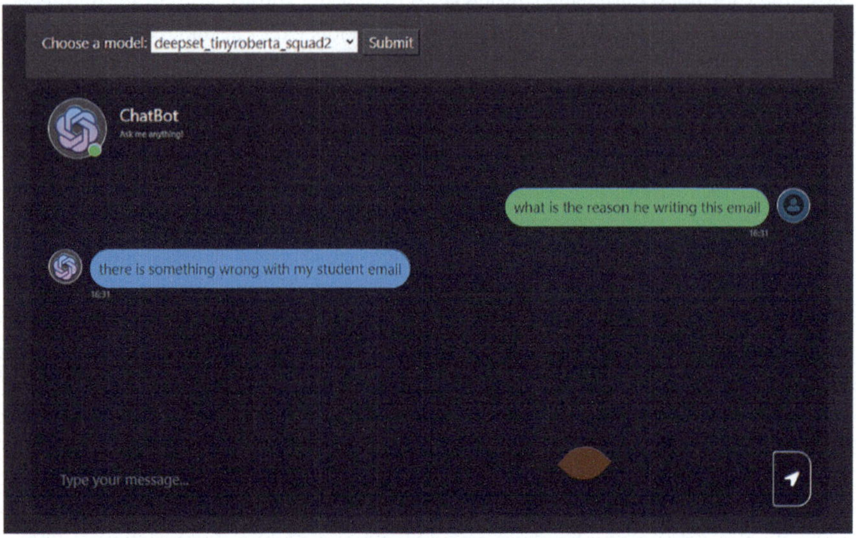

Fig. 1. The interactive GUI for the QA proposed solution.

Figure 2 and Table 3 is showing the performance of various models in answering all questions based on ROUGE score. The results show that TinyBERT performs the best across all ROUGE scores, especially in capturing key phrases (ROUGE-2 = 0.0632, while other models scored 0). This suggests that TinyBERT is the most effective model for this task. Other models, like RoBERTa, BERT, DistilBERT, ELECTRA, and CamemBERT, have lower scores, meaning they struggle more with extracting meaningful phrases. Among them, DistilBERT has a slightly better ROUGE-L score but still falls behind TinyBERT. Overall, TinyBERT is the best choice, while the low ROUGE-2 scores for most models suggest they may have difficulty handling complex phrases.

Table 4 summarises the performance of various models in answering five different questions based on Human Level Performance (HLP). Each model's response to each question is marked as correct (✔) or incorrect (✖). The evaluation includes TinyBERT, RoBERTa, BERT, DistilBERT, ELECTRA, and CamemBERT, each demonstrating varying levels of performance. TinyBERT answered Q2 and Q3 correctly, achieving the highest HLP score of 60%. RoBERTa and BERT both correctly answered Q3 and Q5, each scoring 40%. DistilBERT, ELECTRA, and CamemBERT only answered Q5 correctly, each with a 20% HLP score. These results indicate that TinyBERT outperformed

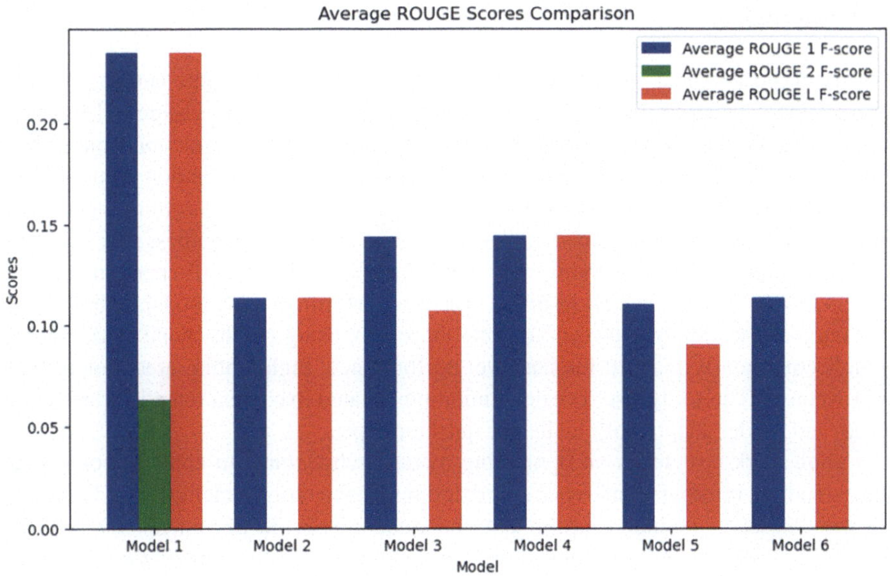

Fig. 2. Average Performance of different QA model in a glance

Table 3. Average Performance of different QA model

Model	TinyBERT	RoBERTa	BERT	DistilBERT	ELECTRA	CamemBERT
ROUGE 1	0.2349	0.1133	0.1439	0.1447	0.1106	0.1133
ROUGE 2	0.0632	0	0	0	0	0
ROUGE L	0.2349	0.1133	0.1072	0.1447	0.0906	0.1133

Table 4. Result for every model based on Human Level Performance (HLP)

Model	TinyBERT	RoBERTa	BERT	DistilBERT	ELECTRA	CamemBERT
Q1	✖	✖	✖	✖	✖	✖
Q2	✔	✖	✖	✖	✖	✖
Q3	✔	✔	✔	✖	✖	✖
Q4	✖	✖	✖	✖	✖	✖
Q5	✔	✔	✔	✔	✔	✔
Performance	60%	40%	40%	20%	20%	20%

the other models in this evaluation. However, they also highlight that despite TinyBERT's superior performance, there remains a significant gap in achieving consistent human-level accuracy across different models and questions.

5 Conclusion

This paper presented an interactive chat application powered by a large language model for question answering (QA) using advanced natural language processing (NLP) techniques. The literature review emphasized the importance of text extraction from PDFs, text preprocessing, and various QA system architectures and evaluation metrics. The proposed methodology utilized a dataset from MUET reading question papers and implemented multiple pre-trained models to assess their QA capabilities.

The evaluation results demonstrated that TinyBERT achieved 60% performance on Human Level Performance (HLP) in QA tasks, outperforming other models in balancing efficiency and accuracy. However, despite these promising results, none of the tested models consistently reached human-level performance, highlighting areas for further improvement. The study also revealed limitations related to context retention, inference speed, and model adaptability to diverse question types.

Future work will focus on optimizing model architectures to enhance contextual understanding, improving inference speed through quantization techniques, and expanding the dataset to include multilingual and domain-specific documents for broader applicability. Additionally, exploring hybrid approaches such as integrating retrieval-augmented generation (RAG) with fine-tuned models may further improve accuracy and robustness in real-world scenarios. These advancements will contribute to the development of a more reliable and scalable AI-driven QA system, reinforcing its potential for professional and academic applications.

Acknowledgement. The authors extend their gratitude to the Tunku Abdul Rahman University of Management and Technology (TAR UMT), for their financial support and provision of resources necessary to carry out this project.

References

1. Abdi, A., Idris, N., Ahmad, Z.: QAPD: an ontology-based question answering system in the physics domain. Soft. Comput. **22**(1), 213–230 (2016)
2. Adhikari, N.S., Agarwal, S.: A comparative study of PDF parsing tools across diverse document categories. arXiv (2024)
3. Ansari, A., Maknojia, M., Shaikh, A.: Intelligent question answering system based on artificial neural network. In: IEEE International Conference on Emerging Technology (ICETech), pp. 1–6. IEEE (2016)
4. Chali, Y., Hasan, S.A., Joty, S.: Improving graph-based random walks for complex question answering using syntactic, shallow semantic and extended string subsequence kernels. Inf. Process. Manage. **47**(6), 843–855 (2011)
5. Chen, A., Stanovsky, G., Singh, S., Gardner, M.: Evaluating question answering evaluation. arXiv (2019)
6. Kakade, C., Kadam, N., Kaira, V., Kewalya, R.: Enhancing sign language interpretation with multiheaded CNN, hand landmarks and large language model (LLM). In: IEEE International Conference on Future Machine Learning and Data Science (FMLDS), pp. 527–532. IEEE (2024)
7. Dwivedi, S., Singh, V.: Research and reviews in question answering system. Procedia Technol. **10**, 417–424 (2013)

8. Qais, E., Veena, M.N.: TxtPrePro: text data preprocessing using Streamlit technique for text analytics process. In: International Conference on Network, Multimedia and Information Technology (NMITCON), pp. 1–6. IEEE (2023)
9. Harshavardhan, S., Vimal, T., Kartik, R.A., Shanthi, M., Ramesh, K.: A GUI-based application for PDF processing tools using Python & CustomTkinter. IJRASET (2023)
10. Hickman, L., Thapa, S., Tay, L., Cao, M., Srinivasan, P.: Text preprocessing for text mining in organizational research: review and recommendations. Organ. Res. Methods **25**(1), 114–146 (2020)
11. Barpute, J.V., Kshirsagar, V., Salunke, S., Vedpathak, S., Varma, K., Solanki, V.: PDF fusion: revolutionizing document management with chat interfaces and voice recognition. In: International Conference on Electronics and Sustainable Communication Systems (ICESC), pp. 1774–1783. IEEE (2024)
12. Kathuria, A., Gupta, A., Singla, R.K.: A review of tools and techniques for preprocessing of textual data. Adv. Intell. Syst. Comput. 407–422 (2020)
13. Lee, Y., et al.: QASA: Advanced question answering on scientific articles. ScienceDirect **17** (2023)
14. Lende, S.P., Raghuwanshi, M.M.: Question answering system on education acts using NLP techniques. In: IEEE International Conference on StartUp (2016)
15. Sharma, Y., Gupta, S.: Deep learning approaches for question answering system. Procedia Comput. Sci. **132**, 785–794 (2018)

Optimizing Job Shop Scheduling: A Comparative Analysis of Metaheuristic Approaches

Arulmoli Arumugam[1], Wan Nor Munirah Ariffin[1(✉)], Emy Aizat Azimi[2], Erni Puspanantasari Putri Machap[3], and Logenthiran Machap[1]

[1] Computational Intelligence and Data Science Focus Group (CIDaS), Faculty of Applied Sciences and Technology, Universiti Tun Hussein Onn Malaysia (Pagoh Campus), 84600 Muar, Johor, Malaysia
normunirah@uthm.edu.my

[2] Universiti Keusahawanan Koperasi Malaysia, 103 Jalan Templer, 46700 Petaling Jaya, Selangor, Malaysia

[3] Department of Industrial Engineering, University of 17 Agustus 1945 Surabaya, Surabaya, Indonesia

Abstract. The Job Shop Scheduling Problem (JSSP) is a classic NP-hard optimization problem encountered in manufacturing and industrial processes. It involves scheduling a set of jobs, each with a sequence of operations, onto a finite number of machines while optimizing specific objectives such as minimizing makespan or total tardiness. Given the computational complexity of JSSP, metaheuristic algorithms have emerged as powerful tools for finding near-optimal solutions. This paper provides a comparative analysis of several prominent metaheuristic algorithms, including Genetic Algorithms (GA), Simulated Annealing (SA), Tabu Search (TS), Ant Colony Optimization (ACO), and Particle Swarm Optimization (PSO). The study evaluates these algorithms based on performance criteria such as runtime, convergence, and scalability. The findings highlight the strengths and limitations of each approach, providing insights into their suitability for different JSSP instances. Furthermore, the paper discusses recent hybrid metaheuristic approaches and outlines future research directions for improving JSSP optimization.

Keywords: Job Shop Scheduling Problem (JSSP) · Metaheuristics · Optimization · Scheduling Algorithms

1 Introduction

The Job Shop Scheduling Problem (JSSP) is a critical combinatorial optimization challenge that arises in manufacturing, production planning, and industrial operations. The goal of JSSP is to assign jobs to a set of machines while optimizing objectives such as minimizing makespan or total tardiness. Due to its NP-hard nature, solving the JSSP exactly for large instances is computationally intractable, prompting the development of numerous heuristic and metaheuristic approaches to address the problem more efficiently (Pinedo, 2016).

P. Chang et al. (Eds.): SCIEMATHIC 2024, SPPHY 427, pp. 326–335, 2025.
https://doi.org/10.1007/978-981-96-7679-8_25

Traditional exact methods, such as branch-and-bound, dynamic programming, and integer programming, have proven effective for small problem instances but become impractical for larger instances due to exponential computational complexity. As a result, metaheuristic algorithms, including Genetic Algorithms (GA), Simulated Annealing (SA), Tabu Search (TS), Ant Colony Optimization (ACO), and Particle Swarm Optimization (PSO), have been widely adopted for solving large-scale JSSPs. Each metaheuristic exhibits unique characteristics and trade-offs in terms of computational efficiency, exploration-exploitation balance, and convergence behavior.

This paper aims to provide a comparative analysis of these metaheuristics, emphasizing their strengths, weaknesses, and adaptability to different JSSP scenarios. The study also explores the integration of hybrid metaheuristics, which combine multiple optimization techniques to enhance performance. Although this study does not involve experimental testing, it synthesizes findings from a broad range of research papers to offer a balanced view of algorithm performance and applicability. Additionally, the study will investigate hybrid approaches, which combine multiple metaheuristics, to demonstrate their potential in enhancing performance and overcoming the limitations of individual algorithms (Talbi, 2009).

The significance of this research lies in its potential to shed light on how different metaheuristic algorithms behave when applied to complex job shop scheduling problems. By examining the performance of these algorithms across various studies, the paper will help identify the most effective methods for specific JSSP scenarios. This is particularly important for industries where optimizing production schedules can lead to significant cost savings, improved resource utilization, and enhanced operational efficiency. Furthermore, the research provides insights into emerging trends in hybrid metaheuristics, which could lay the foundation for future research in developing more robust and scalable scheduling solutions (Wang & Huang, 2020).

The paper is structured as follows. Section 2 reviews the existing literature on JSSP and metaheuristic algorithms, providing an in-depth discussion of each algorithm's underlying principles. Section 3 presents a comparative analysis of these algorithms based on previous research, highlighting their advantages, limitations, and application domains. Section 4 discusses recent developments in hybrid metaheuristics and their potential to improve performance in JSSP optimization. Finally, Sect. 5 concludes the study by summarizing key findings, identifying the most effective algorithms for different problem scenarios, and suggesting directions for future research.

2 Previous Work

The transition from exact optimization methods to metaheuristic approaches has been widely studied in JSSP literature. Early approaches to solving JSSP involved exact methods such as branch-and-bound, dynamic programming, and integer programming. These techniques are effective for small instances where computational complexity is manageable. Branch-and-bound systematically explores possible schedules, pruning infeasible solutions. Dynamic programming breaks down the problem into smaller sub-problems to solve sequentially. Integer programming models the problem mathematically using binary variables, where constraints and objectives are encoded into a solvable system.

Yoon et. al (2016) introduced a mathematical model can effectively develop job rotation schedules in automotive assembly lines, reducing cumulative workload and potentially reducing work-related musculoskeletal disorders without additional engineering work.

However, these methods exhibit poor scalability due to the exponential growth in search space as problem size increases (Brucker & Knust, 2012). For larger JSSPs, exact methods become computationally intractable, which led to the exploration of heuristic and metaheuristic approaches. Metaheuristics have become the preferred choice for large-scale JSSP instances due to their ability to balance solution quality and computational efficiency (Pellerin, 2020). These algorithms are designed to explore vast search spaces effectively and are not guaranteed to find the optimal solution but can often achieve near-optimal results in a reasonable time frame.

Over the years, a variety of methods have been proposed to solve JSSP. Traditional exact approaches, such as branch-and-bound, dynamic programming, and integer programming, have been shown to be effective for small problem instances. However, as problem sizes increase, these methods become impractical due to the exponential growth in computational time required to find optimal solution (Brucker & Knust, 2012). In response, metaheuristic algorithms have emerged as a preferred solution for large-scale JSSPs, balancing computational efficiency and solution quality. Prominent among these are Genetic Algorithms (GA), Simulated Annealing (SA), Tabu Search (TS), Ant Colony Optimization (ACO), and Particle Swarm Optimization (PSO). Each of these methods offers distinct advantages. For instance, GAs mimic the principles of natural evolution, evolving better schedules through crossover and mutation to explore large solution spaces effectively (Holland, 1975). Similarly, SA applies a probabilistic technique to avoid local optima by accepting worse solutions during the early stages of the search, mimicking the physical annealing process Kirkpatrick et. Al, 1983). TS, on the other hand, uses memory structures (the tabu list) to prevent cycling and guide the search toward unexplored areas of the solution space (Glover, 1989). ACO is inspired by the foraging behavior of ants, using pheromone trails to influence solution construction, while PSO is based on the social behavior of bird flocks, iteratively improving solutions by adjusting particle positions based on personal and social experiences (Dorigio & Di Caro, 1999; Kennedy, Eberhart, 1995).

Despite significant advancements in metaheuristic algorithm development, no single approach consistently outperforms others across all instances of the JSSP. Each algorithm exhibits strengths and weaknesses that depend on problem characteristics such as problem size, job complexity, machine configurations, and algorithmic parameter settings. Therefore, there is a pressing need for a thorough comparative analysis of these algorithms to identify the conditions under which specific methods excel. This will provide practitioners with valuable insights into selecting the most appropriate algorithm for their particular scheduling problems (Zobolas et. Al, 2008).

2.1 Genetic Algorithms

Genetic Algorithms (GA), inspired by the process of natural selection and evolution, are population-based algorithms that operate by generating multiple potential solutions (chromosomes) and evolving them over generations. GAs use key genetic operators—crossover, mutation, and selection—to generate new offspring solutions from existing ones, with the aim of improving solution quality iteratively (Holland, 1975). Crossover combines features of two parent solutions, while mutation introduces random alterations to explore new areas of the search space. GAs are known for their robustness and adaptability to a wide range of optimization problems, including JSSP, but they require careful parameter tuning and can be computationally expensive for larger instances.

2.2 Simulated Annealing

Simulated Annealing (SA) is a probabilistic algorithm inspired by the annealing process in metallurgy, where metals are heated and slowly cooled to achieve a more stable crystalline structure (Kirkpatrick et al., 1983). SA explores the solution space by accepting both improving and, under certain conditions, worse solutions to avoid getting stuck in local optima. This is controlled by a temperature parameter that decreases over time, reducing the likelihood of accepting worse solutions as the search progresses. While SA is simple to implement and can effectively escape local optima, its performance depends heavily on the cooling schedule and temperature settings, and it can be slow for larger problems.

2.3 Tabu Search

Tabu Search (TS) is a neighborhood-based algorithm that iteratively explores the solution space by moving from one solution to a neighboring solution. TS employs a tabu list, a memory structure that keeps track of recently visited solutions or attributes to prevent cycling and returning to previous solutions (Glover, 1989). This helps guide the search into new regions of the solution space, avoiding local optima. TS has been widely used for solving combinatorial optimization problems like JSSP due to its effectiveness in handling high-complexity problems. However, the computational cost of maintaining and managing the tabu list can become a drawback, especially in large-scale instances.

2.4 Ant Colony Optimization

Ant Colony Optimization (ACO) is inspired by the foraging behavior of ants, which deposit pheromone trails to communicate paths to food sources. In ACO, artificial ants build solutions by probabilistically selecting components based on pheromone levels, which are updated based on the quality of the solutions found (Dorigo & Di Caro, 1999). ACO has demonstrated success in solving JSSP due to its ability to explore diverse solution spaces and avoid premature convergence. However, for larger JSSP instances, ACO may struggle with scalability and may require hybridization with other techniques to improve solution quality.

2.5 Particle Swarm Optimization

Particle Swarm Optimization (PSO) is a population-based algorithm inspired by the social behavior of birds flocking or fish schooling (Kennedy & Eberhart, 1995). In PSO, particles (candidate solutions) explore the solution space by updating their positions based on their own experience and that of their neighbors. Each particle adjusts its velocity to move towards the best-known position, gradually improving the overall solution quality. PSO is known for its simplicity and ease of implementation but is primarily suited for continuous optimization problems. Adaptations are required to handle the discrete nature of JSSP effectively, and PSO may struggle with large combinatorial search spaces.

2.6 Greedy Randomized Adaptive Search Procedure

GRASP is a multi-start algorithm that builds solutions iteratively in two phases: a construction phase, where a greedy heuristic is used to build an initial solution, and a local search phase, which refines the solution by exploring its neighborhood (Feo & Resende, 1989). GRASP has shown to be effective for quickly generating good-quality solutions for JSSP, but the solutions may not always be optimal, and hybridization with other metaheuristics is often needed for improvement.

Variable Neighborhood Search
VNS systematically changes the neighborhood structure of solutions to avoid local optima by progressively exploring larger neighborhoods (Mladenović & Hansen, 1997). VNS has been successfully applied to JSSP, offering a flexible framework for escaping local optima. However, it may fail to converge to a global optimum without sufficient exploration of diverse neighborhoods.

Artificial Immune Systems
AIS is inspired by the biological immune system and its ability to recognize and respond to pathogens. In optimization, AIS uses mechanisms like clonal selection and immune memory to iteratively improve solutions (Burnet, 1970s). AIS has been applied to JSSP due to its robustness and ability to explore diverse solutions. However, it can struggle with finding optimal solutions in some instances, necessitating further hybridization with other techniques.

Bee-Inspired Algorithms
Bee-inspired algorithms, such as Bee Colony Optimization (BCO), are inspired by the foraging behavior of honeybees and involve the collective effort of bees to explore and exploit food sources. These algorithms have been applied to JSSP but often require hybridization with other algorithms to prevent premature convergence and improve solution quality.

Artificial Neural Networks
Although primarily used for modelling complex, non-linear relationships, ANNs have been applied to JSSP in predictive modelling and optimization. ANNs are known for their ability to generalize from training data but require extensive computational resources and training data for effective use in JSSP.

3 Comparative Insights

Each metaheuristic algorithm offers distinct advantages and challenges when applied to JSSP. While GAs, SA, and ACO are effective in exploring large solution spaces, they can be computationally intensive. TS and VNS excel in guiding the search towards promising regions but may struggle with global optimization. PSO, while simple to implement, requires adaptation for discrete problems like JSSP. As no single algorithm outperforms others across all JSSP instances, hybrid algorithms combining multiple techniques (e.g., GA-TS or SA-ACO) are increasingly being used to overcome the limitations of individual methods and enhance overall performance.

Table 1. The analysis of metaheuristics techniques in scheduling

Method	Advantages	Drawbacks	Applicability
Genetic Algorithms (GA)	i. Robust and adaptable to a variety of problem domains ii. Efficient in exploring large solution spaces	i. Requires careful parameter tuning ii. High computational cost, especially for larger problems	Suitable for large-scale JSSP but may require hybridization to avoid premature convergence and improve solution quality
Simulated Annealing (SA)	i. Effective at escaping local optima ii. Simple and easy to implement iii. Good for high-complexity problems	i. Performance highly dependent on temperature schedule ii. Slow convergence, particularly for large problems	Effective for small-to-medium JSSP, with potential improvements through hybridization for larger problems
Tabu Search (TS)	i. Avoids local optima using memory structures (tabu list) ii. Well-suited for highly complex problems	- High computational cost in managing the tabu list - May require hybridization for improved performance	Best suited for complex, large-scale JSSP, especially when hybridized with other methods
Ant Colony Optimization (ACO)	i. Capable of avoiding premature convergence ii. Effective in exploring diverse solution spaces	- Poor scalability for large problems - High computational demand for larger instances	Suitable for medium-sized JSSP but requires hybridization for large, complex instances
Particle Swarm Optimization (PSO)	i. Simple structure and easy to implement ii. Well-suited for continuous optimization problems	- Inappropriate for discrete combinatorial problems like JSSP without modifications - Susceptible to local optima	Requires adaptations for discrete problems like JSSP; best for smaller instances
Greedy Randomized Adaptive Search Procedure (GRASP)	i. Quickly converges to good-quality solutions ii. Effective in generating initial solutions	- Solutions are not always optimal - Often used in combination with other methods	Suitable for initial solution generation; requires hybridization to improve final solution quality
Variable Neighborhood Search (VNS)	i. Flexible in exploring multiple neighborhoods ii. Effective at escaping local optima	- May fail to reach the global optimum if insufficient neighborhoods are explored	Suitable for large-scale JSSP; hybridization improves solution accuracy
Artificial Immune System (AIS)	i. Robust and capable of exploring diverse solution spaces ii. Adapts to problem-specific constraints	- May struggle to find optimal solutions - Requires hybridization for improved performance	Effective for complex problems but requires hybridization for optimality in large-scale JSSP

(*continued*)

Table 1. (*continued*)

Method	Advantages	Drawbacks	Applicability
Bee-Inspired Algorithms	i. Converges rapidly ii. Highly flexible with fewer parameters to tune iii. Can store memory of solutions	- Premature convergence is a common issue - Less effective compared to Tabu Search	Suitable for medium-sized JSSP; requires hybridization for improved performance
Artificial Neural Networks (ANNs)	i. Excellent at capturing non-linear relationships ii. Capable of learning from data and modeling complex systems	- High computational burden - Requires large training datasets - Empirical nature of model building	Applied to predictive modeling and optimization in JSSP but impractical for large-scale, direct scheduling

Table 1 shows that each metaheuristic algorithm used for solving the Job Shop Scheduling Problem (JSSP) offers unique advantages and faces specific drawbacks, making them suitable for different problem types. Genetic Algorithms (GA) and Simulated Annealing (SA) demonstrate strong adaptability and robustness but are computationally intensive, especially for large problem instances. Tabu Search (TS) is highly effective in handling complex problems due to its memory structures but incurs high computational costs. Ant Colony Optimization (ACO) and Particle Swarm Optimization (PSO) perform well on medium-sized problems, though both struggle with scalability and often require hybridization for large instances. Greedy Randomized Adaptive Search Procedure (GRASP) provides quick convergence and effective initial solutions, but typically needs enhancement from other methods. Similarly, Variable Neighborhood Search (VNS) is versatile and explores multiple neighborhoods, yet it may fail to reach global optima without sufficient exploration. Artificial Immune System (AIS) and Bee-Inspired Algorithms are robust and flexible but prone to premature convergence and suboptimal results, necessitating hybridization. Lastly, Artificial Neural Networks (ANNs) excel at capturing complex, non-linear relationships but are impractical for large-scale JSSP due to their computational demands. Hybrid approaches are often adopted to combine the strengths of different algorithms and improve overall performance.

4 Recent Developments in Hybrid Metaheuristics

Recent advancements in solving the Job Shop Scheduling Problem (JSSP) have led to the growing development and use of hybrid metaheuristic algorithms, which combine the strengths of multiple optimization methods to overcome the limitations of individual approaches. The goal of hybridization is to enhance solution quality, accelerate convergence rates, and improve the algorithm's ability to handle large, complex scheduling problems (Talbi, 2009). Metaheuristic algorithms for multi-mode resource-constrained project scheduling show promise in solving the problem on both existing benchmark datasets and a newly generated dataset (Peteghem & Vanhoucke, 2014).

One prominent development is the combination of Genetic Algorithms (GA) with Local Search (LS) techniques, such as Tabu Search (TS) or Simulated Annealing (SA). While GAs are effective in exploring vast solution spaces, they often struggle with fine-tuning solutions. Integrating GAs with TS or SA allows the algorithm to conduct a more focused, local search once promising solutions are identified. This hybrid approach capitalizes on GAs' global exploration ability and the efficient local optimization provided by TS and SA, resulting in more precise and higher-quality schedules (Zobolas et al., 2008; Brucker & Knust, 2012).

Another significant advancement is the integration of Ant Colony Optimization (ACO) with Particle Swarm Optimization (PSO). ACO excels in generating good solutions early in the optimization process by simulating pheromone trails, but it may suffer from premature convergence. PSO, which relies on the social behavior of particles to adjust their positions iteratively, can help ACO escape local optima by introducing diversity into the search space. This combination has been particularly effective in solving large-scale JSSPs, as it balances exploration and exploitation more efficiently than either algorithm alone (Dorigo & Di Caro, 1999; Kennedy & Eberhart, 1995; Kumar & Jadon, 2014).

Hybrid Simulated Annealing (SA) and Variable Neighborhood Search (VNS) have also emerged as powerful tools for JSSP optimization. SA is known for its ability to escape local optima by accepting worse solutions at the beginning of the search process. By coupling it with VNS, which systematically changes the neighborhood structure of solutions, the hybrid method ensures that the search process does not get stuck in local minima. This hybrid is especially suitable for highly complex and dynamic JSSP environments (Mladenović & Hansen, 1997; Wang & Huang, 2020).

Memetic Algorithms (MA), which combine the population-based approach of GAs with local search methods, have shown promise in solving JSSPs. MAs enhance the performance of traditional GAs by incorporating local search strategies at various stages of the optimization process, improving both exploration and intensification. This hybrid approach has been demonstrated to significantly reduce computational time while maintaining solution quality, making it ideal for large-scale industrial applications (Neri et al., 2012; Talbi, 2009).

Hyperheuristics represent another recent development, focusing on selecting or generating heuristics based on the problem's characteristics. Rather than directly solving JSSP, hyperheuristics manage a pool of low-level heuristics, choosing the most appropriate one for each stage of the search. By integrating multiple metaheuristics, such as GRASP and Bee-Inspired Algorithms, hyperheuristics allow dynamic adaptation to different problem instances, leading to better optimization performance across a wide range of scheduling challenges (Burke et al., 2010; Kaya et al., 2022).

Finally, Machine Learning (ML)-based hybrid metaheuristics are gaining traction in JSSP research. ML techniques are being used to predict the performance of metaheuristics on different problem instances, allowing the hybrid algorithm to dynamically adjust its parameters or select the best approach based on real-time feedback. This data-driven hybridization leads to more efficient use of computational resources and significantly improves the adaptability of metaheuristics to varying JSSP environments (Nazir & Khan, 2021; Park et al., 2021).

In summary, hybrid metaheuristics have shown great potential to improve the performance of JSSP optimization. By combining global exploration with local search capabilities, or integrating data-driven methods, these hybrid algorithms can address the weaknesses of individual approaches, resulting in faster, more accurate, and scalable solutions for complex scheduling problems. These developments open up new possibilities for optimizing production processes in manufacturing and other industries where JSSP plays a critical roles.

5 Conclusion and Future Research Directions

This study has provided a comprehensive review of various metaheuristic algorithms used to solve the Job Shop Scheduling Problem (JSSP), examining their underlying principles, advantages, and limitations. The analysis highlights that no single algorithm consistently outperforms others across all JSSP instances, as the effectiveness of each approach depends on specific problem characteristics, such as problem size, job complexity, and machine configurations.

Among the algorithms discussed, Genetic Algorithms (GA) and Tabu Search (TS) stand out for their strong adaptability and ability to explore large solution spaces, making them suitable for large-scale JSSPs. However, these methods often require hybridization to fine-tune solutions effectively. Simulated Annealing (SA) and Variable Neighborhood Search (VNS) are effective for escaping local optima, particularly in complex scheduling problems, but they may suffer from slow convergence and require strategic parameter tuning. Ant Colony Optimization (ACO) and Particle Swarm Optimization (PSO) perform well on medium-sized JSSP instances, though their scalability can be limited in larger problems, necessitating further hybrid approaches.

The review also underscored the growing trend of hybrid metaheuristics, which combine the strengths of multiple algorithms to enhance performance. Hybrid techniques, such as combining GA with TS or SA, have shown great potential in improving solution quality while maintaining computational efficiency. Emerging methods like Memetic Algorithms (MA) and Machine Learning (ML)-driven hybrid metaheuristics offer promising avenues for optimizing JSSP, leveraging both adaptive learning and robust search techniques to handle more complex scheduling environments.

Future research should focus on further exploring the potential of hybrid metaheuristics, particularly those integrating machine learning and data-driven approaches, to dynamically adapt algorithm parameters and improve performance across a wide range of problem scenarios. Additionally, there is a need for more extensive comparative studies on these hybrid methods across different JSSP benchmarks to provide clearer guidelines for practitioners in selecting the most appropriate methods. Research could also explore the development of real-time scheduling solutions, addressing the need for fast, scalable algorithms capable of adapting to dynamic and uncertain production environments.

Lastly, more focus should be placed on multi-objective optimization within JSSP, as real-world applications often require balancing several objectives, such as minimizing makespan, energy consumption, and resource utilization. Hybrid approaches that can efficiently handle multi-objective problems will be critical in advancing the state-of-the-art in job shop scheduling optimization.

References

Brucker, P., Knust, S.: Complex Scheduling. Springer (2012)

Burke, E.K., et al.: Hyper-heuristics: a survey of the state of the art. J. Oper. Res. Soc. **64**(12), 1695–1724 (2010)

Dorigo, M., Di Caro, G.: The Ant Colony Optimization Metaheuristic. New Ideas in Optimization. McGraw-Hill (1999)

Glover, F.: Tabu search—Part I. ORSA J. Comput. **1**(3), 190–206 (1989)

Holland, J.H.: Adaptation in Natural and Artificial Systems. University of Michigan Press (1975)

Kaya, E., Gorkemli, B., Akay, B., Karaboga, D.: A review of bee-inspired meta-heuristics for combinatorial optimization problems. Eng. Appl. Artif. Intell. **115**, 105311 (2022)

Kennedy, J., Eberhart, R.C.: Particle swarm optimization. In: Proceedings of the IEEE International Conference on Neural Networks, vol. 4, pp. 1942–1948 (1995)

Kirkpatrick, S., Gelatt, C.D., Vecchi, M.P.: Optimization by simulated annealing. Science **220**(4598), 671–680 (1983)

Kumar, S., Jadon, P.: Hybrid ACO/PSO algorithm for job shop scheduling. arXiv preprint arXiv:1407.5931 (2014)

Mladenović, N., Hansen, P.: Variable neighborhood search. Comput. Oper. Res. **24**(11), 1097–1100 (1997)

Nazir, A., Khan, R.A.: Application of machine learning in job shop scheduling optimization. Comput. Secur. **102**, 102164 (2021)

Neri, F., Cotta, C., Moscato, P.: Memetic algorithms and memetic computing optimization: a literature review. Swarm Evol. Comput. **2**, 1–14 (2012)

Park, J., Chun, J., Kim, S.H., Kim, Y., Park, J.: Application of machine learning techniques in manufacturing optimization. Int. J. Prod. Res. **59**(11), 3360–3377 (2021)

Pellerin, R., Perrier, N., Berthaut, F.: A survey of hybrid metaheuristics for the resource-constrained project scheduling problem. Eur. J. Oper. Res. **280**, 395–416 (2020). https://doi.org/10.1016/J.EJOR.2019.01.063

Peteghem, V., Vanhoucke, M.: An experimental investigation of metaheuristics for the multi-mode resource-constrained project scheduling problem on new dataset instances. Eur. J. Oper. Res. **235**, 62–72 (2014). https://doi.org/10.1016/j.ejor.2013.10.012

Pinedo, M.: Scheduling: Theory, Algorithms, and Systems. Springer (2016)

Wang, L., Huang, Z.: Hybrid metaheuristics in solving job shop scheduling problems: a survey. J. Sched. **23**(2), 123–139 (2020)

Wang, Y., Liu, Y., Huang, Z.: Machine learning-enhanced metaheuristic algorithms for job shop scheduling problems: a review. Appl. Sci. **10**(4), 1352 (2020)

Yoon, S., Ko, J., Jung, M.: A model for developing job rotation schedules that eliminate sequential high workloads and minimize between-worker variability in cumulative daily workloads: application to automotive assembly lines. Appl. Ergon. **55**, 8–15 (2016)

Zobolas, G.I., Tarantilis, C.D., Ioannou, G.: Hybrid metaheuristics for the job shop scheduling problem. Eur. J. Oper. Res. **197**(1), 350–361 (2009).Talbi, E. G. (2009). Metaheuristics: From Design to Implementation. Wiley (2008)

Author Index

P. Chang et al. (Eds.): SCIEMATHIC 2024, SPPHY 427, pp. 337–338, 2025.
https://doi.org/10.1007/978-981-96-7679-8

The manufacturer's authorised representative in the EU is Springer
Nature Customer Service Centre GmbH, Europaplatz 3, 69115 Heidelberg,
Germany. If you have any concerns regarding our products, please
contact ProductSafety@springernature.com

Printed and bound by CPI Group (UK) Ltd, Croydon, CR0 4YY
01/10/2026
02225922-0011